SOUTHERN RIVERS

SOUTHERN RIVERS

RESTORING AMERICA'S FRESHWATER BIODIVERSITY

R. SCOT DUNCAN

THE UNIVERSITY OF ALABAMA PRESS   TUSCALOOSA

The University of Alabama Press
Tuscaloosa, Alabama 35487-0380
uapress.ua.edu

Copyright © 2024 by the University of Alabama Press
All rights reserved.

Publication made possible in part by a generous
contribution from Ann T. Forster.

Inquiries about reproducing material from this work should
be addressed to the University of Alabama Press.

Typeface: Garamond

Cover image: Cheoah River, Nantahala National Forest,
North Carolina; courtesy of Alan Cressler

Cover design: Michele Myatt Quinn

Cataloging-in-Publication data is available from the Library of Congress.
ISBN: 978-0-8173-2182-6 (cloth)
ISBN: 978-0-8173-6128-0 (paper)
E-ISBN: 978-0-8173-9484-4

For my wife, Ginger, and our daughters, Lilith and Autumn

Men may dam it and say that they have made a lake, but it will still be a river. It will keep its nature and bide its time, like a caged animal alert for the slightest opening. In time, it will have its way; the dam, like the ancient cliffs, will be carried away piecemeal in the currents.

—Wendell Berry, "The Unforeseen Wilderness," 1972

CONTENTS

Preface   xi
Acknowledgments   xv

PART I

1 The Point   3
2 Welcome to the Anthropocene   9
3 Brimming with Species   13
4 A Simple Dock   22
5 Three Questions   29

PART II

6 Uncle Dallas and the Sturgeon   37
7 Ancient Witnesses   45
8 Industrialization   49
9 Jubilee   59
10 Warts and All   74
11 Unsung Heroes   86
12 Death by Mud   95
13 A Golden Age Begins   105
14 How to Drown a River   114
15 Unlikely Survivor   120
16 Case Closed?   128
17 Hitchhiking Elephantears   136
18 Pearls and Caviar   145

PART III

19 Sacrifice Zone   153
20 Coal's Curse   163
21 Toxic Chemistry   172
22 Quitting Coal   182
23 Hothouse Earth   188
24 A Thirsty Future   198
25 Rising Waters   203
26 Salty Floods   210

PART IV

27 Ecological Free Fall   219
28 Ivan's Wisdom   224

PART V

29 Armor, Adapt, Retreat   233
30 Working with Nature   244
31 Drought   253
32 Reservoir Reservations   262
33 Water from the Rock and Sea   265
34 The Source Within   271
35 Mussel Power   280
36 Back from the Brink   284
37 Restoration Blueprint   291
38 Let 'Em Flow   306
39 Struggling Sturgeon   315
40 A Shad Story   323
41 Safe, Timely, and Effective   329
42 American Eel, Superhero   343

43 Halfway Solutions   350
44 A New Era Begins   354
45 The Road to Removal   360
46 The Future of Hydropower   371
47 Escaping the Flood Trap   380
48 Lake Life   391

PART VI
49 Where Do We Go from Here?   401
50 Homecoming   409

Notes   415
Bibliography   445
Further Readings   481
Index   483

PREFACE

The Southeast is unique for its astonishing diversity of freshwater animals and the exquisite beauty of the rivers in which they dwell. The region's streams range from rocky creeks in the headwater valleys of the Southern Appalachians to the muddy behemoths that grind across the Coastal Plain in their migration to the sea. Maps highlighting the concentration of freshwater taxa on the planet—fishes, turtles, snails, mussels, crayfish, aquatic insects, and others—reveal that the Southeast harbors more aquatic species than most other places on Earth.

The people of the Southeast depend heavily on their rivers. Rivers provide electricity and drinking water. They are used to ship freight and irrigate crops. Rivers are the destination for beloved traditions such as fishing, tubing, and even baptisms. And yet, like with rivers throughout the world, we use southeastern rivers to dispose of treated sewage and industrial waste. When it rains, the region's rivers overflow with pollution from urban and agricultural landscapes. Many southeastern rivers are so polluted their fish are unsafe to eat.

Despite their economic and cultural importance, rivers are invisible to most who live in the Southeast. Mountains, deserts, and oceans command attention, but rivers slink by unnoticed. They sink into the landscape, veiled behind curtains of forest or guarded by No Trespassing signs. Most people only ever see rivers through windshields from bridges. It takes an act of pilgrimage to stand beside a creek or river. This invisibility must end because southeastern rivers are in crisis.

I began writing this book to celebrate the river biodiversity of the Southeast and the herculean efforts by scientists and environmentalists to ensure it survives. While there are formidable challenges to species conservation in the Southeast, I was optimistic about the future of these rivers and their species. In recent years, scientists have made great advancements in the fields of ecology and species conservation, and I wanted to share these inspiring successes.

But as I began my research, my optimism began to wither. I had been unaware of how industrialized southeastern rivers were, and what the consequences were of this degradation. Then I read the latest climate change predictions for the region. Droughts, water shortages, sea level rise, stronger hurricanes, and species extinction—the Southeast is facing an environmental crisis unlike any other in its thirty thousand years of human occupation.

These revelations shook me. I am a native to the region and its waters. I was raised on the shores of an estuary, where river water mingles with the ocean on every tide. During a half century of life in the Southeast, I had grown to treasure the region's biodiversity, landscapes, and people. Now, I was worried about the region's future. That was when this book took on a weightier purpose than I had initially conceived. It became a journey of discovery—a quest for hope about the Southeast's future. I wasn't sure what I would find.

I am pleased—and relieved—to share that I am optimistic again. It is not a shallow everything-will-be-fine optimism. It is a calculated optimism. We have all that we need—in the Southeast and globally—to change how we live with nature and one another, so that humanity and biodiversity enjoy a secure and prosperous future. I am convinced we can do this.

Building a better future begins with understanding the scope of the challenges we face. To that end, in the first half of this book we explore the Southeast's environmental history as it relates to rivers. Stories of challenge, hardship, inequity, and loss provide the context and lessons needed for pulling out of the downward spiral of ecological deterioration and its cultural and economic consequences.

In the second half of the book the journey continues as we discover the technology and ideas needed to manage rivers and water resources better, create a more secure future for ourselves, and revive river ecosystems and biodiversity. Perhaps the most important lesson I share is that the needs of humanity and the rest of Earth's biodiversity are now inseparable. We cannot thrive without biodiversity, and biodiversity cannot survive without us. We live in mutual interdependency.

To apply this lesson to rivers: people the world over need healthy river ecosystems to provide clean drinking and irrigation water and to sustain coastal lands and fisheries. Rivers do their best at providing these and other ecosystem services when their native species are present and their populations are strong. Where rivers have lost their biodiversity, people drink polluted water, harvest less food, and suffer more flooding.

On the other hand, the future of river biodiversity—indeed, the biodiversity of the *entire* planet—now depends on humanity's choices. We have assumed the job of managing the biosphere, that razor-thin layer of the planet supporting all life. Consequently, millions of species now depend on our next moves. Will we destroy the last strongholds of biodiversity, and thereby condemn ourselves and all future generations to ecological poverty and resource scarcity? Or, will we adjust how we live so that humanity and biodiversity thrive side by side?

This time, *right now*, will be one of the most important in the unfolding of the human story. If we want a future of security, hope, and prosperity, we must begin making smart, strategic decisions about how we choose to live with nature. The experts say we have just a few years to get this right.

The work to build a better future will be done locally. It's in our own communities where we will adopt the practices that help us overcome the social and environmental challenges we've created. In the Southeast, much of this work will involve reengineering how we think about and live with rivers. And the sooner we begin, the better.

ACKNOWLEDGMENTS

Behind this book is a talented and generous network of family, friends, and allies without whom *Southern Rivers* would not exist.

I offer my deepest gratitude to my wife, Ginger, and my daughters, Lilith and Autumn. Although most of the thousands of hours of research and writing were tucked into early mornings and late evenings, this project often kept me away from them. In return, they offered nothing but love, patience, and support, and for that I am so thankful.

My brother, Will, my parents, Lucy and Bob, and my dear friend Ann Forster offered encouragement, ideas, and wisdom whenever I reached out. I am especially grateful to Ann for a generous donation to cover publication costs.

During my sabbatical year in Seattle, Lonnie Somer kept my spirits up by taking me on epic birding adventures and talking with me about rivers and conservation during our long drives.

Many readers helped shape this book. Bob Duncan, Lucy Duncan, Will Duncan, Ann Forster, Paul Johnson, Francesca Gross, and Marty Schulman offered helpful feedback on an early draft. The most instrumental reader was my wife, Ginger. A gifted writer and superior editor, her deep review of a late-stage draft led to substantial improvements in focus, continuity, and readability. Reviewers Jim McClintock and Christopher Manganiello offered encouragement and advice that honed the final draft.

Many people shared their expertise for key parts of this project. Top of the list here, by far, is Will Duncan. Will is a first-class conservation biologist and my unfailing mentor in all things aquatic. Special thanks to Paul Johnson for sharing his rich experience, colorful stories, photographs, and stacks of technical reports whenever asked. Others helping me find elusive information or offering key insights include Bill Finch, Ann Forster, Paul Freeman, Francesca Gross, Randy Haddock, Bernie Kuhajda, Jim Lamar, Roger Mangham, Ken Marion, Nathan Markham, Steve Northcutt, Mitch Reid, Angie Shugart, Beth Stewart, Jason Throneberry, Dan Tonsmeire, Benjamin Wells, Mike Wicker, and Charles Yeager. I also want to offer my gratitude to the thousands of scientists, authors, journalists, and advocates on whose labors this book was built. Thanks to them, we have reasons to be hopeful about the future of southeastern rivers.

Ed Brands custom-designed the fabulous map of southeastern watersheds and never complained about my micro-editorial demands.

Praise and gratitude for the photographers! In particular, a very special thanks to Allan Cressler, who shared many stunning images for the project. Too numerous to list here are the dozens of others who kindly donated images on request, or who had posted their images online for use in projects such as this.

Birmingham-Southern College supported the launch of this project with a one-year sabbatical.

Finally, deep appreciation to Claire Lewis Evans and the rest of the production team at the University of Alabama Press for shepherding me through the publication process and creating a work of beauty.

PART I

Our river journey begins at the westernmost tip of Fair Point on the Gulf Breeze Peninsula. This sliver of land is sacred ground to my family and is known to us simply as "The Point." Photo by R. Scot Duncan.

1
The Point

KNOWLEDGE EMPOWERS, BUT RIGHT NOW, it feels like a curse. I've returned to one of my most sacred places. I find footing on the rubble and lift my head. This spot is the exact boundary between Pensacola Bay and Santa Rosa Sound, two estuaries of the Florida Panhandle. The journey —our journey—begins here. A mountain creek or a river shoal would also be logical. But this place, my childhood playground at the terminus of a watershed, it feels right.

I usually come here to enjoy the view and wildlife and am never disappointed. Today I'm encircled by sunlit waters rippling with activity. A Sheepshead crunches on barnacles. A young stingray glides along the bottom. Mullet meander along the margins of the rubble while young male dolphins frolic offshore.

This location is the westernmost tip of the Gulf Breeze Peninsula. On maps it is Fair Point, but in my family, we simply call it "The Point." My parents live nearby on land that's been in the family since the late 1800s. The Point is sacred because when I was a child it was part of my everyday life. I was here often—usually alone—beachcombing, crabbing, fishing, or snorkeling. Behind The Point's beaches was a coastal oak forest and a scattering of old wooden cottages. In those woods I built forts, hunted for treasure, and led epic expeditions.

I break from watching the dolphins and scan the horizon. Landmarks greet me like old friends. To the northeast is the Pensacola Bay Bridge, stretching across three miles of open bay and humming with cars headed to the outer beach. Pensacola's waterfront spans the north horizon. A huge red ship is at port, and to its left is a cluster of large white fuel tanks. With a squint, I can see the oversize American flag at Joe Patti's Seafood next door. Farther west is the low-lying shoreline of Warrington. My dear friend Ann lives there, on land that's been in the Forster family since 1909. On a clear day I can see her house, safely elevated on pilings. Due west from where I stand, at 3.5 miles distance, is Pensacola Naval Air Station, the "Cradle of Naval Aviation." To the south is a horizon free of obstruction save for a narrow line of dunes stretching westward and nearly out of sight. That strip of sand is Santa Rosa Island, and its westernmost end, free of apartments and condominiums, is part of Gulf Islands National Seashore. For me, one who longs for natural landscapes, this portion of the panorama is my favorite.

Today I am not at The Point to relax. I'm here to confront the anxiety that now manifests whenever I visit. When I was a child, The Point was a wonderland. Now, I see this place through the eyes of a middle-aged biologist who has studied ecological change for three decades on four continents. Here at The Point I see threat, instability, and symptoms of the environmental problems I now study. It's during reflective moments like this when my knowledge feels burdensome.

Take, for example, biodiversity loss. When the owner of The Point died in 1998, the land was sold to developers. The forest and cottages were unceremoniously scraped away and replaced with a gated community of large houses, precisely manicured lawns, and exotic tropical plants. Witnessing the loss of my childhood playground was wrenching. Paradise lost for me and my family, paradise found for our new neighbors. In the great scheme of things, the loss of The Point's natural habitats is inconsequential. But centuries of accumulated micro-losses like this are how we've created an extinction crisis. Death by a thousand cuts.

More ominously, The Point itself is disappearing. Today at its tip I stand on concrete rubble. When I was a child in the 1970s, The Point was surrounded by beach. When my father was a child, The Point was a driftwood-strewn sand spit stretching a half mile into the bay. The spit and the beach are gone. The rubble on which I stand was dumped here in the 1940s—a first attempt at preventing more shoreline loss.

The shoreline is receding elsewhere in the estuary. Farther up the peninsula, stately slash pines are toppling into the bay, cluttering beaches with trunks and root tangles. Pleistocene bluffs capped with remnants of the scrub oak ecosystem are being devoured by increasingly hungry storm waves. Waterfront landowners, including my parents, have armored their properties with walls of wood, metal, and rock. These are symptoms of sea level rise, one of the greatest threats humanity now faces.

Signs of water quality decline are also here at The Point. A few yards from where I stand the yellow sand gives way to seagrass meadows. When my parents fished here as newlyweds, they could land a dozen Speckled Trout in an hour. But in my youth, my mother and I would cast here for hours and never get a strike. Today isn't much better. One reason is that the meadows are shrinking. Throughout the Pensacola Bay estuary, seagrasses have declined to less than 10 percent of their extent in 1950.[1] Here at The Point, a once-continuous meadow is now a chain of irregular patches.

Normally, Pensacola Bay is a clearwater estuary with a visibility of six to seven feet. On this day the waters appear fine. But after any heavy rain, bay waters swirl an ugly brown, sometimes for weeks at a time. These sediments block the light needed by seagrasses and plankton, cutting off the base of the estuary's food chain. Higher in the estuary dead zones caused by pollution form each summer. These are expanses of water devoid of the oxygen needed to maintain aquatic life. Some creatures flee. Others sicken or die.

And then there are the toxins: PCBs, dioxins, arsenic, cadmium, and copper. These and other carcinogens were released liberally into the bay until the early 1970s.[2] They persist as legacy pollutants, troublesome toxins that degrade slowly, or not at all. They lurk in the sediment and are resuspended in the waves when strong storms cause turbulence. Wildlife absorb the contaminants, and the toxins grow increasingly concentrated as they ascend the food chain. Popular species of fish and shellfish—including Striped Mullet and Blue Crab—can carry these toxins at levels where frequent consumption increases cancer risk.[3] I fear this problem more than most of the others. When I was a child, mullet and crabs were staples of my family's diet. We'd catch the fish from our dock, clean them, and bait crab traps with the heads. One night's dinner was fried mullet, the next night's dinner was boiled crabs. This is still a beloved and frequent ritual enjoyed by my parents, my wife and children, and my brother and his family. I have eaten thousands of mullet and crabs—my parents have eaten tens of thousands. This is no exaggeration. We've done the math. How much PCB and dioxin are in our bodies, lying in wait like little mutagenic time bombs? How many in the Pensacola region who regularly eat mullet and crab from the bay have been stricken with cancer as a result? How many have died, and who is next?

This is what I mean when I say that knowledge can feel like a curse. Wherever I go, I see symptoms like these that indicate ecological decline and growing environmental insecurity. However, for years I didn't understand the connections between all these problems. They were localized and independent of one another. The toxins were a result of chemical spills long ago. Coastal erosion was caused by changes in the bay's currents due to dredging, bridges, and sea walls. The muddy waters resulted from poorly managed agriculture and construction sites.

Now I know better. The problems in Pensacola Bay are connected through cause and effect. What's more, they are part of a pattern found in nearly all estuaries of the southeastern United States, including Mobile Bay, Apalachicola Bay, Charleston Harbor, Pamlico and Albemarle Sound, and Chesapeake Bay. The litany of ailments is daunting: land loss, dead zones, pollution, collapsed fisheries, invasive species, harmful algal blooms, and toxins in seafood. What connects all these problems? It is how we manage our rivers.

Abundance of fresh water is the Southeast's greatest natural asset. Take away the rains, rivers, aquifers, and humidity, and you would have a dry landscape hostile to most forms of life. Food would be hard to grow. Drinking water would be expensive. Instead, water is abundant. Life in the Southeast prospers with ease. The region overflows with biodiversity. The natural landscape is green and lush. Water is so abundant that most of us who live here never worry whether our taps will run dry or our lawns will stay green.

Rivers concentrate this liquid asset into dependable positions in the landscape. And while we lack the knowledge and technology for complete control of our oceans, the climate, or Earth's shifting crust, rivers can be mastered by human engineering. As a result, rivers are now essential to the southeastern economy. We extract and clean river water, then pipe it to our homes and businesses for drinking, cooking, bathing, and flushing away waste. Pumps slurp water from rivers to irrigate crops, and water livestock and poultry. We hoard river water behind dams and release it to generate electricity. Coal, gas, and nuclear power plants extract river water to use as a coolant or as the steam that spins their turbines. River water is used to rinse coal mined from the Appalachian Mountains, and carry away our stormwater, treated sewage, and industrial waste.

We've also reshaped rivers for our convenience. Creeks have been diverted, piped underground, and paved to guide them through our farmlands and cities. Big rivers have been dredged and straightened for commercial shipping. Dams have been built to transform river valleys into waterfront property and havens for recreation. River floodplains have been lumbered, delta marshes have been ditched and drained, and wetlands have been filled to create dry land for agriculture and urban sprawl. These are all examples of river industrialization. I do not use the term *industrialization* in the traditional sense, to mean a society's transformation from an agricultural to a manufacturing economy. In this book I instead use the term more broadly to indicate a society's use or transformation of nature—including plants, animals, minerals, and landscapes—in the process of building or protecting economic security and wealth. This includes the commodification of nature, but also the resulting generation of waste and wasted ecosystems. My use of the term also includes the modification of nature to provide security, such as the building of a levee to prevent flooding.

All this river industrialization has brought tremendous prosperity and opportunity to the Southeast, but it has also created a nasty tangle of unintended consequences that are costing us in prosperity, health, and human life. Because river systems are the liquid arteries unifying our landscape, problems in rivers do not stay local. Fresh water is increasingly scarce. Urban sprawl is causing dangerous floods. Nutrient pollution feeds algal blooms that threaten human health and damage fisheries. Waves on popular southeastern beaches churn with silt after heavy rains. River swimming holes and coastal beaches are monitored for fecal pathogens from sewage leaks and overflows. State governments routinely warn people—especially children and women of reproductive age—to limit or cease their consumption of popular freshwater and saltwater fishes due to toxins. These and the signs of ecological decline I see at The Point are all blowback from our industrialization of rivers and their watersheds.

We can do better than this. We *must* do better than this. This is why I'm at The Point today, at the very bottom of a watershed. It's the start of a journey, a quest for a better grasp of what's happened to the southeastern environment, especially our rivers. Researchers have completed a tremendous amount of science on the southeastern environment in recent years. They made discoveries in biodiversity, ecology, conservation, hydrology, and climatology. There are even breakthroughs in environmental justice and economics. Not only do I want a deeper understanding of these subjects, but I want synthesis. I want to understand how we got here. I hunger to see how all the parts fit together. But more than anything else, I must know whether there is reason to hope things can get better. Because knowledge shouldn't be a curse.

2
Welcome to the Anthropocene

Just how bad is the environmental crisis in the Southeast and beyond? River industrialization, and the good and bad that come of it, is a global phenomenon centuries in the making. The practice is nearly universal, and by some measures, the Southeast doesn't have it so bad. Shockingly, about 25 percent of the world's rivers are like the Colorado River—they are tapped dry before reaching the coast.[1] Others have been hopelessly polluted, even to the point of extinguishing all but the hardiest life-forms.

What we've done to rivers is one part of planetary industrialization. The rivers, oceans, lands, and atmosphere are now under significant human influence. In recognition of this, scientists have coined a term for the new era—the Anthropocene. The geologists formally proposing this as a new geological age describe it as "a time interval marked by rapid but profound and far-reaching change to Earth's geology, currently driven by various forms of human impact."[2] The concept immediately clicked with me.

The advent of the Anthropocene epoch marks the end of the Holocene epoch, the previous twelve millennia during which Earth's climate was well suited for human needs. Prior to that, during the Pleistocene epoch, humanity endured a series of brutal ice ages. Punishingly cold temperatures kept us in a hunter-gatherer cultural phase. Human culture flourished when the planet broke free of the ice ages and entered the mild and stable Holocene. The benevolent climate fostered our development of agriculture, complex civilizations, and advanced technologies.

The seeds of the Anthropocene were sewn in the Industrial Revolution, which began in the mid-1800s, when we developed modern machines and new forms of producing energy. Through a series of breakthroughs, we massively amplified our ability to shape the world.

And we have been tremendously successful. We number in the billions and have settled across the planet. We produce an abundance of food. Medicine has reduced suffering and mortality. Civil order is largely maintained by law, not force. Education, music, art, literature, and performance have proliferated. Travelers, goods, and ideas crisscross the globe every second of every day. Humankind's collective knowledge is available to billions through the World Wide Web. Sure, we have a long history of violence and oppression, and there's still far too much misery, injustice, and conflict. But life for the average human is far more secure, comfortable, and interesting than at any time in recent millennia.

The result of this success is that our impact on the environment upscaled from the local to the global. No place on the planet is unsullied by human influence. We've converted all the easily inhabitable portions of our planet into managed forest, pasture, cropland, and cities. No less than 75 percent of Earth's land surface has been significantly altered, and two-thirds of the world's oceans have been affected by pollution and the harvest of marine life. Only 15 percent of the world's wetlands remain. Even the most remote locations have lost biodiversity, been subject to resource extraction, or felt the far-reaching influence of our pollution. These trends show no signs of slacking. For example, between 2010 and 2015, humanity cleared an area of forest greater than the size of New Mexico.[3]

The idea that humanity had initiated a new era began surfacing in the late nineteenth century.[4] However, the concept did not receive much attention until the final decades of the twentieth century, when global change scientists began detecting a strong human influence in data from around the world. By then, the global impacts of rapid population growth, the nuclear arms race, and the industrialization of nearly *everything* were starkly obvious even to nonscientists. Since then, the evidence for our global-scale manipulation of planetary systems has mounted further, most dramatically with breakthroughs in understanding human-induced climate change.[5]

Strictly speaking, the reshaping of Earth to our benefit isn't a bad thing. Providing for the security and prosperity of humanity is a noble pursuit. The problem with our newfound power is that our global influence is largely accidental and difficult to regulate, and is destroying many of the very environmental systems that provide a stable and resource-rich environment for humanity. This is why no one is celebrating the Anthropocene's arrival.

At the time of this writing, the International Union of Geological Sciences, the official historians of Grand Time, are still deliberating whether to recognize the Anthropocene as a formal change to the geological timetable for Earth. They must agree on a mark in the global sedimentary record to signal the onset of the new epoch. The most promising markers include plastic fragments, coal ash particles, carbon dioxide concentrations, technofossils, and radioactive fallout from nuclear weapons testing.[6] Geologists must also agree on a date to mark the onset of the epoch. They seem to be homing in on 1950. It will be years before the union comes to a decision, but at this point it doesn't really matter. Scientists everywhere have widely adopted the concept. Just like the Anthropocene itself, there's no going back.

This moment in history, as we recognize the new era, deserves reflection. It was never our intent to wreck the planet. These problems snuck up on us as we were busy finding new ways to survive and prosper. But when we acknowledge our (albeit accidental) role as planetary managers, we also shed our innocence. The recognition of our planetary influence means humanity also must also accept the burden of responsibility for our actions. Ignorance is no longer an excuse for our missteps. From henceforth, our actions—and inactions—are intentional and deliberate.

Beyond our manipulation of rivers and fresh water, how else have we reshaped the planet? In 2009, sustainability scientist Johan Rockström and an interdisciplinary team of twenty-eight other prominent scientists provided a framework for charting humanity's global influence. They identified nine planetary systems on which humanity depends for survival. Some of these systems are familiar, such as biodiversity, climate stability, and fresh water availability. Others, including atmospheric aerosol and stratospheric ozone levels, are unfamiliar to those outside scientific circles. The scientists also proposed boundaries for safe planetary operation. They state that "transgressing one or more planetary boundaries may be deleterious or even catastrophic due to the risk of crossing thresholds that will trigger non-linear, abrupt environmental change within continental- to planetary-scale systems."[7] More simply, if we stay below the boundary, we'll be okay. If we cross the boundary, we risk our own survival.

When the team reviewed the data, they concluded we have exceeded the boundaries of safe operation for four Earth systems: climate stability, biodiversity maintenance, land-system change, and nutrient cycles. Of these four systems the best known is climate stability, and for good reason. Burning coal and other fossil fuels over two centuries has stuffed our atmosphere with heat-trapping greenhouse gasses, especially an incomprehensible amount of carbon dioxide (2,040 ± 310 gigatons, to be exact).[8] This has triggered an alarming cascade of planetary-scale changes that we cannot quickly reverse: rapidly rising atmospheric and oceanic temperatures, shifting rainfall patterns, sea level rise, storm intensification, and ocean acidification. In other words, the Anthropocene may be an age of calamity.

The nine planetary boundaries model, as it's come to be known, caused quite a stir when it was released. Now, over a decade later, it is foundational to understanding the Anthropocene and humanity's new role on the planet. It also highlights that humanity is facing an existential crisis of our own making. We didn't need a pending asteroid impact, alien invasion, or nuclear warfare to bring us to this point. All we needed was to keep feeding ourselves and making more babies, two activities we humans do very well.

How bad is the situation in the Southeast? Well, of the nine essential planetary systems, deterioration of seven is affecting the environmental security of the region's inhabitants in measurable ways: growing freshwater scarcity, climate change, biodiversity loss, nutrient pollution, ocean acidification, land use change, and chemical pollution. All are degrading the rivers and estuaries of the Southeast. Biodiversity loss is a particular concern because southeastern rivers are amid an extinction epidemic.

3
Brimming with Species

I WAS UNAWARE OF THE magnitude of these problems in summer 2002. At the time my wife and I had just moved to Birmingham with our two-year-old daughter, Lilith. Ginger was a freshly minted physician starting residency at University of Alabama at Birmingham. I had accepted a tenure-track position at Birmingham-Southern College, a small undergraduate liberal arts school. My top priorities that summer were finding daycare and developing the courses that I would be teaching in September.

In addition, I faced the challenge of starting a new research program. I am a conservation biologist, and for the previous decade I had studied tropical forest restoration. Continuing tropical research wasn't the right decision for me because as a new father I wanted to stay close to home. Though I wanted to develop a local research program involving my undergraduate students, the biggest obstacle to starting new research was my own ignorance. Despite having been raised just two hundred miles to the south, I knew little about the ecology of the southeastern interior.

I was excited to learn that the network of creeks and rivers draining the Southern Appalachians and the adjacent Coastal Plain harbored an exceptional range of freshwater species. Nearly two-thirds of fish species in the United States reside here, and the region is one of the top hotspots for fish diversity on the planet. These fishes range from colorful minnows in shady mountain creeks to enormous catfishes, gars, and paddlefishes in big muddy rivers.

Freshwater fishes are just the beginning. Amphibian biodiversity is greater in the Southeast than in any other region in the temperate zone of the planet, and no other place on Earth has more salamander species. The Mobile River Delta on the Gulf Coast is one of the two foremost biodiversity hotspots for aquatic turtles in the world.

Headwater mountain streams harbor an exceptional range of freshwater species and provide clean water to the region. Pictured here is the Raven Fork in the Great Smoky Mountains National Park, North Carolina. Courtesy of Alan Cressler.

For several invertebrate taxa—most notably freshwater mussels, snails, and crayfishes—no other place on the planet has as many species. Half the world's crayfish species are found in the region, as are nearly 40 percent of the world's freshwater mussels. Over a quarter of the aquatic species in the region are found nowhere else.[1]

A principal reason why the Southeast has so much aquatic biodiversity is its climate. The region enjoys a generous supply of sunlight and heat due to its low position in the temperate zone. The Gulf stays relatively warm all year long because it absorbs heat during the long summer and is also continuously fed hot tropical water by currents arriving from the Caribbean Sea. This warm water lingers in the region because the Gulf's semi-enclosed configuration prevents mixing with the cooler waters of the Atlantic Ocean. Due to its warmth, water

Southeastern creeks and rivers sparkle with exquisite fishes, but many species are on the brink of extinction. *From the top*, Slackwater Darter, Rainbow Shiner, Scarlet Shiner, and Blackbanded Darter. Blackbanded Darter courtesy of Alan Cressler. All others courtesy of Bernie Kuhajda, Tennessee Aquarium Conservation Institute.

readily evaporates from the Gulf. Winds sweep this water vapor inland to bathe the Southeast in heat and humidity all year long. Storms bring this water down as the rain that sustains the region's creeks and rivers.

Though the climate brings water and warmth to the Southeast, geology crafted a landscape of many rivers. The Southern Appalachians are the geological centerpiece responsible for the region's unprecedented biodiversity. They began building over three hundred million years ago, when the African and South American continents collided with North America. Along the collision zone, Earth's crust buckled, folded, and splintered. Magma surged upward through fractures. The

BRIMMING WITH SPECIES 15

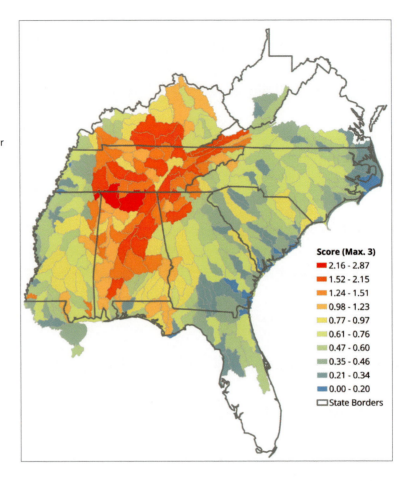

Watersheds of the Southeast scored for their diversity of fish, mussel, and crayfish species. Those with higher counts of total species, localized species, and endangered species received higher scores. Such analyses reveal the watersheds most in need of conservation. The Middle Tennessee River and eastern Mobile River systems received the highest scores in this analysis. Courtesy of Duncan Elkins, the River Basin Center, University of Georgia.

chaotic pile of rubble that resulted was the Appalachian Mountains. In their heyday they were as tall as the Andes of South America.

Earth was in a tropical climate phase for much of the past hundred million years. With no glaciers and polar ice, sea level was much higher than today. However, when the climate began cooling, 2.5 million years ago at the beginning of the Pleistocene, water evaporating from the oceans fell and accumulated as ice at high latitudes and elevations. As sea level dropped, the shallow seafloors of the Atlantic and Gulf of Mexico were revealed and became today's Coastal Plain. The steep creeks of the ancient mountains coalesced on these plains and became the sprawling river systems defining today's southeastern lowlands.

Just as the Southern Appalachians are mother to southeastern rivers, they are also the evolutionary engines producing much of the region's aquatic biodiversity. Most new animal species emerge when populations are isolated for long periods of time and evolve new and unique characteristics. Without the Southern Appalachians, the Southeast would be dominated by a few large, connected river systems. Instead, the mountains fracture the region into a mosaic of distinct watersheds. Some connect directly to the coast, while others are nested within larger basins. These watersheds provided the isolation necessary for hundreds of unique aquatic species to arise.

Consequently, many southeastern aquatic species live only in a single watershed. Species with such restricted geographic ranges are known as endemics. Endemics are of interest to biologists because of their unique characteristics and because of the clues they offer to the evolutionary and geological past. Endemism in the Mobile and the Tennessee River basins is particularly high, but endemic species can be found throughout the region, even in many of the small Coastal Plain watersheds. Most endemics are fishes, mussels, and snails because they cannot break isolation by overland dispersal into adjacent watersheds. Turtles, amphibians, crayfishes, and aquatic insects have respectable amounts of diversity in the Southeast, but because these taxa can survive out of water, they often have larger ranges.

An additional cause of the diversification of the Southeast's aquatic fauna is that the rivers of the region differ in topography, underlying

Left: Though overlooked and underappreciated, freshwater snails keep stream bottoms tidy. This, the Smooth Hornsnail, is found only in Alabama. While its populations are secure, many other southeastern snails are endangered and dozens are extinct. Courtesy of the Alabama Aquatic Biodiversity Center.

Right: The Southeast is home to more crayfish species than any other region on Earth. They vary in size, habitat, and color. And some of them, like this Peninsula Crayfish, are naturally blue! Courtesy of Alan Cressler.

All headwater streams in the Southeast find their way to the lowlands of the Coastal Plain. Most merge with others to form the region's major rivers. Pictured here is the Apalachicola River as seen from the Garden of Eden Trail in the Apalachicola Bluffs and Ravines Preserve, owned by the Nature Conservancy. Courtesy of Alan Cressler.

geology, water chemistry, and climate. This has pushed species to evolve differently in rivers across the region. Some rivers drain limestone topography, yielding alkaline, blue-tinted waters. Big Coastal Plain rivers run brown and heavy with sediment. Many nearshore rivers carry acidic, tea-colored swamp water. These differences have nudged aquatic species to evolve in different ways in different types of rivers.

͜

As I settled into my role as a professor in Birmingham, I discovered there was no shortage of research possibilities. Dozens of ecosystems and thousands of species were nearby, and only a handful had been carefully studied. My initial excitement about this was tempered by something else I learned—the region is in the grips of an extinction epidemic. Dozens of species—especially aquatic animals—were already extinct, surviving only as fading museum specimens and dry descriptions in scientific manuscripts. Poised to join them are hundreds of other river species facing extinction. There are more endangered species in the Southeast than in any other region of the developed world.

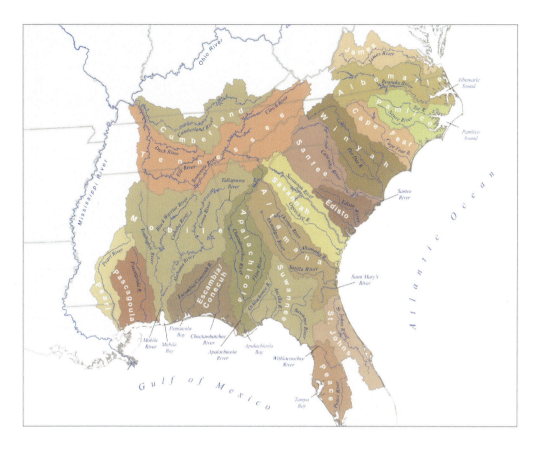

For example, the number of imperiled freshwater fishes in the region has more than doubled in recent decades and now includes nearly a third of the region's fishes.[2] Many freshwater species have retreated to a handful of locations, and some consist of just one precariously small population. A single disaster, perhaps a chemical spill or drought, could push them into oblivion. And in every case of extinction or endangerment in the region, our management of rivers is to blame.

The Southeast is not alone. The extinction wave sweeping the region is just one of dozens across the planet. Currently, the global extinction rate is somewhere between tens and hundreds of times greater than at any other time in the past ten million years. No corner of the planet, no remote mountain chain nor far-flung island, has been spared extinctions at the hand of humanity. And the crisis is far from

The Southeast has a rich diversity of watersheds thanks to the Appalachian Mountains. Courtesy of Ed Brands.

over. A recent United Nations report warns that one million species are at risk of extinction this century if we do not change how we live on the planet.[3] The beginning of the Anthropocene is on track to become one of Earth's greatest mass extinction events. Though these problems are global in scale, every extinction is driven by local events. What happens next to the southeastern aquatic fauna is entirely up to its residents.

Why is biodiversity maintenance an essential planetary system? Why is extinction something we should worry about? In recent decades ecologists have uncovered profound connections between the species diversity of ecosystems and the degree to which these ecosystems provide resources we need. Ecosystem services, as they are called, include the provisioning of oxygen, clean air and water, soil, erosion control, new medicines, disease control, food, fiber, and renewable fuels. Whether we realize it or not, each of us relies on these ecosystem services every day. They are essential to the security, economy, and culture of even the most highly developed societies. An advantage of ecosystem services over human-engineered systems is that ecosystems do not need reminders, encouragement, praise, or payment to do their thing. Plus, they are self-repairing and self-sustaining when we provide them with a healthy environment.

Rivers offer multiple ecosystem services to residents of the Southeast. They provide water for drinking, irrigation, and industrial use. River microbes and animals cleanse the water before extraction. Rivers carry away our waste and provide mechanical energy for spinning turbines at hydropower plants. River water is used as steam and coolant in fossil fuel and nuclear power plants. Floods deliver nutrients and silt to floodplains, thereby benefiting agriculture and forestry. The water and nutrients reaching the coast sustain commercial fisheries for oysters, fishes, and shellfishes. The sand rivers bring to the coast builds barrier islands that protect harbors and settlements from oceanic waves and storms. The silt delivered by rivers to estuaries builds marshes that shelter the young of seafood species and absorb storm surges when hurricanes arrive. Less essential but still important are the cultural ecosystem services that southeastern rivers provide, and the economic benefits they bring: recreational boating, sport fishing, waterfowl hunting, scenic beauty, and inspiration for literature, music, and art. If rivers ceased providing their ecosystem services, and we were forced to design and build replacement systems, the expense would cripple our economy. As for the cultural ecosystem services rivers provide—they are irreplaceable.

Extinction is a problem because biodiversity is key to sustaining many of the vital ecosystem services rivers provide. Ecosystems with more native species, and healthy populations of them, provide more and higher quality ecosystem services than do ecosystems where biodiversity has been diminished, or replaced by non-native species. We can lose some species and still retain ecosystem services. Every southeastern ecosystem has lost at least a few in the past few centuries, but these ecosystems still provide many services. However, the more species we lose, the less we get in return because no single species provides a full range of ecosystem services. Maintaining more species also preserves resiliency in ecosystem function. When one species declines another can partially take over its role. For those familiar with financial investment, the value of biodiversity to support ecosystem services is like the value of a diverse stock portfolio to buffer one's investment against market volatility.

So, there are many good reasons for us to protect ecosystems and the native biodiversity within them. We have no good alternative, because we lack the knowledge, technology, and wealth to replicate what nature does. Ecosystems give us these services, but they come with a warning. William Ruckelshaus, the first administrator of the US Environmental Protection Agency, said, "Nature provides a free lunch, so long as we control our appetite."[4] One of the great uncertainties of the Anthropocene is whether we can tame our hunger.

4

A Simple Dock

SOME WITHIN THE SPECIES CONSERVATION community—from environmentalists to scientists—have bristled against the notion of using the ecosystem services concept to defend the existence of species. The idea that species must have utility to humans for their survival to be justified has frustrated those whose conservation ethic is that other species have just as much right to exist as we humans do. I am inclined to agree, but the issue isn't that simple.

The origins of my own conservation ethic are rooted where our journey began, on the shores of Pensacola Bay. Gulf Breeze is the small residential city on the peninsula by the same name. My mother, Lucy Duncan, taught in local elementary schools. My father enforced labor laws for the US Department of Labor. Despite my parents' middle-class incomes, we lived a life of privilege by having the bay as our backyard.

Our family's activities tracked the daily and seasonal rhythms of the bay. A simple four-foot-wide dock was the center of family activity. The water was clear and shallow, and a daily ritual was to walk the dock to check the incoming weather and see what was swimming or drifting by. Summer brought swimming and schools of mullet that Dad would catch with a cast net. The house often smelled of fried fish and crab boil. Fall tides would bring tropical fishes to the shallows, and we'd snorkel in search of them. Winter chilled the bay, and most fishes left for deeper waters. Cold fronts would arrive, and for a few days the bay would churn with waves whipped frothy white by winds strong enough to keep us off the dock. Spring on the bay was marked by schools of young minnows and the first chilly swim of the season.

My younger brother, Will, and I were the sixth generation of Duncans to live on Fair Point. Much of the Gulf Breeze Peninsula was homesteaded by my third great-uncle James Duncan in the 1870s, after

the family sold its farm in Fern Creek, Kentucky. I'm sad to say that these Duncans were former enslavers and had supported the Confederacy. James's brother Nelson—my direct ancestor—served in the Confederate Army with a horse artillery unit and was captured and imprisoned. His eldest brother, Tom, was allegedly a Confederate spy. After the war the Duncans sold the farm and headed south in search of new beginnings.

At the time, there were still patches of wilderness in the Southeast. The Duncans found one on the shores of Pensacola Bay. They built a large house overlooking a bayou. Fish, blue crabs, and oysters from the bay were everyday meals. They grew vegetables as best they could in the sandy soil and tried to keep their chickens and goats away from the alligators. They later relocated the family house farther down the peninsula, at the base of Fair Point. My great grandfather Addison Duncan Sr. was the captain of the *Santa Rosa*, a US survey vessel charting the intracoastal waterway. After he died in Apalachicola on a survey trip, his wife, Mary Ann McElheran Duncan, sold off the land a few acres at a time for living expenses. By the time I was born, Gulf Breeze

The years I spent as a child on this dock in Pensacola Bay shaped my choice to become a conservation biologist. Having survived more hurricanes and rebuilds than I can recall, the dock remains a centerpiece of activity for our family. Courtesy of Lucy Duncan.

had grown into a city and only a few lots remained in the family—all on the north shore of the peninsula; I was raised on one of them, in a brick house built by my grandfather Addison Duncan Jr. My grandparents and great aunt lived just steps away. Today, nearly fifty years later, my parents still live in the house where I was raised, and still spend a portion of each day on the dock.

Will and I became biologists because our family found endless joy, adventure, and fascination in nature. Birdwatching became Bob and Lucy's chief pastime just before I was born. They became hooked when noticing the abundance and diversity of birds accumulating in their yard and neighborhood every spring and fall. Fair Point is a stopover site for birds migrating across or around the Gulf of Mexico. Migrations at places like Fair Point can be spectacular. Hundreds of colorful birds—from hawks to orioles to buntings—can be seen in a day. It wasn't inevitable that my brother and I became birders, but we did.

My parents' passion for birding dovetailed with their conservation ethic and admiration of science. They've published scientific papers documenting new ornithological occurrences in the region and written books on the bird migration of the Florida Panhandle. They both,

After the Civil War, the Duncans left Kentucky and settled in Gulf Breeze on the shore of Pensacola Bay, Florida. The abundance of fish, blue crabs, and oysters kept the family well fed. Courtesy of Bob and Lucy Duncan.

but especially Lucy, commit considerable time to teaching birding classes, leading field trips, and advocating for conservation initiatives in the area.

This was a remarkable upbringing for my brother and me, but there were darker undercurrents. My parents began birding in the late 1960s and early 1970s, during a time when many Americans became concerned about the environment. At the time it was undeniable, even to everyday people, that unchecked pollution resulting from full-on industrialization was a grave threat to nature and the public. Some of the environmental events of those years are now icons of environmental history: the DDT crisis, Santa Barbara oil spill, Love Canal toxic waste disaster, and fires on the Cuyahoga River. Concern about pollution led to the passage in the 1970s of strong federal laws to protect the environment and public health.

The multitudes of migrant birds, such as this Hooded Warbler, passing through my parents' neighborhood inspired them to become birders and conservationists. Courtesy of Charles Grisham.

Bob and Lucy were active in local environmental issues, for example protesting the discharge of pollution in the lower Escambia River, which was decimating Pensacola Bay's fish populations and wildlife. They were outspoken advocates for the establishment of the Gulf Islands National Seashore on Santa Rosa Island, which was staunchly opposed by locals who stood to gain from developing the land or who resisted federal ownership of local lands. Environmental issues of national or local flavor were often the subject of conversation between my parents and their friends, whether on the trail or at the dining room table. These experiences shaped my emerging worldview and conservation ethic. I developed a strong sense of morality regarding nature. I regarded other species, even ecosystems, as perpetual underdogs, struggling to survive against overwhelming odds. I learned to cherish what wild nature was left, never assuming it would be there when I returned. In coastal Florida, it often wasn't.

Ronald Reagan was elected president in 1980, when I was ten. It was the beginning of an intense conservative backlash against the environmental laws enacted during the 1970s. In the conservative South, environmentalists learned to be cautious of when, where, and with whom they shared their views. Best I could tell, I was the only kid on the planet besides my brother who cared about nature. This was before the internet, and I had no way to connect with other youth with similar views and interests. As a teenager, I became aware that my personal philosophy was antithetical to the superficiality and profligate consumerism that was culturally dominant in the 1980s.

Not until I enrolled at Eckerd College in Saint Petersburg, Florida, did I finally meet others—including my future wife, Ginger—who shared a fascination with nature, a love of science, and a conservation ethic. These friends were from other regions of the United States and other countries where environmentalism was surviving. During college I pursued experiences that set me further on the path to becoming a biologist. I spent three months in Antarctica as a science intern through a program sponsored by the Boy Scouts of America and the National Science Foundation. My contributions to the research were minimal, but the scientists I met were inspired, dedicated, brilliant, and fun-loving people, and what they were doing was important. I also saw that field science was physically and mentally challenging, and sometimes downright dangerous. All of this was appealing to my nineteen-year-old sensibilities.

When I returned to college in the spring, I immediately declared biology as my new major. Over the next few years, I worked with several of Eckerd's biology faculty. I studied coastal mangrove swamps in Costa Rica and captured and tagged sea turtles in Panama. I fell completely in love with the tropics. The forests were amazing, the cultures were fascinating, and the birding was extraordinary. Witnessing rampant tropical deforestation in Central America convinced me to study tropical conservation biology. After graduation, I studied strategies for restoring tropical forests in Uganda under the guidance of Colin Chapman at the University of Florida, while Ginger went to medical school. Several years later my wife and I headed to Birmingham with four advanced degrees and a two-year-old daughter.

It's been a half century now, and my conservation ethic is as strong as ever. I derive intense joy from seeking, watching, and interacting with other species. Sometimes this is purely for entertainment, but at other times, observing nature is my version of a spiritual experience. For me, to observe the seemingly infinite complexity of nature is to witness what must surely be one of the most extraordinary marvels of the entire universe. My understanding of biological evolution has enhanced my appreciation of species. While the lives of individual creatures come and go, a species is like a living work of art, shaped by tens of millions of years of evolutionary brushstrokes and always being updated. When such a living masterpiece is lost to extinction, it's gone forever. We can preserve glimpses of it via descriptions, pictures, videos, and museum specimens, but all are hollow surrogates compared to the real thing. Evolution of new species is exceedingly rare on the human time scale, and due to the vagaries of evolution, a new species can never truly replace one that's been lost. Finally, I value biodiversity because it is proving to be scarce in a universe that beyond Earth's thin biosphere is vast, cold, and desolate. As Edward O. Wilson warned us, a future without biodiversity would be an age of loneliness for humanity.[1] For all these reasons, I believe species are too precious to be extinguished out of inconvenience, ignorance, or greed. And at the very least, since they've nowhere else to go, sharing Earth with other species is the right thing to do.

I know my philosophical musings about the value of biodiversity don't have mass appeal. A few share similar beliefs, but most find them poorly suited for their worldviews. I'm okay with that. I teach about biodiversity to my college students, and we often discuss different ways of valuing species. Some students have a conservation ethic derived from a love of the outdoors, especially from a heritage of hunting and fishing. Others value nature through their religion. I value all these approaches. Humanity is too culturally diverse for there to be a one-size-fits-all reason to value biodiversity. Besides, an extinction crisis is no time to fuss with your allies about ideological flavors.

It is practical to protect biodiversity to preserve ecosystem services, especially given how essential they are to humanity's capacity to survive and prosper. This is an expedient line of reasoning I use at every opportunity. But ultimately, this argument alone will not be enough. As I mentioned previously, most ecosystems in the world have already lost many species to extinction, yet we enjoy valuable services from them every day. Some might logically argue that losing a few more species is not a big deal. Thus, if we are to share the planet in perpetuity with the ten million or more species also living here on Earth, we will need other compelling incentives for keeping biodiversity around, especially when we are tempted to destroy nature for short-term gain. There's value in viewing other species as tools. There's virtue in respecting them as neighbors.

5
Three Questions

My introduction to river ecology and conservation began immediately after my arrival in Birmingham. In 2002, while my wife and I were still unpacking, my brother, Will, arrived to help us move in. Will is an alumnus of the biology program at Birmingham-Southern College, the same department in which I was to begin teaching in a few weeks. One muggy July afternoon he and I headed down to the nearby Cahaba River for a break. Will had spoken often of the river. It was in the Cahaba where he got hooked on river biology, under the tutelage of professor Dan Holliman and a couple other notable river scientists in the area. After his graduation, and while I had been wandering through the tropical forests of East Africa, Will had been studying southeastern river ecology. At the time of his visit, he was a graduate student at the University of Georgia.

After picking our way along a neglected trail, we emerged at the river's edge. I was immediately taken with its beauty. The river was bounded by steep hillsides ringing with birdsong and cloaked with Appalachian broadleaf forest as lush as any I had seen in Africa. The river was more of a creek at this position, high in its headwaters. The water was clear and sparkled with the flash of minnows. There were gurgling shoals, swift runs, and deep pools. Trees leaned in to block the hot summer sun. It was an ecological landscape very different from the placid rivers of Florida, and profoundly different from any ecosystem I'd seen before. I was smitten. It was an instant river crush.

Will schooled me on how to read a stream as we walked the riverbank and waded in the shallows. Different habitats, and species to expect in each. Where to find mayflies and stone flies, and how to tell the difference between them. Over there is where the bass and sunfish hang out, but here is where you will find the minnows. This is a cut bank, and that is a gravel bar, and this is why they are symptoms of too much urban stormwater. On and on it went. By the time we headed home, I was beginning to comprehend how significant southeastern streams are to biodiversity conservation, and how fragile they were. The depth of Will's knowledge also impressed me that day, and he and I still laugh about how it took time for me to adjust to my little brother as my teacher. Will later earned a doctorate in ecology studying streams under Judy Meyer and is now a biologist and project administrator with the US Fish and Wildlife Service. He's still the first person I reach out to when I need help understanding rivers.

Within a few years of moving to Birmingham, I began working with other biologists to study the ecology of local streams, and some of their endangered species. In particular, my students and I have helped study two small, endangered fishes whose entire range has been enveloped by the greater Birmingham metropolitan area—the Watercress Darter and the Vermilion Darter. Because darters are so sensitive to stream degradation, many are classified as endangered (nearing extinction) or threatened (nearing endangerment) by the US Fish and Wildlife Service.

Within a year of my moving to Birmingham, Will introduced me to Randy Haddock, chief scientist for the Cahaba River Society. The society is a nonprofit whose mission is to protect and restore the Cahaba River. Over the next few years, Randy mentored me on stream ecology and conservation, recommending research papers and books, showing me the best places to take students, and teaching me to find and identify different stream species. Randy also spent hours explaining to me and my students such topics as sediment flow in streams, or the ways rivers change seasonally. In 2006, after my research program was up and running, I was asked by Beth Stewart, executive director of the Cahaba River Society, to serve on its board of directors. The next seven years were eye opening for me as I was led by Beth and Randy on a deep dive into the national and local water policies affecting drinking water and biodiversity.

Organizations like the Cahaba River Society exist because we didn't fix enough environmental problems with the passage of environmental policy reforms in the 1970s. These laws did reverse and prevent many natural resource catastrophes over the subsequent half century, and the country's population and its landscapes are healthier and safer as a result. However, these laws have significant problems with their scope and enforcement. Water quality in many rivers improved for a while, but then progress halted. Other rivers have seen little improvement, and a few are in worse shape than before the reforms. A handful of imperiled species have been rescued from extinction, but the roster of endangered species grows longer by the year. In sum, the ecological river revival of the 1970s stalled, and one response has been the proliferation of advocacy groups like the Cahaba River Society. River Network, a national organization dedicated to helping groups like the society with resources and networking opportunities had nearly 6,600 member organizations across the United States as of mid-2020.

Meanwhile, daunting new threats to people and biodiversity are emerging across the country due to the changing climate. A multidecadal drought in the western United States is intensifying water shortages and wildfires and threatening human health. Melting permafrost in Alaska is speeding erosion, undermining infrastructure, and releasing methane, a potent greenhouse gas. Heavy rainstorms have increased in the Northeast. In the Midwest rising temperatures are undermining agriculture, and greater precipitation is causing more flooding. Nature is rewriting the rules, and our playbook is out of date.

As for the Southeast, it is at more risk from climate change than any other region in the United States, according to a 2017 study published by a team of scientists, economists, and policy experts.[1] They forecasted the economic impacts of climate change if global carbon dioxide levels continue to rise at their present rate. To the surprise of many, they found that the South can expect the greatest economic losses. Higher temperatures will create dangerously hot summers, intensify droughts, and boost energy demand. Storms, including hurricanes, will produce stronger winds and drop more rain. Along the coast, sea level rise will cause flooding and land loss. And if all that weren't enough, freshwater availability for the region will decline.

These are ominous predictions. Not just for those of us who live in the South, but for the entire United States, given how important the region is to the national economy. The prediction about declining water supply is particularly troubling for anyone concerned about southeastern freshwater ecosystems. Because rivers are the region's main source of fresh water and are essential to our economy and environmental security, they are destined to take center stage in the Southeast in the coming decades. If we want a secure and sustainable way of living in the Anthropocene—the new geological age in which humanity is the dominant force shaping Earth's environments and systems—we must learn how to manage our rivers more wisely. We must finish the river revival that began in the 1970s, while also preparing for the uncertainty and new challenges the Anthropocene is bringing. We can do this and do it well—and the sooner the better.

When I learned how much climate change will influence the Southeast, three questions arose for me that will shape the rest of our journey to understand the southeastern river crisis. First, *how did we get here?* Despite my having taught environmental science and river ecology to scores of students over the years, there is much I still don't understand, and I know this information can help us find solutions to the problems the region faces. For example, what were rivers like before they were industrialized? How did our pollution problems arise? Why are our regulations not working? And why are so many species endangered?

The second question is vital to the future prosperity of the Southeast. *Will we have enough water?* Predictions are that water supplies in the region will decline in the coming decades due to climate change. How can this be true when the region receives so much rainfall? What can we do to ensure that we have enough clean drinking water and water for other uses such as power production, industry, navigation, and irrigation? What water will be left in rivers for other, less essential uses? Rivers occupy a treasured place in southeastern culture, inspiring art, music, and literature. What will become of this if our rivers lose their appeal? And what about river recreation? What will happen to swimming holes and rope swings; creek wading and skipping stones; duck blinds and fly fishing; tubing, paddling, and whitewater thrills; bass boats on the river at dawn; and fish fries?

Our third question is *can southeastern river biodiversity survive the Anthropocene?* Even without the prospects of climate change, the ecological forecast for the region is grim. The human population is growing, and demand for natural resources—especially water and undeveloped land—is rising. A calamitous future caused by climate change and a clumsy adjustment to it could cause the extinction of the unique species that give southeastern rivers so much of their delightful character. We must avoid a bleak future of empty rivers.

To be clear, I am not worried about common, widespread species. Largemouth Bass, Bluegill, and Channel Catfish—these adaptable species are generalists. They survive in a broad range of habitats and are doing well in the Anthropocene. They are the ecological workhorses we can count on to help ecosystems function at a baseline level, even when environmental conditions have deteriorated. Instead, the species we are at risk of losing are our rare and endemic species: Alligator Gar, Alabama Lampmussel, Anthony's Riversnail, plus hundreds more known only to experts. Some of these species—the specialists—have precise ecological needs and are now rare in the industrialized landscape. A few have been overfished or overhunted. Many are naturally rare or have lost ground to agriculture and urban sprawl. Collectively, they transform an ordinary ecosystem into one bursting with local flavor and discovery. Without this biodiversity a creek is just a ditch, a forest is just a stand of trees, and a meadow is just a lawn.

As resource scarcity, pollution, and climate change become greater threats to our lives, we must avoid a future where we are forced to choose between using resources to sustain biodiversity or sustain ourselves. What would the priority be on the eve of Birmingham, Atlanta, or Charlotte running out of water? Would we spare local fishes and mussels from extirpation by leaving water in the river and beginning to ration water? Not likely. In a situation like that there's little hope that biodiversity would win, until there's a shift in our mindset about how to value other species. Conservation history is littered with the memory of species that stood in the way of our convenience, need, and greed.

Our next step in the journey is to answer that first question: *how did we get here?* How did we get to the point where we intentionally pollute with mud, manure, and chemicals the very rivers from which we extract our drinking water? The situation seems absurd. If we are going to finish the river revival and prepare for challenges of the Anthropocene, then we need to understand how we created these problems.

PART II

6

Uncle Dallas and the Sturgeon

It's a chilly morning in March 1924 on Pensacola Bay. My great-great-uncle Dallas, born in 1876, is rowing along the north shore of Fair Point. The last winds of a cold front have subsided, and the bay is glassy. Dallas has rowed this route hundreds of times. His destination is just beyond The Point, where the yellow waters of the sand spit yield to the bay's depths. It's one of the area's best fishing spots. Migrating schools of Striped Mullet, Red Drum, and Speckled Trout often pass by. Dallas set several long nets there yesterday. Buoyed with floats and with a weighted baseline, their grid-like mesh hangs like a semi-transparent wall. In the winter Dallas fishes with smaller nets for mullet. But not today. With a twelve-inch mesh woven with sturdy cotton line, Dallas's nets are made for bigger fish. Dallas Duncan is fishing for sturgeon.

Dallas glances northward across three miles of open water to the familiar sights of Pensacola. A pall of smoke hangs low over the city—the product of heating and cooking fires. The docks are busy with the typical medley of vessels. Small snapper boats and large three-masted schooners are at anchor awaiting wind. Several modern steel-hulled cargo ships powered by coal-fed steam engines are off-loading tropical hardwoods or taking aboard coal or lumber. A tugboat helps steer a large oceanic steamer sporting tall masts.[1] Not long ago, the port was far busier, with the export of naval stores and lumber from the longleaf forests, but most of the pine in the area has been cut.[2] Commercial fishing for Red Snapper has declined too, also due to overharvest.[3]

After twenty minutes Dallas reaches his line of nets. He stows the oars, rolls up his sleeves, and grabs the float line at one end. The bay's waters are dark, so Dallas pulls the mesh up, feeling the net tension and looking for fishes, tangles, and debris. Gill nets do not discriminate. They capture anything big enough to get entangled. Dallas keeps a club handy for captured sharks that haven't yet drowned, and pliers to snap off the venomous barbs that ensnare rays. Cutting the net is always a last resort. He'll dump any dead bycatch farther out in the bay, since carcasses would attract sharks.

After a few minutes, Dallas feels the mesh tighten. It tugs, and tugs again—something big is down there, and it's still alive. Hauling the net upward, he soon sees the brown, knobby back of a Gulf Sturgeon tangled and askew. It's a big one, just over six feet long. A fish this size is strong, but the sturgeon is exhausted from fighting all night and after a few panicked thrusts it gives up. With a practiced motion Dallas hauls the massive fish over the gunwale and onto the floor of the boat. Working from the outer strands inward, he gradually removes the net from around the sturgeon's massive head and fins. The net removed, the exhausted sturgeon rests quietly on the bottom of the boat. Dallas throws a wet tarp over it to keep the creature cool and alive until he's back at the house.

Having netted hundreds of sturgeon before, Dallas finds the fish to be an unremarkable sight. But most of us would find it odd looking and unlike any creature we've ever seen. Sturgeon have a distinctly primitive look because their back and flanks are lined with knobby, bony plates, called scutes. Their bodies are long and tapered, and broadest behind the head. They have a fleshy, upturned snout dotted with tiny electroreceptors helping them sense their way through dark waters. A flattened belly helps them cruise just above the bottom. The mouth, set back from the snout, sports four whisker-like barbels that lightly drag along the bottom. Through taste and touch, the barbels guide sturgeon to the burrows of shrimp and worms. The mouth is protrusible so it can extend over a burrow entrance and slurp up hapless prey.

There are about twenty-five species of sturgeon in the world, and all live in the northern hemisphere.[4] Eight species dwell in North America. Some species can grow to massive proportions. The Beluga Sturgeon of Europe can grow nearly twenty-six feet long and weigh 1.4 US tons.[5] Dallas's Gulf Sturgeon is a subspecies of the Atlantic Sturgeon, which is restricted to the Gulf of Mexico and the Atlantic Coast of the United States and Canada. Both subspecies can weigh hundreds

Millions of the primitive-looking Gulf Sturgeon (*pictured*) and the Atlantic Sturgeon once migrated up southeastern rivers to spawn in headwater streams. Today these sturgeon are among the region's most imperiled fishes. Photo by Ryan Hagerty, US Fish and Wildlife Service.

of pounds. While Atlantic Sturgeon are larger on average, maximum lengths of both subspecies range from fourteen to fifteen feet.[6]

Dallas might find this information interesting at another time, but not now. He's estimating how much the fish might bring at the docks. Though males and females are nearly indistinguishable, Dallas suspects this sturgeon is a female ripe with roe because of her girth. Eggs can make up to 25 percent of a female's mass when she's ready to spawn. Dallas guesses she's nearly ninety-five pounds, and will sell for about nine dollars, more if she's loaded with roe. If she's carrying eggs, she was heading for one of the bay's rivers. The Gulf and Atlantic Sturgeon are anadromous, meaning they migrate from marine waters into fresh water to spawn.

Ripe female sturgeon are becoming harder for Dallas to catch. Most of the sturgeon he nets in Pensacola Bay these days are small, averaging about fifty pounds. When he began fishing the bay, the average sturgeon he'd catch was twice that size.[7]

By the end of the morning, Dallas has one more sturgeon, a four-footer. It's small, but he'll keep it. It will fetch about two to three dollars —money he could use. After one last glance to make sure the nets are sitting well, Dallas rows homeward. He'll tie a rope around the tail of each fish and tie the other end to a stake driven deep into the sand in the shallows in front of the house. This will keep the sturgeon alive while he runs the nets for a few more days. Then he'll load the captives into the *Trilby*, the family's sloop, and sail them across the bay to the fishmongers at the docks.

Gulf and Atlantic Sturgeon were abundant when the first Europeans arrived in the Southeast. When the Duncans settled on Fair Point in the late nineteenth century, the vast sturgeon populations supported an industry employing thousands. Today, over a century later, both subspecies are facing extinction. Most who live in the Southeast—even avid fishers—have never seen one. Despite having spent thousands of hours on Pensacola Bay, I've glimpsed a Gulf Sturgeon only once, and for just a split second, when it leaped from the depths. How could a widespread creature become so endangered? Was overfishing by my Uncle Dallas and others like him to blame?

As anadromous fish, Gulf and Atlantic Sturgeon know the terrain of the region's rivers, estuaries, and nearby ocean. They've also seen a lot of history. Fossils date back tens of millions of years, and archaeologists find their bony scutes in middens of ancient settlements. And because they were commercially valuable in the modern era, sturgeon have left behind a paper trail of landing data that biologists have used to chart the rapid rise and fall of one of America's most profitable fisheries.

Much of the reason for the sturgeon's grave situation is that females produce copious amounts of eggs for spawning. Like other bony fishes, sturgeon have external fertilization. Males and females expel sex cells (sperm and eggs) while swimming close to each other. An egg then awaits a swimming sperm to find and fertilize it, and this initiates the growth of a young sturgeon. Young sturgeon would not last long in the predator-rich environment of an estuary, so ripe sturgeon migrate into rivers to spawn. The migration occurs in early spring, when river waters are high, cool, and well oxygenated, all of which are good conditions for big fish navigating up small rivers.

Along the way, a male must find a fertile female and trail her in the dark river waters until they reach the right habitat for spawning. It's not known who makes the first move, but the ritual leading up to spawning involves a male rubbing his snout against the female's flanks to stimulate her release of eggs. These antics get animated, and both sexes show significant abrasions afterward.[8] After spawning, most adult sturgeon remain in the river to rest, then migrate back to the estuary in autumn.

A large female Gulf or Atlantic Sturgeon can produce over a million eggs in a season. The goal of this is to overwhelm predators of eggs and freshly hatched sturgeon, even though there are fewer predators in freshwater streams than saltwater environments. Overproduction of offspring ensures that at least a few young survive. However, such heavy reproductive investments in egg production come at a cost. Females need about three to four years of reconditioning and growth before they can spawn again. It's a strenuous cycle, but one that has worked for anadromous sturgeon throughout the world for millions of years.

The voluminous production of eggs as an insurance strategy became a liability in recent centuries as humans began dominating the waterways of the world. Salt-cured sturgeon eggs are known as caviar. It's considered a delicacy and is priced accordingly. Because a large ripe female can yield several pounds of eggs at a time, sturgeon have been sought by fishers ever since caviar consumption arose in ancient Greece.[9]

The behavior of Atlantic and Gulf sturgeon during the spawning season makes them easy to catch. They form loose schools, squeeze together into narrow rivers, and show up on schedule each spring. It's no wonder that Native American communities across the Southeast looked forward to the sturgeon runs every spring.

In contrast, sturgeon were rarely eaten by early European colonists or the first Americans, as journalist and sturgeon historian Inga Saffron explains. She suggests the fish's primitive looks, bottom-feeding lifestyle, and oily flesh repulsed them. In the eighteenth century, several businesses in New Jersey exported salted sturgeon meat to Caribbean plantation owners, who fed it to enslaved people tending the sugarcane fields. By the mid-nineteenth century, fishmongers were selling smoked sturgeon in New York to impoverished immigrants, who had less finicky palates than did immigrants of British ancestry.

The import of sturgeon stimulated other markets. Some boiled heads to extract lamp oil or sold scraps to hog farmers. Others extracted isinglass from swim bladders and sold it for binding, jellification, and clarifying beer. Despite these enterprises, fishing pressure on the Atlantic Sturgeon was light, and probably sustainable.[10] But this was not to last. Sturgeon were too abundant and exploitable for industry to overlook them forever.

The grace period for sturgeon came to an end in 1870, when Bendix Blohm, a German immigrant, discovered a curing method for sturgeon eggs. Word of Blohm's impressive financial success at canning and selling American caviar spread rapidly. Within a few years, the wealthy of the northeastern United States and Europe were buying vast amounts of caviar. Wall upon wall of net greeted migrating sturgeon headed into northeastern rivers to spawn. In 1890, the year of peak harvest, the industry produced 3,527 US tons of sturgeon products.[11]

The Atlantic Sturgeon's life history strategy made it impossible for them to survive intense fishing pressure. Like most long-lived, large-bodied animals, females do not reach sexual maturity until late in life. This is about age ten in the Southeast and age sixteen to twenty-eight farther north, where waters are colder. Thus, the population couldn't replace a captured female for one to three decades. The female's infrequent reproduction also hampered population resiliency. But the feature of their biology that made them most vulnerable to intense fishing pressure was that male and female Gulf and Atlantic Sturgeon only spawn in the rivers in which they hatched. Due to this loyalty to natal rivers, intensive fishing in a river or estuary decimated populations within a few short years, with there being no hope of recruiting from nearby populations.

For these reasons, and the absence of harvest regulations, the sturgeon boom was short lived. Sturgeon landings in the Northeast peaked first. Then as yields declined, fishers shifted to rivers farther south—first Virginia, then the Carolinas, then Georgia and Florida. As landings waned, fishing pressure intensified. This is a common and perverse outcome of unregulated markets because scarcity drives up both desirability and price of the product. In river after river, the declines continued until there were so few sturgeon left that fishing was unprofitable. In 1910, just twenty years after peak US harvest, landings of Atlantic Sturgeon declined to 331 US tons. By 1920, it was a mere 28 US tons.[12] Atlantic Sturgeon were facing extinction after just fifty years of fishing.

Commercial fisheries for the Gulf Sturgeon began a bit later and never approached the levels of harvest of the East Coast. The first commercial sturgeon operation began in 1887 in Tampa Bay, but fishers depleted the population after just three years. A more enduring fishery began in 1896 at the mouth of the Suwannee River, in the Big Bend region of Florida. During the first full year of harvest, fishers landed four and a half tons of sturgeon.[13] Commercial fishing then spread across the Florida Panhandle and all the way to the Pearl River, on the border between Mississippi and Louisiana.[14]

Dallas Duncan would have fished for sturgeon in Pensacola Bay in the period between 1901 and 1928.[15] During this time, sturgeon landings in Pensacola peaked and then plummeted as they did everywhere. Dallas's son Finley was born in 1914 and spent his boyhood on Fair Point playing with my grandfather George Addison Duncan (born 1910), and helping his father fish. Late in life, a historian interviewed Finley about fishing with his father. Back then, they considered fifteen to twenty sturgeon to be a good week's haul.[16] In 1928 Dallas quit fishing in Pensacola Bay because the sturgeon had become scarce. He moved his family east to fish in Choctawhatchee Bay, which still had a sizable population. Finley didn't become a professional fisher. It's a tough life, and the income isn't steady. And by the time he was an adult, there were few Gulf Sturgeon left to catch.

Overfishing is clearly to blame for the precipitous decline of the Gulf and Atlantic Sturgeon, and aspects of sturgeon life history hinder their recovery. But some sturgeon experts believe that overfishing isn't fully responsible for their current endangerment. At least a few sturgeon survived in most rivers of the Gulf and Atlantic Coasts. These small populations have had eighty to one hundred years to recover since the fisheries for them collapsed. That's more than enough time to rebound, yet both subspecies are imperiled.

If overfishing is only an accomplice, what else could be causing sturgeon endangerment? We can rule out most of the usual suspects that cause species to decline. Sturgeon were not a naturally rare species, otherwise commercial fishing would never have been possible. Nor can we blame a sensitivity to environmental change. Fossil sturgeon date to over three hundred million years ago.[17] The oldest fossil from North America dates to between seventy-two and eighty-six million years ago, during the dinosaur heyday.[18] Sturgeon have endured climate swings and environmental catastrophes that would have pummeled modern civilization back to the Stone Age.

We can also dismiss limited fertility as a suspect for perpetuating endangerment. A pair of spawning sturgeon produce enough offspring to restore an entire population, were they all to survive. A narrow geographic range can be a problem for species survival, but fishers netted sturgeon in rivers from Texas to Tampa, Florida, and from North Florida to Labrador in northern Canada. Do sturgeon have ecological preferences that are too narrow to allow recovery? Nope. The sturgeon's broad diet allows them to feed wherever they wander in marine waters. Plus, their ability to inhabit rivers, estuaries, and the ocean demonstrates an impressive range of tolerances.

So, we have a mystery to solve: why are the Gulf and Atlantic Sturgeon still endangered? Perhaps we will find more clues to what happened as the sturgeon take us on a tour of the era before the European conquest of North America.

7
Ancient Witnesses

Sturgeon have witnessed spectacular changes to the Southeast and its rivers. They endured the Cretaceous, a time when crocodiles and gharials dozens of feet in length hunted them. Sturgeon survived a planetwide catastrophe sixty-five million years ago, when a six-mile-wide asteroid struck the Yucatan, wiped out most life in the Southeast, and then darkened Earth's skies long enough to cause the extinction of dinosaurs. Sturgeon watched from the depths as ecosystems recovered and were repopulated by mammals and birds. More recently, sturgeon endured Earth's plunge into the frigid climate of the Pleistocene and a series of ice ages lasting 2.5 million years. During this time, sea level repeatedly dropped to four hundred feet below its present position, and long, bitterly cold winters and short summers were the norm for millennia.[1]

About fourteen thousand years ago, near the end of the Pleistocene's final ice age, sturgeon witnessed the arrival of the Paleo-Indians to the Southeast. This moment forever altered the algorithms governing species survival in the region. Paleo-Indians were skilled hunters. Whether they hunted sturgeon is unknown, but if they did, the fish is fortunate to have survived. The new residents had mastered oral communication, cooperation, abstract thinking, teaching, and rapid learning, and they created tools faster than prey could evolve countermeasures. North American wildlife had no experience with predators such as these. During the final millennia of the Pleistocene, many large animals across the continent and in the Southeast—including mammoths, mastodons, ground sloths, and other megafauna—went extinct. While a slowly warming climate may have played a minor role, most evidence suggests overhunting did them in. Gulf and Atlantic Sturgeon are some of the only megafaunal animals that survived.

Rivers were a focal point of activity from the first moments of human presence in the Southeast. Archaeologists have discovered most Paleo-Indian artifacts in or near rivers, especially along the Tennessee, Cumberland, and Savannah, and the rivers of the Big Bend region of Florida. These nomadic hunters favored encampments by rivers. Game trails across shoals would have been natural ambush sites, and large rivers would have helped corral prey on game drives.

By about 11,450 BP (before the present), Native American culture shifted into a new cultural phase in the Southeast—the Archaic period. The declines in megafaunal populations helped force the transition. The Archaic people were less nomadic and exploited local resources in evermore sophisticated ways.[2] Archaeologists have recovered canoes, net sinkers, and fishhooks from seasonal river encampments. They have found evidence of small-scale cultivation in floodplains. Archaic people built weirs of rocks and logs in rivers to corral migrating fishes. Mounds of discarded shells along riverbanks reveal that they consumed great quantities of mussels.[3] Riverside trails and the rivers themselves connected societies separated by thousands of miles.

Rivers were also essential to Native American life during the next cultural phase, the Woodland period (3200–1000 BP). Permanent villages, pottery, elaborate burial sites, and small-scale farming mark the beginning of this period. Creeks and rivers were the focus of daily life and seasonal rhythms as before. In the Late Woodland period, there was a decline in ceremonial life and a shift to hunting smaller game. Many archaeologists believe Late Woodland populations had exceeded their carrying capacity—the maximum population size that the environment can sustain with current technologies and cultural practices. This shift may have been triggered by the Little Ice Age, a prolonged period of cooler temperatures that reduced ecological and agricultural productivity.[4] But instead of regressing into a lifeway resembling the Archaic period, Woodland culture was rescued by a grass from Mexico.

Maize had arrived via trade routes hundreds of years earlier. Woodland people often planted it alongside squashes, beans, and sunflowers. Archaeologists suspect that when food insecurity reached a threshold, Late Woodland villages cleared large tracts of forest for maize

cultivation. The shift to an agrarian lifestyle was rapid where it occurred. People still cultivated other foods and harvested from the wild, but maize was the dietary mainstay.[5] This transformation initiated the Mississippian cultural period (1000–1600 CE).

Rivers were indispensable to Mississippian culture. Farmers planted on floodplains, where millennia of seasonal floods had built up thick layers of fertile soil. They cleared forest for miles in every direction for planting. A hierarchical ruling class emerged due to the need for organized plantings and communal grain storage.[6] The elite lived atop tall, flat-topped mounds built with surplus labor by subjects and enslaved people. Villages coalesced into chiefdoms, some of which were regionally united under a centralized power held together with alliances and allegiances. Rivalries and wars were common.

Native Americans constructed weirs of rock, logs, and branches to corral and capture migrating fish such as shad and sturgeon. The remains of these weirs, such as this one on the Etowah River in Georgia, survive in some southeastern rivers. Courtesy of Alan Cressler.

Despite the high culture, social stratification, and emergence of big agriculture, people still harvested from the river. Fishes, including sturgeon, and river mussels supplied protein that maize could not. Mussel shells were used as spoons, bowls, and hoes. Artisans used ground mussel shell to temper pottery before firing. Mississippian people valued pearls found in mussels and used them to make jewelry and decorate clothing.

As described by early European colonists, Native Americans of this period fished for sturgeon in a wide variety of ways using practices developed millennia earlier. In the river shallows fishers would ambush migrating sturgeon with nets, spears, harpoons, and clubs.[7] Some caught sturgeon by slipping a noose around the base of the tail, then pulling the fish to the bank after much struggle. Others trapped them with elaborate weirs of rock and wood spanning the river. Having learned that sturgeon are attracted to light at night, some practiced torch fishing. In the dark of night, one fisher guided the canoe, another held a torch over the water, and a third stood ready with a harpoon. Migrating sturgeon were so abundant that fishers dried or smoked surplus meat to preserve it. They extracted oil from carcasses and used it for torches and cooking.

Though the Southeast's first peoples used rivers for fourteen thousand years, sturgeon survived just fine. They are hardy, resilient animals that endured millions of years of predation, harsh environmental change, and considerable hunting pressure in recent millennia. Thus, what keeps the sturgeon endangered today must be a new threat, something arising since the arrival of Europeans to the southeastern shore.

8
Industrialization

THE FIVE CENTURIES SINCE THE arrival of Europeans to North America account for less than 4 percent of human history in the Southeast. Yet, as sturgeon can attest, during this time the region has undergone its most radical ecological transformation since that asteroid struck the Yucatan sixty-five million years ago and nearly extinguished all terrestrial life in this hemisphere. How has this transformation affected southeastern rivers, and what aspect of it can account for the sturgeon's ongoing endangerment?

Even though Native Americans had actively managed the Southeast for millennia, Europeans took land management to a whole new level. Historians have offered us many perspectives on the contrast and conflict between Native American and European cultures. Suffice it here to say that Europeans—arriving with a mindset and technology honed for exploitation—were decisively successful at imposing their will on the land and its people. The goal was commercial exploitation of resources. The strategy was the use and transformation of nature to build and protect economic security and wealth. Today, after centuries of this, we have thoroughly reshaped the mountains, rivers, and coastlines of the Southeast.

Colonization of the Southeast and other regions of the world by Europeans initiated a three-phase process of landscape industrialization: (1) claim the land and subjugate, expel, or eliminate its occupants; (2) harvest and export readily marketable native commodities (e.g., plants, animals, and minerals); and (3) create a landscape for continual production of income and wealth, chiefly through agriculture and mining. The process was not preplanned; it simply emerged as cultures and their institutions discovered ways to exploit the rest of the world to build wealth. Nor is this to say that the process is formulaic, where one phase must follow another. Instead, the industrialization process is fluid and, arguably, cyclic. There are many places in the Southeast and around the world where those with more power and privilege exploit those with less. And some still harvest native commodities for commercial sale, though most of the southeastern landscape is in agriculture.

In the Southeast, the Spanish commenced the first phase of dominating the Native peoples, and they did it with zeal. For a quarter century after Columbus's first trip to the New World, they probed the region, mapped coastlines, and attempted settlements. Skirmishes with Native Americans were frequent. To their credit, resistance by local tribes was often fierce, and contributed to the failure of several expeditions. But Spanish hunger for gold, glory, and evangelization was fierce. In 1539 an expedition into the southeastern interior was launched that changed everything. In command was Hernando de Soto, a tried-and-true conquistador who had cut his teeth as a brutal tactician when subduing the Native peoples of Nicaragua and Peru. De Soto's mission was to secure Spain's claim to the region and acquire gold and silver by any means necessary.[1]

De Soto's entourage included over six hundred men, hundreds of warhorses, packhorses, and mules, and some surprisingly influential hogs. Beginning near Tampa, Florida, the army ranged as far north as North Carolina and Tennessee, and as far west as Arkansas and East Texas. The men spent most of their time among riverside Mississippian chiefdoms, where wealth and power were concentrated.[2]

De Soto and his men were vicious. They arrived at new villages with kidnapped and mutilated prisoners to intimidate chiefs into providing supplies and valuables. Armed resistance was frequent, but wood, bone, and stone weapons were no match for Spanish armor and weaponry, including swords, lances, crossbows, muskets, cannons, and armor-wearing attack dogs. The Spanish often enslaved the conquered. Resistors were captured and mutilated, often by amputating noses, chins, and hands. Soldiers enslaved young women for sexual exploitation, but not before priests baptized them to protect the souls of the rapists.[3]

De Soto died of illness after four years of marauding. Fearing his death would embolden the Native populations, the survivors slipped his weighted body into the Mississippi River under the cover of darkness. Warfare and disease had halved the expedition's ranks, and much of their weaponry and supplies were gone. Skirmishes with enraged and increasingly emboldened Native people were frequent, and the Southeast lacked the gold and silver the Spanish sought. The expedition's survivors left the region, eventually escaping to the Spanish stronghold of Mexico City.

Though the expedition produced no silver or gold, and the dead leader was fed to the fish, De Soto's misadventure was wildly successful in two unfortunate ways. First, American culture has glorified De Soto for centuries, a fate he would have loved. This veneration persists even though modern historians established his brutality decades ago. In *The Gulf: The Making of an American Sea*, historian Jack E. Davis inventoried some of what has been named in De Soto's honor.[4] The list includes schools, city parks, national forests, national parks, streets, a car, a fort, and a naval vessel. Alas, I must add one more to the list. It's our very own Gulf Sturgeon. Its full name is *Acipenser oxyrhinchus desotoi*, the last part of which is the subspecies name and a nod to the brute.

De Soto's expedition was also successful because it paved the way for future invasions and settlements by Europeans in the Southeast. They didn't achieve this with superior Spanish weaponry. Instead, they spread infection. Native American immune systems were not prepared for the diseases unwittingly unleashed on them by the Spanish, including smallpox, influenza, and measles.[5] This is why hogs play a surprising role in this story. Swine carry many human diseases, and many escaped into the wild during the expedition. Long after De Soto's genocidal rampage ended, hog hunting by Native Americans led to the spread of diseases throughout tribes. Millions of Native people died in the following decades, and the Mississippian culture collapsed without a large labor force to grow maize. This effectively softened up the region for the Europeans arriving later.

After De Soto, Spain maintained a fort at Saint Augustine, but otherwise lost interest in the Southeast. Though still tormented by disease outbreaks, Native American populations recovered somewhat during the reprieve, adopting lifeways like those of the Late Woodland period, but with more reliance on agriculture. They integrated new traditions with old ones. Bands scattered by the Mississippian collapse reorganized into powerful new tribes—Chickasaw, Creek, Choctaw, and Cherokee.

Had Spanish exploits elsewhere been similarly unsuccessful, the history of North and South America would have been quite different. But Spain had conquered and colonized much of the Caribbean, Mesoamerica, and South America, and the gold and silver shipped back brought the empire wealth and power. Furthermore, Spanish use of enslaved Africans made cultivation of sugarcane and other cash crops highly profitable. The English, Dutch, Portuguese, and French tried to emulate Spain's success, often with similar ruthlessness. Initial attempts by them to colonize the Southeast were unsuccessful, the French failing on Parris Island and the English failing at Roanoke. Starvation, hostile Native populations, disease, storms, and general ineptitude were to blame.

The 1607 establishment of the Jamestown colony by the English was a game changer for the Southeast. While the Spanish injected the region with plagues that transformed and decimated the inhabitants, the English brought the first cash crops, whose cultivation reshaped the landscape. Jamestown was purely a business venture to enrich financiers of the Virginia Company. Initially the goals of the Jamestowners were the same as those of the Spanish—find precious metals. The first 104 colonists were to erect a fort and gather the gold and silver they assumed was lying about the countryside. They also planned find the Northwest Passage—the fabled waterway to the Pacific that would expedite trade with China and bring great fortunes to those controlling its use. They built the town near the mouth of the James River, on acreage they assumed was unused by the locals.

The colony got off to a rocky start, nearly failing several times. Many of the first colonists were wealthy upper-class businessmen and house servants lacking useful skills such as farming, fishing, or carpentry. Many colonists spent too much time searching for gold, and not enough time growing food. The few crops planted failed due to a drought gripping the region. Malnutrition and disease took hold.

Health declined further because their well was tainted with salt water. Local tribes, unified under the powerful leader Powhatan, periodically brought food to the colonists to trade for axes, knives, glass beads, and copper. These exchanges likely saved the colony from complete failure. But by the end of 1607, two of every three colonists had died. The next year was just as dismal. Though new colonists arrived, and the newcomers possessed more appropriate skills, they brought insufficient food supplies to end the ongoing famine. By the next winter, the fragile peace with the Powhatan people deteriorated as colonists demanded more food than the chief was willing to spare.[6]

It was then the Atlantic Sturgeon saved the colony. The colonists had seen the Native people feasting on sturgeon, and the colony's famous chronicler, Captain John Smith, wrote that they had watched the Native people net sixty-eight sturgeon at once. When the sturgeon began migrating up the James River in spring 1609, colonists waded into the cold shallows, hoping to skewer migrating sturgeon with swords or bludgeon them with skillets.[7] This might seem ridiculous given how rare sturgeon are today, but historical accounts from Jamestown and later settlements report that throngs of sturgeon would fill Atlantic Coast rivers during migration. Fortunately for the colonists, skillets and swords were enough. According to Smith, they landed eighty-seven sturgeon, enough to even feed the dogs.[8] Archaeologists have recovered thousands of sturgeon scutes in a Jamestown midden dating back to the famine.[9] Sturgeon saved Jamestown.

The drought continued through summer 1609 and into the winter. During this time, colonists remained inside the stockade, fearing Powhatan's warriors. Fields were untended. Rivers were unfished. The winter of 1609–10 became known as the Starving Time. Some ate dogs, rats, or leather. There was at least one case of cannibalism.[10] Nearly three of every four colonists perished before reinforcements and supplies arrived the following May.[11]

The fate of the southeastern landscape and its rivers changed forever when, in 1612, colonist John Rolfe successfully cultivated specimens of a new strain of tobacco. By then, Jamestown had survived its most desperate period, but its long-term fate was in question. Colonists had little to show their financiers for their efforts. Nary a nugget of gold or silver had been found, and they had not seriously tried to find a route to China. Some Jamestowners attempted to ship caviar from sturgeon back to England, but the barrels of salted roe that arrived were rancid. Preoccupied with simply surviving, they didn't try again.

Tobacco was native to the New World and popular among Native Americans. The first European explorers had been bewildered at the sight of Native Americans—all genders and ages—inhaling its smoke, but they had quickly grown to appreciate the plant's stimulating and addictive qualities. Tobacco had become green gold for the Spanish, who delivered ships laden with it from Caribbean colonies to eager markets throughout Europe in the late sixteenth and early seventeenth centuries. Rolfe and other colonists discovered that Virginia soil—and importantly, English soil—could grow lots of tobacco. What's more, one of the Virginia cultivars was exceedingly popular, being much sweeter and stronger than Spanish tobacco. Jamestown colonists were soon growing and exporting millions of pounds of it to England and reaping the financial rewards. Investors financed more shipments of supplies and settlers to work the fields.[12] Upon the realization that this was a full-on invasion, resistance from the Powhatan people flared up for a time, but the crush of settlers and disease was too much. The colony had been rescued again, this time by tobacco.

Inspired by the success in Virginia, more colonists and investors came to the Southeast. The Carolina colony was founded in 1663, with Charles Town (near the confluence of the Cooper, Wando, and Ashley Rivers) established as its seat seven years later, and the colony of Georgia was initiated in 1733 at Savannah, on the banks of a river by the same name. After the Revolutionary War, the rest of the Southeast came under US rule in treaties with Spain in 1798 (Mississippi and Alabama) and 1821 (Florida). Whether the land was under English, British, or American rule, what ensued during the 350 years after Jamestown was a socio-ecological transformation of the Southeast involving the three phases of landscape industrialization (claiming, harvesting, and transforming the land).

Subjugation, killing, and eviction of Native Americans in the Southeast occurred for nearly 250 years after the British founded Jamestown. The history of this genocide is complex and involved maneuvers to pit tribes against one another, frequent wars, disease outbreaks, overt racism, and broken treaties. By the mid-1800s most surviving Native Americans in the Southeast were farmers of the Cherokee, Chickasaw, Choctaw, Muscogee (Creek), and Seminole tribes. Many had adopted cultural practices brought to the region by the populations of Euro-Americans and enslaved Africans. But neither cultural assimilation nor treaties of the day were enough to protect them and their land rights. After the passage of the Indian Removal Act in 1830, federal troops and local militias forced the five tribes of the Southeast to move hundreds of miles west to less valuable land in what is now Oklahoma. Only a few evaded relocation and remained. The vacated lands were then opened to new settlers and the next phases of industrialization.

From the beginning, colonists harvested en masse species that supplied marketable quantities of meat or other products. As technologies gradually improved, so did the speed of this second phase of landscape industrialization. Axes were replaced with crosscut saws. Muskets were exchanged for shotguns and rifles. Handlines, spears, and weirs were abandoned for enormous nets woven with factory-made line. Initially, only a few plants or animals could be gathered at once. With time, an entire school, flock, forest, or herd could be harvested in one go.

Hunters gunned down hundreds of thousands of colorful birds, such as the now-extinct Carolina Parakeet, and egrets and herons with elegant plumes for the feather trade. Fur trappers nearly exterminated beavers, and their disappearance led to the loss of the wetland meadow ecosystem. In the 1800s, the development of rail systems and refrigeration incentivized unchecked hunting of wild game because spoilage was no longer a problem and hungry markets were waiting in large coastal cities. Urban palates could enjoy wild game and seafood every night. For many southeastern species, the pressure was too much. Sturgeon and other valuable fish species became scarce. Commercial hunters completely eradicated the Passenger Pigeon, formerly the most abundant bird species on the continent. Gunners destroyed the southeastern elk and bison populations, and almost eradicated deer and turkey.

The commercial exploitation of wild plants and animals was a sideshow compared to the wholesale replacement of natural ecosystems with agriculture. Settlers converted the fertile soils of lowlands first. Poor immigrants arriving later farmed the steeper rocky uplands. Farmers drained or filled wetlands to create more cropland. Communities organized to suppress the seasonal wildfires that had sustained prairies and woodlands. Governments paid bounties for the skins of cougar, bobcat, bear, and wolf to protect livestock. By the early twentieth century, nearly every shred of virgin habitat in the Southeast was under cultivation or had been logged at least once. The Southeast became a land of row crops, timber plantations, and pasture, much like the landscape the European colonists had left behind. Goats, sheep, cattle, and hogs became the region's most common large animals. Chickens became its most abundant birds. The agricultural phase of industrial transformation led to the extinction of many terrestrial species and the eradication of entire ecosystems.

The greatest tragedy of this history is the human suffering it caused. This was not just the near extermination of Native Americans while taking their lands, but also the enslavement of Africans to farm the crops grown on those lands. Tobacco cultivation required intensive, back-breaking work, and landowners struggled to recruit labor to maintain and expand production. Initially, workers arrived as indentured servants, poor English who were unable to afford the voyage or the start-up costs of farming. But Europeans were unreliable because of diseases brought to the New World by early traders of enslaved Africans, including yellow fever and strains of malaria.[13] Some tobacco farmers used enslaved Native Americans, but this practice ended about 1715 because nearby Native populations had dwindled due to war, enslavement, and disease.

Many West Africans had acquired immunity to yellow fever, and full or partial genetic immunity to multiple forms of malaria. Though the mass importation of enslaved Africans to the Southeast for agricultural labor may have been inevitable, African resistance to these diseases encouraged the practice. The widespread use of enslaved Africans began in the Carolina colony in the 1670s and quickly spread to the other southern colonies.[14] Thus began several hundred years of one of the most wrenching and tumultuous chapters in all of human history, one that continues to shape the culture of the Southeast, the United States, and the world.

Rivers helped the transformation of the Southeast and were, themselves, industrialized in the process. River deltas provided the footholds for the first settlements, and estuaries initially sustained settlers with abundant food and sheltered their fleets. Rivers provided passage inland to explore the interior and trade with Native American villages. Floodplains became fertile farms. Riverside villages arose, and from them radiated road networks that enabled inland expansion.

The Fall Line, the geological boundary between the Coastal Plain and the Appalachian uplands, shaped the settlement of the interior. Below it, rivers were deeper and gently glided across soft sediments. Above the line, rivers were shallower, steeper, and rockier. Below the Fall Line nineteenth-century paddle steamers transported people and goods, but only small barges and flatboats could navigate the rocky shoals above the line, and only then during periods of high water. Fall Line positions on major rivers became hubs for dozens of major settlements in the Southeast due to the constant loading and unloading of passengers and goods. The rivers' momentum as they tumbled through the Fall Line was tapped to power mills and factories. Columbia arose on the Congaree, Macon on the Ocmulgee, Tuscaloosa on the Black Warrior.[15]

Rivers were reshaped as landscape industrialization continued. Logs and rocks were cleared from rivers to improve navigation. Headwater creeks were dammed to power mills for grinding grain and sawing timber. Longleaf pine woodlands and cypress swamps alongside rivers were logged, and log rafts were floated to downstream lumber mills. Rivers were dredged to deepen and straighten channels. Large, expensive canals with diversion dams and boat locks were dug to allow vessels to bypass falls and shoals, and to connect one river to other rivers and water bodies. Riverside villages became cities. Commercial fisheries developed for river and estuarine fishes. Bay oysters were shipped to restaurants of major cities. River water was extracted for public supplies, irrigation, mining, and manufacturing. Small dams were built to store water, improve navigation, and power cotton gins and factory machinery. And all the while, rivers were used to dispense with sewage and factory waste.[16]

The three hundred years of landscape industrialization—from the advent of tobacco farming in Jamestown to the dawn of the twentieth century—did considerable damage to river biodiversity and ecosystems. This history helps us understand the challenges the Atlantic and Gulf sturgeon and other river species face today, but it doesn't explain why sturgeon are still endangered. We know overfishing caused their decline, but sturgeon populations have had a century to recover after the fisheries collapsed. Because sturgeon were doing fine until after the peak years of sturgeon harvesting, we can narrow our focus to the period after about 1900. Industrialization of the rivers intensified during the twentieth century. Existing threats to biodiversity worsened, and new threats arose. Dozens of species were exterminated, and others became endangered. To understand the scope of how southeastern river ecosystems were transformed and the impact this had on people and aquatic biodiversity, we are going to set aside the sturgeon mystery for a while. We'll begin our exploration of modern river industrialization by investigating a puzzling event witnessed by two young girls late one night on the shores of Mobile Bay.

9
Jubilee

Around two o'clock on a hot, humid August night in 1950, two young girls are shaken awake by their excited mother and hastily stuffed—nightgowns and all—into their day clothes. Judy, age eight, is carried to the car by her father, Richard Rutland, and laid down in the back seat. Her sister, Lucy, age four, is carried to the car by their mother, another Lucy. The younger Lucy—my mother—curls up on the floorboard below her sister, a blanket beneath her and the driveshaft hump for a pillow. Everyone else in the Springhill neighborhood of Mobile, Alabama, is dead asleep. Lucy quickly tosses a few more supplies in the trunk, including snacks for the kids, while Richard cranks up the two-toned Mercury and tells her to hurry. He's impatient because the natural spectacle they are hoping to see won't last but a few hours.

As they motor through the empty streets, Lucy slips back to sleep as only a young child can do. Judy, excited but still groggy, watches the blur of streetlights and drooping live oaks against the dark sky. There's no traffic as Richard guides the car along Government Street. They duck into the Bankhead Tunnel, slip beneath the Mobile River, and emerge onto the US 90 causeway crossing the delta. The elder Lucy glances over her shoulder to check on the children. Richard worries they will arrive too late—the phone call that woke him was already an hour ago. After another thirty minutes they arrive at a park on the bay's eastern shore. Unlike the rest of South Alabama at half past two in the morning, the park is bustling with activity. They are not too late—the jubilee is still in full swing.

Clusters of people along the shore are talking excitedly. The young Rutland girls see silhouettes of people hauling buckets and bending over basins. Others are wading in the knee-deep shallows holding lanterns and flashlights. As the Rutlands approach the water, they see within the circles of illumination that the waters are crowded with animal life. Crabs, shrimp, eels, flounders, catfish, and stingrays are so numerous that they sometimes obscure the yellow sand beneath. The creatures are strangely lethargic, resting on the bottom or swimming lazily. People are swiping dip nets through the water and scooping up blue crabs, sometimes several at a time. A few well-prepared men are spearing flounders with gigs. The animals do not attempt to flee. Tubs, basins, and bushel baskets fill with the seafood and are stashed on the beach under wet burlap sacks.

Richard and his wife join the frenzy while the girls watch from the beach and enjoy the gentleness of the warm night air. After an hour or so, their buckets are full and the Rutlands head home. As they leave, a predawn breeze stirs the shoreline air. The glassy waters then ripple with light wind, and as if a switch is flipped, the surviving bay creatures gradually come to life and slip away to deeper waters. The jubilee is over.

Lucy and Judy shared this story with me several times when I was a child. For them it was an unexpected and wonderful midnight adventure. But being familiar with my bay, Pensacola Bay, and with the habits of marine creatures, I was always a bit confused—I believed my mom and aunt, but the story sounded more like a tall tale, or a magical fantasy. Why would these animals gather at the shoreline for easy harvest by people or other predators? I always asked for an explanation, but none of the adults I knew could offer one. They were as puzzled as I was. With time I gave up trying to explain the inexplicable, and eventually forgot about the story.

One afternoon, some thirty years later and soon after I began teaching at Birmingham-Southern College, a student named Heather from Fairhope, Alabama, was in my office. We were swapping stories about our experiences growing up on the coast, when she mentioned "the jubilee." I sat up immediately. Memories of the family story began to trickle back. "Tell me all you know," I asked. By the end of the conversation, I knew that the jubilee was a bona fide phenomenon, not the stuff of myth or local fiction. Locals watch for the event and really do call their friends in the dead of night to spread the word. But when I asked Heather what causes the jubilee, all she offered up was a shrug.

In my thirties, and with a much better understanding of environmental science, I heard the story much differently than I had as a child. The jubilee sounded more like the symptoms of an environmental disaster. The lethargy of the creatures resembles what happens when aquatic animals run low on oxygen. Most marine and freshwater creatures get the oxygen they need through gills and similar structures designed to absorb oxygen molecules that have dissolved into the water. If oxygen levels drop too low, animals can become lethargic and even die.

Oxygen levels can crash in aquatic environments for several reasons. For example, warm water cannot hold much dissolved oxygen. Or, because water absorbs oxygen through contact with the atmosphere, deep layers of water can also lack oxygen. However, the most common cause for low dissolved oxygen levels that I knew of at the time was excess nutrients in a river or bay. And in the long history of human interaction with rivers and estuaries, the source of excess nutrients has been feces in one form or another.

After my conversation with Heather, I did some digging on the topic of jubilees. With casual reading online, I learned it is a phenomenon believed to be unique to Mobile Bay, one that has been documented for well over a century. Jubilees occur along the eastern shore on summer nights with just the right combination of weather and tides. They are celebrated locally as a manifestation of nature's abundance and generosity. This was a good start, but I wanted to know what scientists had to say. In particular, I wanted to learn whether there was a link between jubilees and an overabundance of the pollutants that excrement adds to the water. So, I turned to the scientific literature for a deeper dive.

My suspicions were not unwarranted. One hallmark of the Anthropocene—our new geological era—has been our adding too much of two elements to ecosystems: nitrogen and phosphorus.[1] The movement of these nutrients through ecosystems, a process known as nutrient cycling, is one of the nine planetary systems on which humanity depends for survival on Earth, and it is one of the three planetary systems we have pushed into the danger zone.

Both nitrogen and phosphorus are exceptionally important to sustaining life. For example, each protein molecule contains at least one nitrogen atom, and phosphorus is used by nearly all organisms to store and release energy. Both elements are also components of DNA. For these reasons, biology students study diagrams of how these and other nutrients move between the living and nonliving world. Each chart is a unique and dizzying swarm of arrows showing the pathways nutrients take as they migrate among plants, animals, humans, oceans, rivers, rocks, soil, and the atmosphere. These biogeochemical cycles may seem dreadfully boring to many students, but they are essential to life on Earth.

Despite their importance, nitrogen and phosphorus in forms that can be used by organisms (e.g., nitrate and phosphate) have historically been in short supply on Earth. An understanding of this led to important breakthroughs by plant scientists in the nineteenth century. Suspecting that the scarcity in soil of these and other nutrients limited crop growth, they deduced that providing these elements in the correct form and dosage could boost agricultural productivity. Experimentation validated this hypothesis and revealed why long-practiced agricultural methods such as spreading manure from livestock on cropland improved yields. As it turned out, cow pies, horse hockey, ewe berries, and all other forms of livestock dung are loaded with nutrients. These discoveries initiated a radical change in humanity's relationship with land and water.

To understand why, we must venture back to the Industrial Revolution in Europe, a time when human populations were rising fast, and the warnings of Thomas Malthus were threatening to taint the period's record of dramatic technological successes. Malthus, an early social scientist, had in 1798 observed that improvements in food production spurred bursts in population growth, but at rates that outpaced food production. This doomed populations to food scarcity and population corrections in the form of famine, disease, and war. The paradox became known as the Malthusian Trap. Numerous critics, then and now, point to continued technological innovation as an underestimated factor in Malthus's gloomy prognostications. One agricultural innovation his contemporaries pointed to was synthetic fertilizer, which offered the hope of boosting food production output and avoiding population corrections.[2]

Industrial-scale production of synthetic fertilizer rich in nitrogen and phosphorus was limited at the time due to scarcities in raw materials rich in these elements. Bones were one source. Farmers could spread ground-up bone, known as bonemeal, over fields to provide phosphorus, but supplies from slaughterhouses were limited. Demand for fertilizer was unrelenting, and some dealers began procuring human remains from graveyards. The government banned the practice after farmers and the public learned about the source of the new and improved bonemeal.[3]

Near the beginning of the nineteenth century, German scientist and South American explorer Alexander von Humboldt brought to Europe's attention the availability of seabird guano on Peru's Chincha Islands.[4] Merchants were soon exporting enormous amounts of the nutrient-rich guano to Europe and the United States. After guano supplies dwindled later that century, vast deposits of nitrogen-rich saltpeter were discovered in Chile's Atacama Desert.[5] Control of the Peruvian guano and Chilean saltpeter trade was so lucrative and important that their possession featured centrally in two nineteenth-century wars and a major naval battle early in World War I. It's strange but true: navies of the world battled over bird poop.

A dependence on a foreign source of nitrogen was a strategic concern to emerging global powers of the late nineteenth century. Not only was food production vulnerable to supply disruption, but nitric acid derived from processing guano and saltpeter was the primary ingredient for the manufacture of explosives.[6] Governments tasked their scientists to find new sources.

It was known that our atmosphere comprised mainly nitrogen gas, but in a form unusable by plants, animals, and even industry at the time. Chemists sought to harvest this elusive, yet ubiquitous, resource. Early in the twentieth century, German chemists Fritz Haber and Carl Bosch developed the processes for converting atmospheric nitrogen to ammonia on an industrial scale.[7] In combination with other chemical processes already developed, ammonia could be turned into a fertilizer or explosive.

However, the inevitable agricultural revolution would have to wait. The Haber-Bosch process was expensive, delivery systems were undeveloped, and crop-specific applications were unknown. Instead, the process was used extensively in the production of munitions, first by the Germans in World War I and later by industrialized countries more broadly in World War II. When World War II ended in 1945, chemical production facilities in much of Europe and Japan were in shambles, but American plants were intact and quickly began mass producing inexpensive nitrogen fertilizers to meet demand.[8] In combination with other agricultural reforms, food production across the world skyrocketed and initiated the greatest period of population growth in history, one that has yet to stop. The postwar agricultural boom, known as the Green Revolution, has improved nutrition and food security throughout the world. Countless famines and deaths have been averted. Moreover, of the 7.8 billion humans on the planet in 2020, several billion are alive today because of the Haber-Bosch process.[9]

So, what has any of this to do with rivers and estuaries? Nutrients in fertilizer—which is essentially synthetic manure—does not stay in fields. Much of the nitrogen is released by microbes in various molecular forms into the atmosphere as they consume dead plant tissues. This worsens air pollution, and at least some of this nitrogen falls back to Earth in rain. Even more of these nutrients are washed off farm fields through overwatering or by heavy rains. Both escape routes eventually send these nutrients to rivers, lakes, estuaries, and oceans.

One might expect that an influx of nutrients could be a good thing for rivers and other aquatic ecosystems. If nutrients are in short supply, additions should help these ecosystems prosper, right? Nope. The surplus of nutrients in ecosystems—a condition known as eutrophication—creates chaos. Fast-growing plants, especially algae, absorb these nutrients and quickly funnel them into copious amounts of growth. In healthy aquatic ecosystems, algae play an important role by oxygenating the water through photosynthesis, providing habitat, and becoming a food source during decay. But when a surplus of nutrients triggers an algal bloom, these benefits are dwarfed by the problems created. Where currents slow, algae form floating mats on the surface, blocking light for organisms below. Algae grow on the stream bottom, filling

Nutrient pollution in freshwater and marine ecosystems often triggers devastating algal blooms. Photo by Tom Archer, courtesy of Michigan Sea Grant.

this zone with long tangles of growth. Where currents prevent mats from forming, floating algae, a chief component of plankton, transform clear waters into a green broth.

All this is bad, but the worst ecological damage caused by algal blooms is a drop in dissolved oxygen in the water. Individual algal cells are short lived, and soon after a bloom begins there's an overabundance of dead algae in the water. As microbes such as bacteria consume the dead algae, they soak up dissolved oxygen. This causes waters to become hypoxic, the condition of having very low levels of dissolved oxygen. Chronically hypoxic ecosystems support only a handful of species that can tolerate these conditions. As hypoxic waters spread, mobile animals will attempt to flee affected areas, but slow-moving creatures and immobile species are often overcome. If this doesn't kill them, it can impair activity, growth, and reproduction. It is this ecological cascade of cause and effect that was on my mind as my student reminded me about the jubilee in my office that afternoon.

The story of excrement and rivers continues, and it implicates each of us. Many of the nutrients added to cropland through fertilizing end up in our food. But just as nitrogen, phosphorus, and other nutrients don't stay put in farmland, neither do they stay in us. Which brings us to the point in the story where we have to ask: where does it all go?

The nutrients in livestock manure are also present in human urine and feces, and they are in much higher concentrations due to our omnivorous diets. The way we process this waste has caused eutrophication and other problems in rivers throughout the world. Where population densities are low, leaving our waste in the bushes or behind trees, or even using a privy or latrine, has little ecological impact. Instead, it is the concentration of human waste from densely populated areas that creates serious problems for our ecosystems and us.

Most people in modern times know that contact with or accidental consumption (*shudder*!) of fecal matter can cause the transmission of diseases, some of which can be deadly. But even before modern science established the link between human waste and public health, much effort was invested in removing sewage from urban areas. Early sewer systems were built in ancient Greece, Mesopotamia, and later throughout the Roman Empire. However, after the collapse of the Roman Empire, sewer systems fell into disrepair and it became customary in urban areas to dump human waste into the street. This created both aesthetic and public health problems. A solution that emerged was to build cesspools beneath homes. These pits collected waste and allowed drainage into the surrounding soil. While this improved street conditions, cesspools sometimes leaked into nearby sources of drinking water, and the accumulation of dangerous gasses sometimes suffocated sleeping families.[10]

Diseases had always swept through urban areas, but the rapid growth of urban populations during the Industrial Revolution of the nineteenth century caused the pace of disease outbreaks to quicken. For years it was widely believed that poisonous fumes—or miasmas—emanating from rotting waste caused many human diseases. During the latter half of the nineteenth century, scientists began discovering that microscopic organisms were causing diseases. This new paradigm for understanding disease came to be known as germ theory. As research advanced, it became clear that pathogens (bacteria, parasites, and viruses) were responsible for the outbreaks plaguing cities. Infections of the human gut were among the deadliest. Many pathogens have evolved to exploit the human gastrointestinal track because it is a microbial paradise. It's moist, food rich, and maintained at incubating temperature. Because paradise can be lost if the host mounts a successful immune response or dies, pathogens hop from the gut of one host to the next by inducing the voluminous production of infectious fluids from the victim (also known as diarrhea). A mere droplet of this fecal matter can infect a new host, and so the fecal-oral route of disease transmission has allowed plagues to torment humanity for thousands of years. Cholera and typhoid are among the deadliest, but dozens of other pathogens can cause illness or death, especially if victims are already immunocompromised.

As the link between human waste, disease, and public health became clear, demand grew for a safer method of treating sewage in urban settings. Sewer systems were built to collect waste and whisk it away. Though this waste was sometimes used as manure in fields, the most practical solution was usually to divert sewage into streams.

The high volumes of human waste soon being dumped into creeks and rivers created health problems for people downstream, who relied on these rivers for drinking water, irrigation, fishing, and recreation. Receiving waters near large cities were overwhelmed with sewage, and the problem worsened when the flush toilet became popular.[11] The Thames River in London became so infamously fouled that in the summer of 1858 a heat wave triggered a smell so vile that leaders gagging in the nearby Parliament agreed to pay for an overhaul of the city's sewer system.[12]

Modern sewage treatment plants were developed in the late nineteenth and early twentieth century to address the problem of sewage in rivers. These plants remove debris and fine particles, use microbes to break down dissolved organic molecules, and then disinfect what's left. The process removes some of the nutrients, but typical wastewater has substantial concentrations of nitrogen- and phosphorus-rich molecules. Treated wastewater is sometimes used for irrigation or sprayed over fields, but usually it is discharged into a nearby river, lake, or estuary, and this can cause eutrophication. There are methods for eliminating nitrogen and phosphorus from wastewater, but these processes add extra expense that few communities are willing to pay.

Maintaining the network of pipes that bring waste to treatment plants is an ongoing challenge for communities across the country. The US Environmental Protection Agency estimates that untreated wastewater spills from these networks and into the neighborhoods and creeks occur between twenty-three thousand and seventy-five thousand times each year. Known as sanitary sewer overflows, these failures are often caused by blocked or broken pipes, loss of power to lift stations, or improper design. Sewer overflows have increased in recent years as cities have grown without installing necessary upgrades or increasing capacity.[13]

The amount of treated wastewater in southeastern rivers is impressive, especially for rivers flowing near large metropolitan areas. For example, the City of Atlanta treats 170 million gallons of wastewater per day while serving the needs of 1.5 million people living or working in the entire metropolitan area. An additional 2.5 million are served by wastewater systems operated by other entities.[14] On a typical day, this wastewater accounts for about 1.4 percent of the water in the nearby Chattahoochee River.[15] The fraction of wastewater in urban rivers rises substantially during dry periods and droughts as water inputs decline. Altogether, there are over forty treatment plants on the Chattahoochee between Lake Lanier (near Atlanta) and the Alabama border, one hundred miles away.[16] One *percent*, even 5 percent, of a watershed's streamflow may not seem like much, but this isn't ordinary water. It is loaded with nutrients and contains complex chemicals, such as pharmaceutical compounds, that pass unscathed through our sewage treatment plants. All this contributes to eutrophication and other problems all the way to the Gulf of Mexico.

There's one more connection between excrement and rivers that we must explore. During the second half of the twentieth century, innovations in agricultural output and livestock production led to the mass production of inexpensive beef, pork, and chicken. While the negative human health consequences of increased meat consumption are well documented—more heart disease, cancer, and diabetes—what is less appreciated is the impact this dietary shift has had on rivers.[17]

At present, most meat is produced by raising livestock at high densities in facilities known as concentrated animal feeding operations, or CAFOs. CAFOs create a staggering amount of animal waste. One estimate from the end of last century is that 133 million tons of dry manure were produced annually by livestock, a value thirteen times more than that of human waste produced.[18] The livestock waste produced in America could cover Rhode Island with twelve inches of manure each year.[19] CAFO waste is not just laden with nutrients; it has high concentrations of dangerous bacteria, toxins, growth hormones, and antibiotics.[20]

The Southeast has a lot of CAFOs, especially for poultry and swine. In 2016, six of the top ten chicken-producing states were in the region. Georgia ranked first, producing over 8 billion pounds of chicken that year. Swine farming is concentrated in only a handful of states, and North Carolina is one of them, ranking second in the United States and having an inventory approaching 10 million. North Carolina ranks first for raising turkeys, and grew 1.2 billion pounds of gobblers in 2016.[21] Visualizing what a billion pounds of meat looks like isn't easy. According to the Centers for Disease Control and Prevention, the average US male weighs 198 pounds. Thus, a billion pounds of meat is the equivalent of 5.05 million average American men.[22] That's a lot of meat.

The manure produced by CAFOs can be dry or liquid. Dried manure, often a product of chicken CAFOs, is gathered and spread on fields as fertilizer. Swine waste is usually in liquid form. It is temporarily stored in lagoons, then later sprayed as a fertilizer on cropland.[23] When everything goes as planned, both are reasonable uses of the animal waste. Soil microbes break down the organic components of the waste, harmful bacteria die, and crops uptake the nutrients. But problems are common. When microbes and crops cannot keep up with inputs, excess waste seeps into the groundwater or washes into streams when it rains. Some manure lagoons leak into the groundwater and contaminate local wells, streams, and wetlands. Problems are exacerbated in low areas like floodplains, where the water table is near the surface.[24]

And then there are the hurricanes. North Carolina is hit by more hurricanes than any other state on the East Coast aside from Florida.[25]

Many of the state's CAFOs are built on river floodplains and are vulnerable to catastrophic flooding when hurricanes strike. In 1999, when Hurricanes Dennis and Floyd hit in quick succession, 2.5 million poultry and 30,000 hogs drowned. Fifty hog manure lagoons were flooded, and six were breached.[26] For weeks after both events, the bodies—rivers, estuaries, and the Pamlico Sound—receiving these waters were under extreme eutrophication stress. Nutrient levels spiked, and oxygen levels plummeted.[27] Despite efforts to minimize the risk of future such disasters, problems still continue. Flooding from Hurricanes Matthew (2016) and Florence (2018) drowned 8,300 hogs and 5.2 million chickens, and dozens of hog manure lagoons spilled into floodwaters.[28]

The blowback from nutrient pollution in the Southeast and elsewhere is manifest in many ways, including harmful algal blooms caused by species of cyanobacteria or dinoflagellate.[29] Some blooms simply discolor the water, form a layer of scum, produce foul odors, and/or initiate hypoxia. The worst blooms produce toxins that harm wildlife and people. Fish kills caused by harmful algal blooms in midsummer and early fall are now a regular occurrence in the Southeast. The triggers are not fully understood, but warm water and an overabundance of nutrients make rivers and lakes particularly vulnerable. Because blooms tend to occur in slow-moving waters, rivers with at least a moderate current are less vulnerable. However, the reservoirs and estuaries that rivers flow into can provide perfect conditions for harmful algal blooms.[30]

Such harmful algal blooms are on the rise, compromising public health and the economy. Blooms cost the US tourism industry $1 billion a year. Beaches, rivers, and other waterways must be closed to prevent respiratory distress from inhaling airborne toxins produced by the blooms. People, pets, or livestock can die from drinking poisoned waters. Fish and shellfish (e.g., oysters, clams) accumulate toxins produced by the algae, and eating them damages the liver, kidneys, and nervous system. The closure of fisheries due to harmful algal blooms costs tens of millions of dollars a year in the United States. Municipal drinking water supplies can become unusable during a bloom, and the threat of harmful algal blooms increases the costs of monitoring and water purification.[31]

As scary as harmful algal blooms are, nutrient pollution of rivers is triggering an even more sinister problem—oceanic dead zones. Dead zones are portions of coastal waters made hypoxic or anoxic (lacking oxygen entirely) through eutrophication: rivers bring excess nutrients to coastal waters, causing plankton to grow, die quickly, and then sink to the bottom. Microbes digest the dead plankton and deoxygenate the water near the seafloor. The problem is worst when seas are calm because there are no waves to bring these deep waters into contact with the atmosphere, where oxygen could be absorbed.

Dead zones are a global phenomenon. As Robert J. Diaz and Rutger Rosenberg reported in 2008, anthropogenic (man-made) dead zones began appearing in coastal waters of the world soon after the use of synthetic fertilizers became widespread after World War II. Since the 1960s, their frequency in coastal waters has doubled each decade.[32]

Diaz and Rosenberg found that nearly every coastal body of water along the Southeast's Atlantic and Gulf Coasts experiences dead zones, including Chesapeake Bay, Pamlico Sound, Charleston Harbor, the Savannah and Saint Johns Rivers, and Apalachicola, Pensacola, and Mobile Bays. Some experience a dead zone every few years or multiple times during the year. Others are perpetually hypoxic. The Mississippi River's dead zone is one of the world's worst. In 2019 it attained a size larger than the combined area of Connecticut, Delaware, and Rhode Island.[33]

Chronic hypoxia in any aquatic environment changes the ecosystem in ways that diminish biodiversity and impair ecosystem services. Larger, long-lived, specialized species more sensitive to oxygen levels are replaced by small, short-lived, opportunistic species that can tolerate hypoxia. This species turnover disrupts food webs and can cause the collapse of fisheries. Scientists have estimated that in affected areas of the Gulf of Mexico, four US tons of small creatures are killed by hypoxia for every square mile. If given the chance, the marine community may still recover in chronically hypoxic areas—but it will take decades.[34]

Dead zones, harmful algal blooms, sewer overflows—these are just some of the problems we have created for ourselves and biodiversity by manipulating the abundance and movement of nutrients through ecosystems. So far, I've only described the problems for the Southeast. Over a third of Earth's land surface is used for grazing or farming, and there are nearly eight billion humans defecating each day, on average. That's a lot of fertilizer, and a lot of excrement. It's no wonder, then, that the nutrient cycle is one of the nine essential planetary systems we have most dangerously disrupted.

So, what did I learn about the jubilees in Mobile Bay when I dug into the scientific literature? Are these hypoxic events caused by eutrophication? The answer is murky. The first marine biologists to examine jubilees established that they are hypoxic events. The animals crowding the shoreline are fleeing hypoxic waters. They also concluded that jubilees are a natural event, one that was documented long before the widespread use of fertilizer in the region.[35] Jubilees arise in the summer when winds are light or nonexistent and strong high tides bring hypoxic salt water high into the bay. Because of their salt concentration, these marine waters are heavier than the river water entering the bay. If waves and winds do not mix these layers, the marine water stays near the bottom and becomes more hypoxic as microbes in the mud digest organic debris on the bottom of the bay.

The precise conditions for a jubilee develop when a strong tide brings these hypoxic waters into the narrow section of the upper bay. A light easterly wind then pushes surface water westward. This pulls the hypoxic marine water into the shallows of the eastern shoreline. High bluffs along the eastern shore prevent those winds from oxygenating the hypoxic waters. Bottom-dwelling animals swimming ahead of the hypoxic water become trapped against the shoreline until conditions relent. Jubilees are nocturnal because during the day plankton oxygenate the water via photosynthesis and override hypoxia caused by marine water. The configuration of tides, geography, and climate make Mobile Bay one of the few locations in the world where such an event occurs naturally.[36] So is the jubilee worthy of a guilt-free celebration for seafood enthusiasts?

Maybe not. . . . Marine scientists later studying the shape of Mobile Bay and its water quality concluded that our tinkering with the bay has augmented hypoxic conditions. The port of Mobile is a major hub for commerce and shipbuilding, and since the 1870s portions have been dredged to allow the passage of large ships. The bay averages 10 feet in depth, but now has a ship channel 46 feet deep and over 393 feet wide. The channel has become a freeway for hypoxic waters to enter from the Gulf and reach high in the bay, where jubilees occur. Some of the spoil from this dredging is deposited farther down the bay, where it prevents the mixing of this water with more oxygenated waters.[37] There has also been a rise in excessive nutrients from fertilizer runoff and treated sewage entering the bay, and these changes enhance the occurrence of hypoxic conditions.[38] Large areas of hypoxic dead zones now occur regularly throughout Mobile Bay in summer.

So, the jubilee is a natural event, but one that we have augmented in the process of industrializing the Mobile Bay estuary and the rivers that feed it. Several scientists and longtime bay observers believe jubilees are increasing in frequency and intensity, but because they are difficult to monitor, there's no hard data on this.[39] Regardless, perhaps it's time to view jubilees as symptoms of an ecosystem in distress, and less a manifestation of nature's abundance and a chance to easily harvest oxygen-starved animals. The jubilee attended by my mother and her sister in 1950 was one of the last before a new era began of intensified river pollution and bay modification. Although the era of unchecked environmental manipulation during their youth eventually came to a close, things got a lot worse before they got better.

10

Warts and All

IN THE DECADES AFTER JUDY and Lucy witnessed a jubilee on the shores of Mobile Bay, US environmental health entered a free fall. The nation was in a phase of unprecedented industrial expansion and population growth. Factories, farms, and fertility were creating more waste than ever. During the 1950s, the United States produced over forty-four million pounds of organic waste each day.[1] Most was dumped untreated into rivers, lakes, and estuaries—waters still used for drinking, fishing, irrigation, and recreation.[2] The mantra was *dilution is the solution to pollution*. Ecosystems can assimilate some pollution, but not at this scale. Contamination caused eutrophication, algal blooms, and fish kills. Lake Erie, one of the Great Lakes, was so polluted that it was declared dead. Enough oil spilled into US waters each year to fill over six Olympic-sized swimming pools. The ecological ruin was so pervasive that average Americans were outraged and demanded action from the government. Congress was aware of the issue. Progressive legislators had tried to fix the problem for decades. The trouble was that water quality issues ignited fierce debates between proponents of states' rights and those who advocated for a federal approach.

Water quality problems were nothing new. Pollution arose as a problem in the 1800s as the nation's industrial sector grew. But, like many other issues, it was viewed by US law as a local problem. Legally, one could pollute public waters up to the point that further pollution would harm someone else's property or the public good. The work of defining and enforcing such laws fell to states, and what evolved was a nightmarish tangle of judicial rulings.

By the early twentieth century each state had independently developed an agency to oversee water quality issues. But most state pollution laws were weak, and the will to enforce them was weaker. Legislators avoided passing strong laws because they might encourage businesses to move to other states, an outcome that would torpedo the reelection bid of any politician supporting such a measure. Furthermore, politicians opposing strong state environmental laws enjoyed generous support from the business community. Lacking a mandate and incentive to change, states purposefully kept weak environmental laws. If an enforced action against polluters was necessary, the preferred approach was gentle encouragement of voluntary compliance.[3]

During the first half of the twentieth century, the only federal law superseding state control of water quality was the Refuse Act of 1899, banning the dumping of debris in rivers and harbors to protect navigation. If they wanted, states could fill their rivers to their banks with sewage and industrial pollution.[4] And many of them did. Author and journalist Upton Sinclair described some of this in his iconic 1906 book *The Jungle*, an exposé on the US meatpacking industry. Sinclair featured Chicago's Bubbly Creek, a branch of the Chicago River that was infamous for flowing red and chunky, with blood and animal body parts. The creek's bubbles were from the methane and hydrogen sulfide gas arising from the toxic stew.[5]

Thanks to Sinclair and others, widespread recognition of this problem in the early twentieth century led some to call for federal intervention. However, this would represent a radical departure from the current balance of power between states and the federal government. At the time, the federal government was far weaker than it is today, and states enjoyed their autonomy. Nevertheless, progressives proposed over one hundred bills to address water quality in the first decades of the twentieth century. Every one of them failed. Not only were opponents uneasy about ceding sovereignty of their waters to federal oversight, but there was also disagreement on how to regulate pollution. Congress dropped the issue for a while to focus on the Great Depression and then World War II.[6]

In 1948, Congress finally passed a federal water quality law. However, the Federal Water Pollution Control Act was a victory for states' rights and an epic fail for those pushing for a federal approach. The compromises made by progressives to garner enough votes left the federal government with little authority. Most waters remained under state oversight, and the federal role was simply to help states enforce their own policies. Washington would regulate interstate waters, but this was a mere 15 percent of US waters. Even here, the law limited federal intervention to a narrow range of circumstances. Enforcement action required the consent of the state from which the pollution arose. Not surprisingly, states did not volunteer for punishment.[7]

Meanwhile, pollution of rivers, lakes, and estuaries worsened during the postwar economic boom. Pressure on Congress grew. In 1956, Congress tried to fix the existing legislation with the Water Pollution Control Act Amendments, but again, states' rights advocates dictated the terms of the agreement. The amendments made federal funding available to encourage—but not require—cities and states to build wastewater treatment plants. The act also tried to fix the regulation of interstate waters by allowing the governor of an affected state to call for a conference of stakeholders. But polluters still had to volunteer to curtail their activities. In 1961, Congress extended the law to all navigable US waters, but the legislation still relied on voluntary pollution abatement and was predictably ineffective.[8]

In the 1960s, proponents of a federal solution in Congress gained more support as national water quality continued to decline. The result was the Water Quality Act of 1965. This law was a bold strategy shift. The federal government now required states to set water quality standards for their interstate waters and develop and pursue plans to meet these standards. Enforcement action against polluters could require them to reduce their waste discharge. It appeared that the federal government would finally have the means to regulate pollution, though just in interstate waters. However, the act proved to be ineffective. One reason is that the law required states to prove a direct link between polluters and water quality declines. This was nearly impossible given the scientific and technological limitations of the day. Furthermore, state legislatures protecting their industry ensured that standards were weak and their agencies to enforce them were underfunded and understaffed.[9] On the bright side, the 1965 act required the development of water quality standards and acknowledged the benefits of maintaining water quality acceptable for swimming and fishing.[10] It was incremental progress.

Television helped tilt the power dynamics in favor of federal intervention. As TV became affordable and popular with the growing middle class, media coverage of the environmental crisis expanded. Beamed into living rooms from coast to coast were images of rotting fish and raw sewage flowing through the nation's rivers. Several events became emblematic of the crisis. Families at the dinner table watched aghast in 1969 as a hundred thousand barrels of oil gushed from a well blowout and coated the Santa Barbara coast in California.[11] Footage of fouled beaches and oil-soaked birds struggling to live caused people to question whether unchecked economic growth was worth this price. Later that year the Cuyahoga River caught fire in Cleveland, Ohio. Flames five stories high burned off industrial pollution and torched a bridge. Images of the fire (which was just the latest in a series of Cuyahoga fires) became symbolic of the failure of state and federal environmental protection.[12]

By the late 1960s the public was frightened and furious about pollution. Voters, both Republicans and Democrats, now understood that air and water pollution was not just destroying the environment but also threatening their health and their children's future. Rachel Carson's *Silent Spring* in 1962 had justifiably stoked fears of industrial pollution. This book brought unprecedented attention to toxic industrial pollutants and their role in failing ecosystems. By the late 1960s, industry was introducing over five hundred new chemical compounds into commercial markets each year, with little to no testing of their impact on human or environmental health. Toxins infiltrated the drinking water and food supply. The US Public Health Service in 1969 found that 36 percent of tap water samples collected across the country were unsafe due to chemical or bacteriological contamination.[13] Another survey found that 99 percent of fish had DDT levels beyond FDA limits.[14] Carson can be credited for helping Americans understand the connection between pollution and their survival.

The states' rights defense for lax environmental laws and industry protectionism collapsed in 1970. While he's mostly remembered for the Watergate scandal and his subsequent impeachment and resignation, Republican president Richard Nixon made Congress focus on developing federal environmental policy. In his State of the Union address of 1970, Nixon declared that "the 1970s absolutely must be the years when America pays its debt to the past by reclaiming the purity of its air, its waters and our living environment. It is virtually now or never." Nixon wasn't what anyone would call an environmentalist. This was a bid for the support of environmental voters.[15] So, after decades of foot-dragging and bumbling bipartisan efforts, Congress passed more environmental laws in the early 1970s than at any other time in US history. This included the Clean Air Act (1970), National Environmental Policy Act (1970), Marine Protection, Research, and Sanctuaries Act (1972), Endangered Species Act (1973), and Safe Drinking Water Act (1974).

Rivers and other waters of the United States got their relief in the form of the Federal Water Pollution Control Act Amendments of 1972, better known as the Clean Water Act (CWA). The CWA is the foundation for modern water pollution policy. Its ambitious goal is to restore and maintain the chemical, physical, and biological integrity of the nation's waters. This act marked the beginning of the river revival.

The CWA was a rewrite of the 1948 law to give the federal government clear authority to combat pollution, and the muscle to get it done. Its policy development and enforcement fell to the newly formed Environmental Protection Agency (EPA), while the US Army Corps of Engineers was to oversee dredging and filling activities. Through the act, Congress directed the EPA to have all waters swimmable and fishable by 1983, and to eliminate all pollution discharge into navigable waters by 1985. Though the deadlines have long passed, these still are the overarching goals of the CWA.[16]

Under the CWA pollution is illegal and violators are subject to fines and imprisonment. This applies to all "waters of the United States." No longer are state waters beyond the reach of the federal government. States can keep some autonomy by opting out of direct federal oversight—and most do—but they must ensure their programs meet federal standards. States failing to do this can lose their right to opt out.[17]

The CWA states that polluters must obtain a permit for discharging waste into US waters, and that this discharge must not exceed limits set to maintain or improve water quality for the receiving body of water. To promote improvements, waters may not lapse into a more degraded condition, and wherever feasible permit holders must use the best technology available for limiting pollution. Importantly, citizens can sue violators or sue the government if either fails the law, a stipulation that has caused the proliferation of environmental watchdog groups. The act also made available funding to increase the rate of municipal wastewater treatment plant construction.[18]

The Clean Water Act of 1972 and its later amendments are tremendously powerful. And not surprisingly, they are controversial. One initial area of contention was whether the goals of the CWA were too ambitious. Early on, many states argued that achieving fishable or swimmable status for some waters was technologically unfeasible. Some argued that the economic costs of achieving pollution-free waters would outweigh the benefits.[19]

The most contentious area of dispute has been the interpretation of "waters of the United States" or WOTUS. The legitimacy of the CWA, dating back to the 1948 initial version, relies on the commerce clause of the US Constitution, which gives Congress authority to regulate trade among states, with foreign powers, and with Native American tribes. Prior to 1972, the Federal Water Pollution Control Act of 1948 and subsequent amendments applied only to "navigable waters," those waters capable of facilitating interstate commerce in a traditional sense. In the 1972 amendment, Congress did not clarify the meaning of WOTUS, and controversy over its interpretation has continued unabated ever since.[20]

The key point of contention is whether the law should apply to nonnavigable streams and isolated water bodies. The EPA's policies of the 1970s protected these waters because they supply or potentially supply surface water or groundwater to navigable waterways. This broad policy application protected ecosystems and safeguarded water supplies but limited what private landowners could do on their lands, including the filling of shallow, isolated wetlands. The is a big deal in the Southeast because, compared to other regions of the United States, a disproportionate number of creeks and rivers flow through private land.[21]

Because the interpretation of the CWA legislation falls to the executive branch, the reach of WOTUS regulation has waxed and waned over the years, depending on the party that controlled the White House and the outcomes of federal court rulings on lawsuits against these regulations.[22] However, inconsistency in the interpretation of WOTUS was significantly diminished in May 2023, when the Supreme Court ruled in *Sackett v. EPA* that wetlands can only be protected by the CWA if they share a surface connection to navigable bodies of water.[23] This ruling is a huge victory for property rights advocates, including those in the agriculture, forestry, and mining industries who have long wanted less federal oversight. And because this ruling removes federal protection of vast areas of wetlands and other ecosystems that had enjoyed federal protection for over half a century, this is a major setback to protecting water quality and availability, regulating flooding, and protecting biodiversity.

Despite this new setback, the CWA has led to far better pollution regulation than what would have been possible through self-rule by states. One measure of the CWA's impact is the rate of wetland loss. Wetlands convey surface water into underground reserves, store floodwaters, and filter water of pollutants. Coastal wetlands are nurseries for commercially and recreationally important finfish and shellfish. Wetlands also provide a place for hunting, fishing, and ecotourism—activities contributing to local economies. Prior to 1972, we destroyed wetlands at a rate of about 703 square miles per year in the contiguous United States. After that year, the rate declined sharply, reaching about 22 square miles per year in the 2000s, the last period for which data is available.[24] This is one of the CWA's most important successes.

What about pollution, the primary target of the CWA? While data collection on water conditions was spotty prior to the 1970s, the consensus is that our waters are cleaner and safer. In 2018, economists David Keiser and Joseph Shapiro published an analysis of several massive data sets. They found significant reductions from 1972 to 2001 in acidity, fecal coliforms (gut bacteria), lead, mercury, and other pollutants, though declines plateaued after about 1990. Key to these advances was a tenfold increase in funding for new municipal wastewater treatment plants. The act also required that all plants provide secondary treatment, where microbes are used to remove organic matter than contributes to eutrophication. In 1972, 28 percent of Americans were served by wastewater treatment plants supplying only primary treatment that removed large particles. By 2012 this value had dropped to just 1 percent. Over these same forty years, the number of Americans served by plants with tertiary treatment (advanced treatment to remove nitrogen, phosphorus, or other contaminants) jumped from 4 percent to 40 percent. Consequently, organic waste, nutrients, and fecal coliform loads in rivers have declined, and dissolved oxygen levels have increased.[25]

All this is good news for rivers, people, and biodiversity, but we are a long way from declaring victory in cleaning US waters. Under the CWA, states must monitor the quality of their waters. Waterways are classified into designated use categories, and standards are set accordingly. Some designated use categories have ambitious standards—such as waters suitable for wildlife or public water supply. Other categories have lower standards for purposes such as agricultural or industrial use. States monitor a subset of their waters every year and report their findings to the EPA. In 2017, states reported that 55 percent of their collective river and creek miles were impaired, meaning that their water quality did not meet the designated use standards. The chief pollutants causing impairment are pathogens (e.g., harmful bacteria), sediment, and nutrients, in that order. Mercury, lead, and other metals are another major cause of impairment, as are polychlorinated biphenyls (PCBs), a class of now-banned manufacturing chemicals.[26]

Nutrient levels in southeastern rivers are still high. Total nitrogen levels are far above what the EPA considers normal in 88–99 percent of streams, depending on the subregion. The highest overloads—at ten times the normal amount—are in the interior plains and hilly sections of the Southeast, where agriculture is concentrated.[27] A majority of southeastern streams, 53–95 percent depending on the subregion, carry phosphorus loads exceeding normal levels, and rivers near the coast have phosphorus levels four to five times above normal. With this much contamination, it's not surprising that dead zones arise each summer in southeastern coastal waters.

The EPA conducts its own periodic surveys of US waters. Its most recent (2017) report to Congress revealed that 46 percent of river and creek miles and a third of coastal waters and wetlands are in poor biological condition.[28] A fifth of lakes and reservoirs are hypereutrophic. Whether we judge by the measurements of the states or the EPA, we still have a long way to go before we meet the goals of the CWA.

Another alarming sign that the CWA isn't fully working is the abundance in our environment of hazardous chemicals that the EPA doesn't regulate. Our sewage, for example, is full of pharmaceuticals, personal care products, and other chemicals flushed down toilets or poured down sinks. Many wastewater treatment plants do not remove these chemicals, and they wind up in our rivers. Scientists have little to no understanding of how most of these chemicals affect the people and wildlife that consume them or have contact with them.[29]

Other underregulated groups of chemicals include pesticides, herbicides, flame retardants, and the tens of thousands of other chemicals that US residents are exposed to regularly. One group of these chemicals, per- and polyfluoroalkyl substances (PFAS), includes thousands of varieties used in food packaging and common household products. These chemicals are now detectable in nearly all surface waters of the United States. Scientists have found that PFAS can damage the kidney, the liver, and the reproductive and immune systems. A 2020 study estimated that between eighteen and eighty million US residents are exposed to particularly dangerous forms of PFAS (perfluorooctanoic acid [PFOA] and perfluorooctane sulfonate [PFOS]) through their public drinking water supply, and that these chemicals are detectable in nearly all public water systems in the country. What's more, two of the top states for PFAS exposure through drinking water are Alabama and North Carolina. Research studies, public concern, and lawsuits are proliferating as we learn more about PFAS.[30]

Why do we still have poor water quality a half century after the passage of the Clean Water Act? Policy experts say the act was flawed from the beginning. The law's failure hasn't been in addressing sewage or industrial pollution. The EPA began regulating these so-called point sources of pollution long ago. The more vexing problem is "nonpoint source" pollution—sources too numerous or too dispersed to be controlled at a specific location. Congress was aware of this problem as it drafted the 1972 amendment, but there was neither certainty about the magnitude of nonpoint source pollution, nor obvious mechanisms for its control. In the end, Congress did not require regulation of nonpoint source pollution, though it instructed the EPA to study the problem, authorize funding to seek solutions, and oblige states to report on the extent of nonpoint source pollution.[31]

A decade after the passage of the CWA, advances in monitoring and watershed science revealed that nonpoint source pollution was the main cause for the continued impairment of US waters.[32] The last major update to the CWA, the Water Quality Act of 1987, included the first federal attempt to regulate nonpoint source pollution. But instead of tackling the problem head-on, Congress punted. It required states to develop their own programs to address nonpoint source pollution, though Congress offered funding for control activities and demonstration projects.[33]

Over thirty years after the 1987 update to the CWA, nonpoint source pollution continues to be the biggest obstacle to cleaning up US waters. While the cost of eliminating it would be high, so too is the cost of continuing to allow poor water quality. A 2009 study by Walter Dodds and colleagues estimated that eutrophication caused by nitrogen and phosphorus pollution cost the US economy $2.2 billion annually through loss of recreational opportunities, costs of preparing drinking water and protecting endangered species, and declines in real estate value of waterfront homes.[34]

Agriculture is by far the biggest contributor to nonpoint source stream pollution.[35] Runoff from fields and livestock operations carries vast amounts of nutrients, pesticides, and sediment into streams, and none of it is regulated under the CWA. In 2017, the EPA reported that 133,164 stream miles in the United States were impaired due to agriculture, a cumulative distance more than five times Earth's circumference.[36] Until we can limit the loss of nutrients from agriculture, eutrophication and all its ill effects will plague US waters.

The news on nonpoint source pollution isn't all bad. Despite continued growth in population size and agricultural output, nitrate concentrations in US rivers peaked around 1980 and have remained steady thereafter.[37] Phosphate pollution in US rivers has declined slightly since the early 1960s.[38] In the Southeast, nitrogen declined from the early 1970s through the mid-1990s and then stabilized, and phosphate concentrations declined until the mid-2000s.[39] One reason for this modest success is that today's farmers now achieve higher yields than in the past without the same degree of fertilizer use. Another is that more wastewater treatment plants include tertiary treatment to reduce discharges of nitrogen and phosphorus into receiving waters.[40] And finally, due to many state bans on phosphate in detergents, industry voluntarily stopped putting phosphates in laundry detergent in 1994, and in automatic dishwasher detergent in 2010.[41]

Another favorable trend for southeastern river water quality—but bad news for farmers—has been a decline in agricultural output in the Southeast during the past century as food production became concentrated in the Midwest. Since the 1920s, cropland in the Southeast has dropped by over twenty million acres, an area about the size of South Carolina.[42] Most of this abandoned cropland has been replaced by forest or tree farms. The shift prevented agricultural pollution in the Southeast from being much worse, as would have otherwise happened.[43]

Despite its flaws, the Clean Water Act has helped America's waters and its people and biodiversity. We enjoy much safer drinking water and cleaner waterways. Each of us has friends and family who are healthier, or alive, because the act reduced their exposure to toxins and pathogens. This is worth celebrating.

What is the future of water quality policy in the United States? For sure, Americans will continue to debate the role of federalism, the interpretation of "waters of the United States," and the cost of pollution abatement programs. Just as they have in the past, the outcomes of these policy disputes will have real consequences for the health of streams and everyone who depends on them.

We can expect continued progress on controlling nonpoint source nutrient pollution. The EPA offers grants and initiatives to encourage farming reforms to reduce agricultural runoff. Many states have stepped up by banning practices such as the application of fertilizer or manure when the ground is frozen and cannot absorb nutrients into the soil. And both federal and state agencies promote research into farming practices such as having year-round ground cover to reduce erosion and keep nutrients on site.[44]

Is it too much to hope for a complete rewrite of the CWA to fix its major deficits? Much has changed since 1972. Our population has grown by 56 percent, our economy has globalized, and a rapidly changing climate is forcing us to rethink assumptions about our role in managing the environment. Scientists—from hydrologists to economists—have made sweeping advances in our understanding of policy tools, watersheds, streams, water quality, aquatic ecosystems, and biodiversity. We have a new arsenal of powerful technologies for monitoring and managing waterways. Policy experts have proposed new ways of managing pollution. For example, instead of continuing the command-and-control approach of enforcing the CWA, we could adopt market-based approaches such as pollution taxes and tradeable permits.[45] Or perhaps we should manage our water resources at the watershed level instead of dividing management of them among states.[46] Now would be a prudent time for Congress to update the act. But for this to happen, we need to elect more lawmakers who will transcend entrenched partisanship to solve complex problems. Until then we will live with the Clean Water Act as is—warts and all.

11
Unsung Heroes

THE CLEAN WATER ACT HAS been a massive undertaking to clean up American waters. Expenditures on pollution abatement have totaled well over $1 trillion since 1972, averaging about $100 per citizen per year. Yet as we learned, we still struggle with nonpoint source pollution, and it threatens our health, our economy, and our ecosystems. Adopting new agricultural practices is part of the solution, but due to technological and economic limitations of what can be done on the farm, we need more options. Some biologists have suggested that one solution to our pollution problem has been with us all along.

River mussels are living water filters. They are clam-like mollusks that nestle into the bottom of creeks, rivers, and lakes. They spend their days taking in water, filtering out digestible material, and ejecting indigestible bits onto the stream bottom. Mussels remove algae, phytoplankton, and silt, and return clean water to the river. In the process, mussels remove nutrients that cause eutrophication, plus pathogens and complex pollutants such as pharmaceuticals.[1] Journalists have described mussels as natural water treatment plants, purification systems, and nutrient capacitors.[2] They are the unsung heroes of the southeastern river.

Mussels could be a natural, low-cost solution to keeping our rivers clean, but they have become rare in most rivers. When the first Europeans arrived, mussels paved the bottoms of North American large creeks and rivers. Historical accounts describe dozens to hundreds per square yard. And it was a diverse assemblage of species. Big and small, thick and thin, plain and ornate—there was a mussel for every occasion. About 300 species inhabit North America, 270 of which occupy the Southeast—more species than any other region on Earth. Historically, a typical river of the region harbored more species than can be found on entire continents.

Unfortunately, river industrialization has not been easy on mussels. They are the most imperiled group of animals in North America, and possibly the entire world. Dozens of species are extinct. Populations of most surviving species have declined, and many are close to disappearing completely.[3] Alabama has 186 species, more than any other state in the United States, but 24 are extinct as of 2019. Another 56 are on the US endangered species list, and 10 are listed as threatened. Nearly 50 percent of the state's mussels are gone or at risk of disappearing.[4] Of Georgia's 98 mussel species, 12 are extinct, 11 are federally endangered, and 4 are threatened. Of Tennessee's 130 species, 9 are extinct, 42 are endangered, and 2 are threatened.[5]

Conservationists have suggested that restoring mussel populations would help solve our river pollution problem. It's an alluring solution—one that's good for us, and good for ecosystems and biodiversity. However, there's a hitch. Biologists don't fully understand why mussels disappeared. It's a mystery like the sturgeon's situation: populations crashed long ago but have had enough time to recover. If we want a mussel revival, we need to identify and overcome whatever caused them to disappear from our rivers. And perhaps the explanation for mussel decline will also reveal why sturgeon are still imperiled.

Mussels are living water filters that improve water quality in rivers. Though the Southeast is home to more mussel species than anywhere else on Earth, many of them are on the cusp of extinction. Courtesy of the Alabama Aquatic Biodiversity Center.

Biologists have tried to solve the mystery of mussel decline since the passage of the Endangered Species Act (1973) and the birth of conservation biology in the early 1970s. It's not been easy. By the time they began investigating the decline, many mussel species were extinct, or their populations had already crashed. Others were disappearing faster than biologists could study them, let alone track down the cause. Some malacologists (biologists who study mollusks) have investigated how mussels respond to environmental changes caused by humans. Others have scrutinized the accounts written by the first American malacologists to reconstruct the population sizes and ranges of species and look for clues about the disappearances. Some biologists have modeled prehistoric mussel populations by working with archaeologists studying Native American shell middens. One of the conclusions from these efforts is that troubles for North American mussels began in the late nineteenth century. Ironically, it was due to a mussel's ability to defend itself.

Mussels have durable shells (also known as valves), but occasionally a sand grain or rock fragment will slip between the valves and irritate their squishy insides. If not ejected, the mussel will coat the offending particle with mother-of-pearl, the iridescent material mussels use to build their shells. Mother-of-pearl varies in color. A gleaming white is typical, but other colors include pinks, blues, and purples. A pearl forms as the mussel adds layers to the particle. River pearls come in an infinite variety of shapes, determined by the shape of the invading particle.[6]

Pearls are nature's river bling. They've enchanted humans for millennia. Shiny, colorful, and smooth, they are unlike most other natural objects. Small and durable, they are easily adaptable as jewelry and clothing ornamentation, and their rarity increases their value. Native Americans used river pearls as jewelry and stockpiled them as wealth. Archaeologists have recovered large caches of them from Mississippian era burials. One famous necklace from a dig on the Tennessee River has 1,300 pearls. When Hernando de Soto did not find gold among the Native Americans, his thugs stole their pearls. Archaeologists don't know whether Native Americans sought pearls deliberately or gathered them only when harvesting mussels for food. Regardless, after European diseases decimated Native American populations, freshwater pearls and mussels were ignored for three centuries.[7] Settlers had no use for the shells and found mussel meat to be tough and unpalatable.

All hell broke loose for mussels in 1857, when carpenter Jacob Quackenbush discovered a "magnificent pink pearl" near Paterson, New Jersey. Quackenbush sold the pearl to Tiffany and Company for $1,500—a small fortune at the time. As news spread, pearl fever swept across the nation. Each mussel was a lottery ticket, and rivers were giving them out for free. Though about only one in ten thousand mussels carried a marketable pearl, the enticement of easy money called people of all backgrounds to the river.[8] Each summer when waters were warm and low, thousands rolled up their sleeves and britches, gathered their skirts, and waded onto sandbars and shoals seeking treasure.

Pearl rushes waxed and waned over time and by region throughout the rest of the century.[9] Many of the pearls harvested crossed the Atlantic. Some still dangle from the ears of the wealthy or encircle their necks. A few are in museums. War and tragedy have claimed others. As for the pearl that started it all, it was sold to Eugénie de Montijo, wife of Napoleon III, the last emperor of France. The current location of the pearl—now known as the Queen Pearl in honor of Empress Eugénie—is a mystery unto itself.

Mussel populations declined where pearl fever struck. Pearl hunters killed thousands just to find one pearl.[10] Piles of discarded mussel shells mounted on riverbanks. One mound could have more shells than most contemporary malacologists will see in their lives. In a survey at the very end of the nineteenth century, most mussel harvesters reported partial or complete exhaustion of mussel beds.[11] Several of the nation's few malacologists advocated for protection, but it was to no avail. A much more profitable industry using mussels was launching.

Remember Bendix Blohm, the German immigrant who first produced caviar from Atlantic Sturgeon eggs? Johann Boepple was another German immigrant who altered southeastern rivers forever. Boepple was the first to successfully convert freshwater mussel shells into buttons. He founded the first shell button factory in 1891 on the banks of the Mississippi River in Iowa. Within ten years there were seventy button factories in the Mississippi River Valley.[12]

Whereas pearling was artisanal and haphazard as to where and when it arose, button production from mussel shell became a true industry, and it was systemic in its harvest. Buyers preferred the big, thick-shelled species from large rivers. Some shellers gathered mussels as a side gig, but thousands made their living by harvesting mussels and processing the valves for shipment. This required a nomadic life, and shellers and their families roamed the rivers in houseboats, and at times lived together in temporary sheller camps.[13]

Initially, shellers harvested mussels using long-handled tongs or rakes. Later the preferred method was fishing for them by dragging a

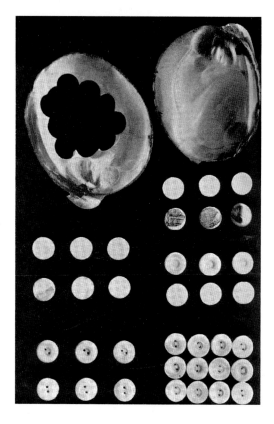

German immigrant Johann Boepple forever altered the ecology of North American rivers when he invented a process for making buttons from thick-shelled mussels. Photo by George F. Kunz, 1898. University of Washington Freshwater and Marine Image Bank/Wikimedia Commons.

row of metal hooks across the bottom of the river. Each looked something like a treble hook with a small metal sphere at the tip of each branch. Because each hook looked like a bird's foot, the whole assembly was known as a crowfoot dredge. Shellers tugged dredges behind boats while drifting with the current. When a hook tip fell between the open valves of a mussel, the mollusk would clamp down and hold on until hauled onto the boat and pulled off by the sheller. Barges transported mussels to factories, where shells were scrubbed, soaked, lathed, sliced, carved, drilled, and polished into buttons that were sorted, boxed, labeled, and shipped throughout the United States. Thousands were employed, and button factories were the economic engine for many towns.[14]

The numbers of mussels harvested for the industry is staggering given their rarity today. Photos from the era show barges fifty feet long and twenty feet wide feet piled five feet high with shell. At the peak of harvest, in 1912, fifty thousand US tons of shell were extracted from

Shellers captured mussels by drifting with the current and dragging along the bottom multipronged hooks tipped with knobs. When a knob fell into an open mussel, the creature sealed its shell and the sheller would haul the mussel aboard. Photo by Robert E. Coker, 1921. University of Washington Freshwater and Marine Image Bank/Wikimedia Commons.

streams across North America, mainly in the Upper South and Midwest. That's about five hundred million mussels in just one year.[15] The piles of riddled shell outside mussel factories resembled the largest of the Native American mussel middens. One represented a millennia's worth of foraging for survival. The other represented a year's worth of industrial production.

Not surprisingly, this level of harvest could not be sustained, and the industry began showing the classic signs of collapse. When shellers depleted the expansive mussel beds in the large rivers, they moved into smaller rivers with smaller beds that were quickly exhausted. As large specimens became rare, manufacturers began buying smaller, younger mussels and undercutting the ability of populations to rebound. As preferred species disappeared, manufacturers bought inferior, smaller species, thus putting pressure on new species in ever-smaller rivers. By 1920, the pearl button industry was in collapse due to dwindling supply. Merchants began importing cheap buttons from abroad. The Great Depression took hold. Plastic buttons delivered the killing blow after World War II. The last pearl buttons were made in 1965.[16]

One might hope this marked the end of mussel harvesting from American rivers, and that mussels would get a much-deserved break. But that was not to be. As the pearl button industry faded, Japan began importing mussel shell from the United States for its nascent cultured pearl industry. Bits of mussel shell were carved into small spheres called seed pearls, then inserted into a species of pearl-forming oyster. These oysters formed perfectly spherical pearls in far less time than pearl formation happened in the wild. As demand for cultured pearls rose during the latter half of the twentieth century, shell from the Southeast, especially Tennessee, became prized. State governments began regulating the harvest to protect endangered species and to ensure shellers harvested mussels sustainably.[17]

Then came the Zebra Mussel invasion of the Great Lakes. Biologists watched helplessly as this small mollusk from Eurasia swept across northern states. The invaders crowded together on every firm structure available and disrupted ecosystems. Their thin shells were useless for the seed pearl industry. Instead, zebra mussels killed native mussels through competition. Sometimes Zebra Mussels settled on native mussels and smothered them. Many feared that the invasion would spread southward and decimate southeastern river mussel populations.

Concerned about their supply of seed pearl, Japanese buyers stockpiled inordinate amounts of shell in the mid-1990s. Demand outpaced supply, and prices skyrocketed. A large Washboard mussel could fetch forty dollars; a bucketful was worth thousands. Illegal harvesting and shell smuggling were rampant. There were murders and unsolved disappearances of shellers and shell buyers in rural towns. Perpetually underfunded and understaffed, conservation law enforcement could do little to stop the harvesting.[18]

Fortunately, the shell harvest frenzy fizzled out. Japanese buyers stockpiled enough shell, disease struck pearl oyster farms, and Chinese competitors developed methods that didn't rely on North American shell. The export of US mussel is now a trickle of what it once was. As for the Zebra Mussels, they spread southward but haven't yet affected southeastern mussel populations.

Pearling and shelling severely depleted mussel populations. Shellers left mussel beds—millions strong—in tatters. This was unfortunate for many reasons, not the least of which was that we lost mussels as a safeguard against water pollution, which was escalating during this time. However, the damage was not enough to cause the declines and disappearances of mussel populations to the levels we see today. Not a single species went extinct because of the pearl, pearl button, or seed pearl industries.[19] Though these industries dealt harmful blows to the continent's mussel fauna, an ecological principle had safeguarded mussels from a full-on precipitous decline.

Pearl hunters and shellers were predators of a sort. And foraging predators—humans included—are subject to ecological rules. When prey become scarce, predators move on to more productive areas. Ecologists call this rule density-dependent foraging, and it applied to harvesting mussels. The first sheller to arrive at a virgin bed could gather mussels with ease. Every pass with the crowfoot dredge brought up a bounty of mussels. But foraging success declined as shellers removed more mussels. When a sheller gathered too few for the effort invested, he would move on to another bed. Surviving mussels could rebuild the population.[20] Density-dependent foraging explains why shellers initially focused most of their attention on large beds with high concentrations of mussels and overlooked or ignored smaller beds. They only moved on to smaller beds after they depleted large ones. But even then, a few mussels always survived. This rule also explains why Native Americans didn't wipe out mussels. Though the shell middens they left behind were voluminous, we know based on chronological sequences and species diversity of mussels in middens dating back five thousand years that Native Americans never eradicated a single species from any river or creek, or caused any mussel extinctions.[21] Finally, density-dependent foraging is why fishers didn't eradicate most sturgeon populations. When sturgeon landings dipped too low, fishers moved on.

Biologists have spent decades investigating the mystery of mussel disappearances. They've deduced several characteristics of what caused their decline. First, because entire populations and species disappeared, the perpetrator was likely a density-*independent* factor. This means that unlike with foraging by pearl hunters and shell harvesters, the likelihood of mussel mortality in an area was not related to mussel abundance. Such a factor killed indiscriminately: young and old, big and small, common and rare. Second, because disappearances were widespread, the cause of mussel decline was a factor affecting a large geographic area. Third, the guilty party would have to be present when the crime was committed. This would be the period between 1930 and 1980, when most mussel species or populations disappeared.

Finally, because surviving mussel species are rare today despite having had enough time to rebuild their populations since shelling was widespread, the factor causing mussel declines is probably still present in our streams. These deductions, in combination with historical detective work and contemporary observations of water quality, led investigators to suspect that a pollutant caused mussel disappearances, one that can be traced all the way back to Jamestown.

12
Death by Mud

TOBACCO—THE CASH CROP THAT SAVED Jamestown and that inspired colonists to take lands from Native Americans and enslave Africans—caused the first pollution problems in North American rivers. It wasn't sewage, pesticides, herbicides, synthetic fertilizers, or cigarette butts: the pollutant was mud.

Mud is made of very small sediment particles more technically known as silt and clay. Healthy rivers transport these and other particles such as sand and pebbles downstream. On their journey to the coast, these sediments sustain important riverine ecosystems. Silt and clay particles are too small to settle in busy river currents and stay in suspension. However, where river currents slow, they will drop to the bottom. Rivers will also drop mud when they rise and spill over their banks onto the floodplain. Because mud particles carry nutrient molecules on them, they promote plant growth wherever floods deposit them. Most silt and clay particles eventually arrive at the coast, where river currents weaken. The mud then settles in swamp and marsh ecosystems or in open waters to become marine muds.

Some pollutants fall into the category of "too much of a good thing." This is true for nutrients, and it's true for silt and clay. Mud pollution, or sediment pollution as scientists call it, can smother habitats, harm wildlife, alter ecological processes, and degrade water quality for our own uses. Clay is smaller than silt. Because of the size and mineral composition of clay particles, they carry a negative electrical charge. As if they were tiny magnets, clays attract positively charged molecules, including certain nutrients, complex organic molecules such as pesticides and herbicides, and heavy metals. Thus, mud pollution can cause problems with eutrophication and toxicity in wetland environments. Mud is one of the most widespread pollutants in

Nearly half of southeastern streams are in poor or fair ecological condition due to erosion and mud pollution. The problem is most noticeable after a heavy rain, but the mud does its worst damage when it settles and smothers stream habitats. Courtesy of Alan Cressler.

US streams. It has caused 43 percent of southeastern streams to be in poor or fair ecological condition.[1]

Sediment pollution is a form of nonpoint source pollution, a widespread problem that (as we learned) is preventing water quality from improving in American rivers. Like other nonpoint source pollutants, sediment pollution arises from numerous or dispersed locations that are difficult to control. It originates where soil is exposed to wind and rain and subsequent erosion. Such places are rare in the natural ecosystems of the eastern United States, where plants hold sediments in place. In the few locations where landslides or animal activity expose soils, plants soon grow and erosion ceases. However, human land use practices expose a lot of dirt, and preventing its erosion is not easy. The most difficult source of sediment pollution to tame is agriculture, and North American problems with sediment pollution began with tobacco farming at Jamestown.

When the English colonists ventured out of the Jamestown stockade to plant tobacco on Powhatan land, they encountered a form of farming they didn't recognize as agriculture. Back home in England, farmers and pastoralists had cleared the countryside of forest centuries before. The land had been in perpetual cultivation or pasture ever since. The forests around Jamestown were in perpetual cultivation of a different type—shifting cultivation. As Charles C. Mann explains in *1493: Uncovering the New World Columbus Created*, the Powhatan would clear a patch of forest and farm it for several years, until soil fertility declined. They would then move on to another area, allowing the forest to grow back. Native people used the second-growth forest for hunting and gathering of forest products. Years later, after the forest restored soil fertility, they would clear and replant the former field. Native peoples had practiced shifting cultivation throughout the Americas for millennia by the time Europeans arrived.[2] Because the land was under constant vegetative cover, sediment pollution from shifting cultivation was negligible.

As English colonists explored the Virginian forests, they didn't see a form of sustainable farming. They saw unused forest ready for conversion to tobacco production.[3] The colonists set on this forest with axes and saws. They hauled away useful wood and burned the rest in place. The hardest job was breaking through the forest soils. Beneath the decaying leaves there was a dark, spongy soil bound together with a network of tree roots. Most roots were iron tough, and some were as thick as a man's torso. Crisscrossing roots from adjacent trees formed a fabric that held soil in place and resisted the shovel, axe, and plow. Eventually steel and muscle bared the soil to the sky, and the ecological transformation was complete.

In the forest rain had always been a good thing. Leaves, branches, and leaf litter intercepted the rain and slowed it down. Water clinging to these surfaces evaporated. Trees sponged up the water that soaked into the soil. The excess became the groundwater that sustained the James River and its tributaries throughout the year. However, in the tobacco ecosystem rainfall's role was both good and bad. When the rain was brief or light, it soaked into the ground and nurtured the crops. But when more rain fell than soils could absorb, water accumulated on the surface and picked up clay and silt particles. Tiny puddles formed, filled, and spilled into one another. Rivulets developed and ran thick with the soil laborers had worked to reveal. This muddy stormwater flowed into the lowest point available in the landscape, usually a stream or wetland.

Agricultural erosion bedeviled the Virginia colonists. Soil fertility declined as stormwater washed away sediment and the nutrients it carried. When tobacco yields declined, farmers planted fields in corn, a less profitable crop. When corn yields declined due to fertility loss, farmers converted fields into pasture.[4] They could ship tobacco and corn to England for profit, but in the age before refrigeration, ranching was not a path to wealth. Thus, unsustainable soil practices drove the continual clearance of forest for new agriculture in Virginia and the other southeastern colonies.

The loss of forests in the region caused creeks and rivers to change in ways that were bad for settlers. The mud flowing into streams filled their channels. New sandbars formed and continually migrated downstream, making river navigation difficult and dangerous. As stream valleys filled with sediment, they lost their capacity to hold water, and flooding arose as a new menace for settlers.[5]

Mud pollution not only reshapes streams; it also affects the creatures in the stream. Imagine a mussel bed in the James River as colonists converted the watershed from forest to agriculture. Mussels snuggle side by side in the firm streambed. Stormwater from the first big summer thunderstorm drops a half inch of mud on the river bottom. Extending the large muscle known as a foot, mussels push their way upward to stay just beneath the streambed surface. A week later, another storm deposits more sediment. Mussels crawl upward, again. More storms, more sediment, more upward migration. However, the new sediments are unconsolidated and unstable. Eventually an intense storm causes a strong flood that sweeps the mussels and the new sediment downstream. Some mussels tumble into proper habitats and nestle into the streambed after the flood. Others use their foot to drag themselves a few feet each day in search of a better place to settle. Some make it, some don't. Attrition of the mussel community has begun.

Mud pollution harms other species, too. In the headwater creeks of the watershed, streams flow across a rocky bottom. The nooks and crannies between rocks support a rich array of invertebrates, including crayfish, caddis flies, stone flies, mayflies, and beetles. When agriculture expands into the area, mud, sand, and fine gravel wash from deforested areas and fill the spaces between the rocks. The stream bottom loses its three-dimensional structure, creek animals lose habitat, and the ecosystem begins to unravel.

In both big rivers and creeks of the Southeast, clay is the worst pollutant of the sediment types. Clays are microscopic and will drift for days and miles until the current slackens. The particles, thin and jagged, transform the water into liquid sandpaper that abrades the fragile gills of mussels, fishes, and other animals. When the James River runs thick with clay, mussels face a cruel dilemma—either go hungry or feed and be injured by clay particles.

Clays and silt suspended in the water block light from penetrating deeply into the stream. This hinders the growth of phytoplankton, a favorite food of mussels. Water with suspended mud also absorbs more sunlight than clear water, and as temperatures rise, dissolved oxygen levels drop.[6] These declines in food and oxygen further weaken mussels and other stream creatures.

Light reduction also hinders the periphyton (algae, diatoms, and cyanobacteria) growing on rocks, logs, and other hard structures on the stream bottom. This diminishes food for snails, which themselves are food for fishes such as suckers and drums. Eventually, mud buries these hard structures, or floods scour them away. The stream ecosystem is rapidly losing structure, biodiversity, and resilience—the ability to recover after disturbance.

In the murky, mud-infused water, fish struggle to find prey, spot predators, or court mates. Those that deposit eggs into rock crevices, including sturgeon, cannot find proper spawning habitats. In healthy streams, fertilized fish eggs sink to the bottom and adhere to rock surfaces. But if a mud layer coats the rocks, spawning is like dropping a glazed donut onto a sandy beach. The egg becomes coated with sediment, does not stick, and rolls downstream to unsuitable habitats or a hungry predator. Even if the egg does survive, the coating prevents the embryo from absorbing enough oxygen from the water.

So, while mud is a natural ingredient in the mixture of sediments migrating through a natural river system, too much of it causes biodiversity to decline. Species disappear, or their populations drop to ecologically irrelevant levels. Generalist species that tolerate poor water quality replace the original fauna. The river is still there, but it is less diverse, less resilient, and less able to offer us its full range of ecosystem services.

The problem of agricultural erosion continues today, over four hundred years after Jamestown. It doesn't matter whether farmers plow land behind a mule or on a tractor. Nor does it matter whether the crop is tobacco, cotton, or genetically modified corn. Stormwater will wash exposed soil into a creek or river. It's a tough situation for both farmers and nearby streams.

For several centuries traditional agricultural practices and resulting erosion were the norm. There was little effort made to keep soil in place. Farmers plowed fields downward from hilltops, thus creating furrows through which rain and mud raced downhill. They cut fields to the edge of streams, leaving no vegetative buffer. So, as agriculture spread across the Southeast, erosion and mud pollution intensified.

By the end of the nineteenth century, agriculturalists had cleared 30–40 percent of the southeastern landscape, and most of this area was plagued by erosion until well into the twentieth century.[7] The worst of the erosion across the area was in the Piedmont, a hilly region extending from east central Alabama to New Jersey and nestled between the Coastal Plain and the Appalachians. Roughly a third of Georgia, South and North Carolina, and Virginia is Piedmont. In the Southeast, settlers deforested the entire region and converted it to cropland, mostly for cotton, one of the region's other cash crops.[8] The combination of a sloped terrain and absence of soil conservation led to copious erosion. In the southern Piedmont, an area about the size

of New York state, erosion took an average of seven inches of soil over two centuries. Some areas lost twelve inches of soil.[9] It is no exaggeration to say that erosion reshaped the region. Pictures from the time show deep gullies cutting through hilly farmland. Some farms became miniature replicas of America's western canyonlands. Due to the loss of fertility, the economy of the region has yet to recover.[10]

Portions of the Mobile River Basin and the Apalachicola River Basin, as well as most of the large rivers of the southeastern Appalachian Atlantic slope, drain through the Piedmont. Their streams received the eroded sediment, and much of it is still in these streams. Scientists refer to these deposits from past eras as legacy sediments. Each major rain event remobilizes and redeposits these sediments during their long ride to the coast. Because they are steep and fast, mountain headwater streams can clear out sediment pollution from an erosion event within a few decades. But in the bigger, slower rivers of the Piedmont and Coastal Plain, it may take six to ten thousand years for all the Piedmont's legacy sediments to reach the coast.[11]

Deforestation and agricultural practices created severe erosion across the Southeast. Deep gullies often formed, such as these, preserved in Providence Canyon State Park, Georgia. Erosion still plagues the Southeast, and the displaced sediment has damaged tens of thousands of miles of the region's streams. Courtesy of Alan Cressler.

Today, sediment is one of the most ubiquitous pollutants in US rivers, responsible for the impairment of several hundred thousand miles of US creeks and rivers. In 2007, 3.4 US tons of soil were lost from cropland per acre per year in the Southeast. Nearly all this sediment enters creeks and rivers, causing habitat and biodiversity loss, eutrophication, and flooding.[12]

Sediment pollution would seem to be a likely suspect for the disappearance and decline of mussels. It has all the characteristics of a density-independent ecological factor that could inflict the damage malacologists have documented: it is ubiquitous and can harm or kill mussels across species, ages, sexes, and sizes. Mud pollution fits the profile nicely, and most mussel experts agree that sediment is the chief culprit in their disappearances.[13] However, a new line of thinking is challenging this conclusion.

In 2012, Wendell R. Haag, a fisheries biologist with the US Forest Service, provided the field of mussel ecology with a much-needed synthesis. His book *North American Freshwater Mussels: Natural History, Ecology, and Conservation* is an exhaustive review of mussel biology as it relates to the mussel conservation crisis.[14] Periodic syntheses are crucial to scientific progress because they can reveal trends and patterns that only emerge when one examines the accumulated work on a subject.

Haag reviewed over seventy years of research on the effect of mud pollution on mussels and found that the evidence linking mussel declines to mud pollution was not as strong as most had assumed. One issue is that river ecology is hard to study using classical experimentation. For example, if we wanted to study how suspended clays affect mussel feeding, we would want to hold constant all other factors that could influence feeding while we manipulate clay concentrations and measure changes in feeding. We would also need a control stream, where all conditions were identical to the experimental stream except for clay concentrations, which we would hold constant. And, we'd want to run the experiment for long enough to simulate what mussels are exposed to during their lifetimes.

Achieving this level of control of real streams is neither practical nor ethical. The difficulties of manipulating sediment levels aside, it's impossible to find enough near-identical streams to use in experiments. These are frequent challenges in science. Similar limitations hinder disciplines as diverse as climatology and cancer biology. Sure, we can conduct controlled laboratory experiments simulating stream environments, but scaling these findings up to a real stream and its full range of ecological relationships is fraught with complications.

Stream scientists have had to rely on the next best thing: comparing two or more similar streams that vary in a way suspected to be important. So they might, say, compare mussel populations between a stream in an agricultural watershed with heavy mud pollution and a stream in a forested watershed lacking sediment pollution. Such studies can readily reveal the impacts of humans on streams, but they fall short of confirming the factors causing those impacts. For example, high rainfall causes a simultaneous increase in current speed, stream depth, and suspended clays, and a decrease in temperature and dissolved oxygen. Thus, if mussel feeding declines after a storm in the agricultural watershed but not in the forested watershed, this is evidence that agriculture is causing the problem, but not of which specific factors are affecting the mussels. Perhaps it is the low oxygen during the flood, or exposure to pesticides on clay particles.

Though there is much correlative evidence suggesting a causal link between mussel disappearances and sediment pollution, Haag concluded that there isn't yet enough direct evidence to be certain of cause and effect. For sure, excessive sediment pollution harms mussels in ways like those I've described; Haag says as much. His point is that the evidence linking mud pollution and mussel disappearances is weaker than biologists have often asserted, and this leaves the door open to other possibilities.

Haag also found evidence potentially exonerating mud pollution as the culprit for mussel disappearances. He cited studies and observations of streams that have high sediment loads but also sustain an abundance and diversity of mussels. Other studies have found that juvenile mussels—those assumed to be most sensitive to sediment pollution—cope with mud pollution just fine. However, the most persuasive finding was that sediment pollution in US streams had waned since the 1930s, precisely during the period that mussel extinctions were on the rise and populations were facing severe declines.

The late nineteenth and early twentieth centuries had seen such massive amounts of erosion from cropland that there was an outcry for soil conservation. The public's demand for action intensified during the 1930s, when much of the United States suffered a multi-year drought, one made infamous by massive dust storms (the dust was airborne farm soil) in the southern plains, a region subsequently dubbed the Dust Bowl. The Soil Conservation Service was established in 1935, and erosion declined with the adoption of soil conservation practices such as hillside terracing and contour plowing. Forest loss in the Southeast slowed during this time, in part due to the Great Depression, and forest cover increased slightly by 1940.[15] Thus, the timing of mussel disappearances simply doesn't match the era of greatest sediment pollution.

If mussel harvesting, eutrophication, and mud pollution are not to blame, what other density-independent factor could have caused mussel disappearances this past century? What altered our rivers so profoundly that across a broad array of ecosystems and watersheds dozens of species and hundreds of their populations are now facing extermination?

13
A Golden Age Begins

Muscle Shoals. For most nonresidents the small city in the northwestern corner of Alabama is famous for its contributions to pop music, particularly the "Muscle Shoals sound." Since the late 1950s, dozens of famous artists, including Aretha Franklin and the Rolling Stones, have recorded some of their greatest hits at sound studios there. For aquatic biologists, however, the city is infamous. Muscle Shoals was where one of the most spectacular features of the eastern North American river landscape vanished, and a new form of ecological destruction began.

Perched on the south bank of the Tennessee River, Muscle Shoals rubs shoulders with the cities of Tuscumbia and Sheffield. Florence, the fourth of the quad cities, sits on the river's north shore. The popularity of the location has everything to do with the river. Native Americans—most recently the Cherokee—occupied the area for thousands of years prior to the arrival of American settlers in the early nineteenth century. Both the Native population and the newcomers enjoyed the abundance of food in the river and forests, and the fertile soils for farming. As a means of transportation, the river aided the trade and sale of goods with communities upstream and downstream. But the Tennessee River provided these amenities for hundreds of miles along its length. What drew so many to settle at Muscle Shoals was a fifty-three-mile expanse of wide and rocky river.

The city of Muscle Shoals was named after this river feature, also known as Muscle Shoals. A shoal is a very shallow section of river, and in upland streams shoals are created when rock resists the river's erosive power. Muscle Shoals (the river feature) owed its existence to the Fort Payne Chert, a dense sedimentary rock that formed over three hundred million years ago.[1] When streams cannot cut downward into a landscape, they push outward, and the chert's obstinacy forced the river to widen up to 1.5 miles. Outcrops of especially stubborn chert created over 120 islands, some large and forested, others small and bare. The combination of a river archipelago and a broad expanse of waters sluicing through rocky chutes and tumbling over small falls created one of the most beautiful spectacles of the preindustrial American river system.[2]

Early settlers to the region were less enthused about the aesthetics of the shoals than they were about the convenience it provided for crossing the Tennessee River. At low water, one could carefully walk across the river, a feat otherwise impossible for hundreds of river miles upstream and for over a thousand river miles south to the Gulf of Mexico. Blurry black-and-white photos show early American settlers crossing the river in mule-drawn wagons, though the sharp slippery rocks and tricky currents made the crossing difficult. Entrepreneurs eventually established ferries to shuttle people and goods across the river at deeper locations above and below the shoals, but the poor and adventurous would still cross the shoals.

The shoals were a hazard for river travelers heading upstream or downstream. Traversing the shoals by large boat at high water might be possible, but it was foolish. Small boats or canoes could make the passage year-round, but navigable channels were narrow, convoluted,

The Tennessee River in North Alabama naturally widened to create the Muscle Shoals, one of the greatest spectacles of the preindustrial American river system. This fifty-three-mile-long aquatic wonderland harbored the highest concentration of mussel species on Earth. Photo by Roland Harper, 1922. The University of Alabama Libraries Special Collections.

106 PART II

and bordered by sharp rock that could quickly sink a vessel. The difficulty of negotiating the fifty-three miles of shoal may explain the origins of the shoal's name. One of two theories is that the "muscle" referred to the muscle power needed to navigate the shoals by boat.[3]

A river this wide and shallow also created an aquatic wonderland, the freshwater equivalent of a coastal tidepool environment, but with a diversity that could rival an Amazonian tropical stream. Rivulets coursed over and through bands of the gnarled chert. The waters were clear and well oxygenated. Snails dotted every wetted surface. Fishes darted through pools and rills. And in the pockets, where the current slackened and gravel accumulated, there were mussels. Tightly packed, valve to valve, and paving the bottom, millions strong. The shoals were crazy thick with mussels. Not just numbers, but species, too. Roughly seventy species are known from Muscle Shoals, representing nearly a quarter of all North American species, and the greatest concentration of mussel species on Earth.[4] The other theory about the origin of the shoal's name is that it references this profusion of mollusks. *Muscle* is an outmoded spelling of *mussel*.

The Fort Payne Chert creating this aquatic paradise also played a role in the shoals' demise. Because of its strength and extent, the rock was to become the foundation on which a dam was built—not just any dam, but the largest one successfully attempted at the time. Completed in 1924, Wilson Dam was a marvel of engineering and innovative federal policy, and it inspired a frenzy of dam building across the United States lasting seventy years and completely transforming the ecology of American rivers. Unfortunately, Wilson Dam destroyed most of the Muscle Shoals, including its snail, fish, and mussel diversity. That which survived was lost several years later, with the construction of the Pickwick Dam downstream. The history of what happened at Muscle Shoals reveals much about the southeastern river crisis and offers clues about the mysterious disappearance of mussels in the twentieth century.

Colonists built the first dam in North America in 1640, to power a gristmill in Scituate, Massachusetts. Like most dams built during the

Wilson Dam on the Tennessee River, under construction in the early 1920s. Photo by War Department, Office of the Chief of Engineers, Nashville District. The National Archives at Atlanta, National Archives identifier 6125743.

next century and a half, it was a small dam on a small creek. Dam operators would store water in a pond behind the dam, then release the water when needed to spin waterwheels. The energy of the spinning waterwheel usually powered a gristmill or sawmill. On occasion, people built dams to store water for other purposes, like to control local flooding.[5] Regardless of their purpose, these early dams produced only local impacts on streams.

With the arrival of the Industrial Revolution in the mid-nineteenth century, entrepreneurs leaned on rivers more heavily. As manufacturing proliferated, industrialists built ever-larger dams to harness the power of ever-larger streams to spin factory turbines, crankshafts, and flywheels. Water stored behind dams was also used as a coolant in manufacturing. Used water—hot and laden with contaminants—was discarded downstream, creating surges in river levels known as factory tides.[6] Though these dams and related activities produced severe local impacts on rivers, nineteenth-century damming didn't cause mussel disappearances like those that swept the eastern United States in the next century. These dams were too small and too few, and mussel populations were healthy when the shell button industry was on the rise in the early twentieth century. However, the construction of large dams on the big rivers of the American landscape is a different matter entirely.

Large American rivers were only in a partial state of industrialization by the turn of the century. In developed areas, big rivers handled much of our pollution in the form of sewage, sediment, and industrial waste. Many had been dredged and cleared of debris to facilitate river transportation. A few large rivers had been dammed with small dams. Despite these modifications, most large American rivers were still flowing relatively untamed as the new century dawned. Their breadth and speed varied with the seasons. They would flood after heavy rains or snowmelt, or slump into a period of slow flow during drought. These natural cycles had governed the ecology of rivers and their valleys for hundreds of thousands of years. But these freedoms were not to last.

By the late nineteenth century, planners, entrepreneurs, and engineers were eyeing rivers and contemplating what could be done by building large dams on them. Big dams on these rivers could control floods or raise water levels to expand transportation networks, but neither of these uses justified the expense and risk of building large dams. Things changed in 1882, when the Appleton Edison Electric Light Company supplied the first electricity to US customers from its dam on the Fox River in Wisconsin. Almost at once the nation began clamoring for electricity and the dams to supply it. Urban dwellers wanted electricity for lighting and streetcars. Manufacturers wanted electricity so they could power factories at sites away from the riverbank. In addition, westward expansion during this time created interest in building large dams to provide power and irrigation for dry land that otherwise could not support agriculture. As it became obvious that hydroelectric dams on large rivers were a gateway to economic growth and profit, wealthy industrialists jockeyed for opportunities to build dams and sell power to the people.[7]

But despite this interest, vexing issues prevented the initiation of large dam projects during the early years of the twentieth century. There was no solid legal framework for making decisions about whether dams should be built. Nor was there legal clarity on how to balance the competing needs of those using the river for commerce and those who wanted to build dams that would limit or block river transportation. Moreover, the issue of building large dams became central in a policy debate raging in the United States over what role, if any, the federal government should have in expensive and influential undertakings such as dam building. Up until this point, the private sector had undertaken most natural resource development initiatives and infrastructure projects.[8]

In 1916, with American entry into the First World War looming, Congress approved funding for the construction of plants to produce ammonium nitrate, the compound used in explosives at the time. The process for doing this, developed by Germans Fritz Haber and Carl Bosch, required an abundance of electricity at levels that only a large hydroelectric dam could produce affordably.[9]

Muscle Shoals entered this history in 1917, when President Woodrow Wilson selected it as the location for two ammonium nitrate plants. Under his leadership, the Department of War cajoled the nascent Alabama Power Company into donating its land at Muscle Shoals to help with the war effort.[10] The shoals were ideal for the project. The Tennessee River supplied a dependably high volume of water, and the chert bedrock provided a strong foundation for the dam. The construction of Wilson Dam, named after President Wilson, was novel in three ways that together ushered in a wave of dam building in the United States and abroad. First, as the largest dam in the world, it proved that technology could harness the power of large rivers. Second, as the first federally funded large dam in US history, it inspired many more federally funded dams.[11] And third, as the first large dam designed for hydropower, flood control, and navigation, it reshaped visions of what dams could do.

Previously, dams were built and operated in different ways depending on their purpose. Flood-control dams frequently released water to maintain upstream capacity for times of heavy rainfall. Navigation and irrigation dams withheld water during wet months for release during the dry season. Small hydropower dams of the day retained water continually and released it in daily pulses to generate electricity on demand. Many engineers were skeptical that a single dam could meet all three needs. The Roosevelt Dam in Arizona partially resolved the issue. Completed by the US Reclamation Service in 1911 and financed by local farmers, the dam provided water storage for irrigation and generated electricity as a secondary priority.[12] Inspired by the Arizona success, engineers designed Wilson Dam to generate electricity for the nitrate plants, regulate floodwaters, and promote navigation by flooding the shoals and providing a boat lock to aid vessels traveling upstream or downstream.

The task of building Wilson Dam fell to the US Army Corps of Engineers. It was the corps' first large multipurpose dam, with many more to come. Construction began in 1918, just months before World War I ended. The corps finished the dam in six years and soon began supplying electricity to the nitrate plants and selling electricity to the Alabama Power Company.[13]

Despite its apparent success, controversy swirled around Wilson Dam after completion. The selling of electricity by the federal government alarmed many in the private sector, especially owners of electric companies. Because tax dollars funded dam construction, the government could sell its electricity at rates lower than the private utilities, thereby undercutting the emerging private electricity market. There was a related debate over whether the federal government should control rivers and other environmental resources instead of leaving those resources to be managed by the private sector and governed by the free market.[14] At one point, Henry Ford, founder of the Ford Motor Company and creator of a mass market for automobiles, offered to buy Wilson Dam. If Congress had accepted the offer, Muscle Shoals—and not Detroit, Michigan—would have become America's motor city.[15]

The collapse of the financial markets and the onset of the Great Depression in 1929 pushed the debate in favor of federal intervention in the economy and natural resource management. The election of Franklin Delano Roosevelt in 1933 ushered in an era of progressive politics, where large public works projects were undertaken to provide employment and promote economic development. Wilson Dam inspired the construction of similar dams across the country. The next five decades became known as the golden age of dam building.

At the peak of this era, 1950–70, about 200 large dams at least forty-nine feet tall were completed each year.[16] Nearly every river in the United States was affected. There are now over 91,468 dams in the continental United States, according to the corps' National Inventory of Dams. Most are publicly owned (local, state, or federal government). About 6,600 are large dams, and about 2,500 are the size of Wilson Dam or larger. At the beginning of the twenty-first century, these dams could store about 1.32 million gallons of water for each US citizen.[17] That's the equivalent of an eleven-mile wall of water measuring one foot wide and three feet tall for each person. Within the contiguous United States, dams are absent on only forty-two rivers measuring over 125 miles long. Dams or streamflow regulated by dams affect over 98 percent of river miles in the United States.[18]

Southeastern states got their fair share of dams. Georgia has the most, with 5,306 dams in the corps' database. It is followed by North Carolina (3,191), Virginia (2,790), South Carolina (2,343), and Alabama (2,273). Florida, Kentucky, and Tennessee have about 1,000–1,200 each.[19] Most of these large dams are in the uplands of the region—especially the Piedmont—where deep river valleys provide rushing waters for spinning turbines, ample reservoir capacity, and firm bedrock for anchoring a dam.[20] There are an additional 59,280 smaller dams in these states that are not in the corps' database but that have been cataloged by a network of state agencies, scientists, and others. Most of these smaller dams were built to create reservoirs for agriculture, fishing, stormwater management, or landscape diversification.[21] To add it all up: the Southeast has at least 78,623 dams, and more are inventoried each year. A mere handful of southeastern rivers are either not impounded or have long free-flowing sections—the Cahaba, Altamaha, Pee Dee, and Suwannee, among them.[22]

By the late 1970s, large federally funded dam projects began to wane for many reasons. For one, all the easy projects had been completed—there simply were few rivers left to dam. But more importantly, there was a shift in how the public perceived dams. Many had lost their unquestioned faith in large, federally funded projects, or they were more sensitive to the human and ecological costs of damming. Fiscal conservatives balked at the price of such projects.[23] Some felt betrayed because many dams failed to provide all the benefits the public had been promised. Meanwhile, environmentalists offered a convincing counternarrative of valuing free-flowing rivers for uses such as tourism, recreation, and wildlife preservation. Empowered by the suite of federal environmental laws passed in the 1970s, environmental organizations became savvy in providing legal resistance to dam projects. Lengthy and costly regulations and legal battles perpetually delayed dam planning and construction. For all these reasons, the building of large dams in the United States came to a near halt during the final decades of the twentieth century.

In our search for an agent responsible for the disappearance of mussels, we have been looking for one that affected a large geographic area, killed off populations independent of their density, and matched the timing of mussel disappearance in the twentieth century. Given its geographic extent and timing, the golden age of dam building may be the perpetrator we've been looking for. If it is, then Wilson Dam on the Muscle Shoals is our villain's first crime scene. And, by drowning the Muscle Shoals, the site of the highest concentration of mussel species and numbers on the planet, it was arguably the most egregious offense in a crime spree that lasted a full fifty years.

14

How to Drown a River

I knew nothing about its history when I first laid eyes on Wilson Dam in December 2015. I had driven north from Birmingham long before dawn with plans to meet up with my friend Charles Grisham. It wasn't an interest in rivers drawing us to the dam; it was birds. Charles lives near Huntsville, Alabama, and I had grown to enjoy his unbridled enthusiasm for tracking down rare species. He is also a ninja with a camera and takes the most beautiful bird photographs I've ever seen.

We were heading to Wilson Dam because a posting on a bird-watching listserv reported that the dam was discharging water and attracting thousands of gulls. The flocks included species Charles had never photographed and I had not seen in several years. Plus, the semester was over, and I was overdue for a birding adventure.

At first sight I noticed that the dam was older and more beautiful than any other I had seen. The few other dams I had visited were much younger and more austere. In contrast, Wilson Dam was built in the neoclassical style of architecture reminiscent of ancient Greece and Rome. It stands 137 feet tall and is composed of 58 stately spillways, each capped with an arch. Understated adornments decorate the masonry of the arches and bridgework above them. The spillways stretch in a phalanx for nearly a mile across the river.[1] The structure is suitably dignified for holding back the mightiest river in the Southeast.

Dam operators were lowering the reservoir to make room for the heavy rains of late winter and spring that were coming. These are the rainiest months for the Southeast, a time when free-flowing streams build in volume and speed, sometimes spilling over into adjacent floodplains. By releasing water now, the impoundment behind the dam would have the capacity to hold these waters and gradually release them downstream. This reduces flood risk and helps ensure a

more predictable flow for barges navigating the river. Dams up and down the Tennessee River and across the eastern United States do the same thing early each winter.

Through each open spillway poured a violent torrent of water that shot outward dozens of feet. The cascades turned the tailwaters below the dam into a chaotic jumble of whitewater, foam, and spray. A mist rolled over the water, chasing the current. The roar of pounding water carried through the chilly air. Even at ninety-one years of age, it was a powerful testament to human engineering.

As we had hoped, there were birds everywhere. The release of water was sucking in river fish from the impoundment, stunning and pulverizing them, then shooting them through the spillway gates. Sadly, this

Wilson Dam releasing reservoir water in December 2015 to increase storage capacity to minimize winter flooding downstream. Photo by R. Scot Duncan.

is a common outcome for fishes trying to migrate downstream past dams. The pummeled fish were easy pickings for the swarms of gulls swirling over the whitewater. Each species of bird sported a distinct pattern of gray, white, and black. Cormorants, jet black with yellow bills and emerald eyes, bobbed in the turbulence between dives to hunt fish. Alongside them were huge American White Pelicans, resembling awkward swans. They seined the water for minnows with long, gaudy-orange beaks and the elastic skin of their throat pouch. All these species were birds of coastal ecosystems that never would have been in North Alabama if not for the dams and the large reservoirs they make. It's another sign of the impact that large dams have had on the region.

With time we found the species we were seeking, plus two Little Gulls, a European species exceptionally rare in the Southeast. Mesmerized by the constant swirl of water and birds, and intent on finding more rarities, we watched the river as long as our schedules would allow.

On the face of it, the scene that day at Wilson Dam did not resemble one of ecological bankruptcy. Water was flowing, birds were feeding, and fish were abundant, even if a few were getting pulverized. So why is it that many river biologists and conservationists bemoan the presence of dams on our rivers? And how, if at all, are dams a threat to mussels? To answer these questions, let's follow what happened to the mussels of Muscle Shoals after Wilson Dam was completed.

Then, as now, most hydroelectric dam operators need a lot of water in reserve so they can generate power whenever electricity is needed (the exception are run-of-river dams, uncommon in the Southeast, which generate power without creating a reservoir). They must stack the water high enough so that the released water packs enough punch to spin the turbines generating electricity. Thus, engineers designed the Wilson Lake reservoir to be much deeper than the shoals it buried. As the reservoir filled, the river's current slowed and came to a near halt. As this segment of the Tennessee River became a lake, changes in depth and current speed launched a cascade of other transformations that extinguished most mussels in the thirty-one miles of shoals above the dam.

The first problems for mussels were that temperature and dissolved oxygen levels plummeted. Beneath many feet of water, sunlight was unable to warm the reservoir bottom. Dissolved oxygen levels dropped because there was no river current to mix deeper water with the warmer, oxygenated surface waters. Low oxygen levels worsened for creatures on the bottom as microbes used up remaining oxygen when digesting the dead plant material washing into the impoundment and settling on the bottom.

Despite the cold and low oxygen conditions, some mussels survived for months after the impoundment filled. Mussels are renowned for being able to wait out tough conditions. However, they must eventually feed, and whenever they tried they met new problems on the lake bottom—mud and starvation. With the slowing of the river's current, mud that washed into the reservoir settled to the bottom. These particles damaged and clogged mussel gills, further eroding their health. Eventually the reservoir buried the rocky shoals above Wilson Dam under twenty feet of mud.[2]

Shoal mussels that inched upward to keep up with the lake bottom faced another problem—undernourishment. Before the dam, the river mussels had relied on an uninterrupted stream of water to bring them sustenance. That food included phytoplankton, whose populations proliferated at the surface of the river. After the dam, there was still plenty of phytoplankton, but it was at the surface and far beyond reach.

Cold, hungry, oxygen-starved, and buried under muck—many of the shoals' mussels didn't survive. Of the sixty-nine mussel species historically known from the shoals, thirty-two of them disappeared after damming.[3] Most of those still present are hardy generalists that can tolerate a wide range of aquatic conditions. They are doing fine in the Tennessee River's new ecosystem. For the remaining species, their small populations are trapped within a few hundred yards of Wilson Dam, where the scouring tailwaters maintain a shallow, silt-free environment.

The impact of dams on river ecosystems extends far downstream. Recall that hydroelectric dams release high-pressure water from deep below the surface. Just as the mussels experienced in the reservoir, this water is much colder than what most southeastern stream species can tolerate.

If another dam downstream doesn't drown the river (as what happened to the remaining twenty-two miles of the Muscle Shoals), it can still lose habitat though sediment starvation. Stream habitats are continuously losing sediments stripped away by the current. However, if a dam traps those sediments, downstream habitats such as sand- and gravel bars will shrink because the river cannot replace lost sediment. Sediment accumulation behind the dam also creates problems for dam operators because the reservoir loses storage capacity as impoundments fill with sediment.

Dams also impose unnatural regularity on the flow of rivers. The volume, speed, timing, temperature, and chemistry of river water varies seasonally in a pattern altogether known as the river's flow regime. In the Southeast, rainfall peaks in late winter and spring and then declines in the summer and fall. Rivers rise and fall accordingly.[4] Most species in a river ecosystem adjust their feeding, migration, and reproduction to coincide with the most favorable seasonal conditions for these activities. Unusually high or low periods of rainfall, such as those caused by tropical storms or droughts, can interrupt average flow regimes in some years. But even these aberrations play a positive role in supporting biodiversity and other river functions. For example, large floods transport sentiments long distances downstream to ecosystems that need them, such as the swamps and marshes of river deltas. Or, a year of little rainfall can improve the survival of young mussels for some species.[5]

Dams end much of this healthy, natural variation sustaining ecosystems and biodiversity.[6] Flood-control dams hold back water during heavy rains, then release it gradually. Navigation dams control the release of water to maintain a consistent range of depths downstream. Irrigation dams retain water during the wet season for extraction during the dry season (when, by the way, it is needed most by river wildlife). Hydroelectric dams stockpile water at night for release during the day, when demand for power peaks. These distortions to the natural flow regime disrupt the life cycles of river species, including mussels, and alter the ecology of the river for dozens to hundreds of miles downstream.[7]

What happened to the Muscle Shoals early this past century would not happen today. The utilitarian view of natural resources that prevailed for most of American history is now balanced with new values arising from the modern environmental movement. The federal environmental laws passed in the 1970s codified many of these values. Soon after they were established, they began impinging on dam planning and construction. The first clash between dams and the new environmental laws took place in the Tennessee River watershed, several hundred miles above Wilson Dam. At the center of the conflict was a small, unassuming fish that many have vilified, and few have championed.

15

Unlikely Survivor

IT'S NOT OFTEN THAT A student research paper affects the future of federal policy, but that's exactly what happened in the mountain foothills of eastern Tennessee in the early 1970s. At issue was the construction of Tellico Dam on the Little Tennessee River. The large dam was under construction when Congress passed powerful federal laws to protect the American environment. This legislation was fresh, untested, and not fully understood by the citizenry or the government.

Tellico was the latest dam built by the Tennessee Valley Authority (TVA), itself an experiment in American governance inspired by the success of Wilson Dam. Established in 1933 as part of Roosevelt's New Deal, the TVA was to have the power of government and the flexibility of a corporation. Its mission was to consolidate and amplify federal investment in the economic development of the Tennessee River region, an area of entrenched poverty spanning seven states. TVA programs included reforesting marginal agricultural lands, promoting soil conservation on farms, and eradicating malaria.[1] However, the TVA's most ambitious goal was to tame the Tennessee River's propensity to flood, improve river navigation, and supply electricity throughout the region. For decades, the dam was the TVA's favorite tool for the job. By 1944 the entire length of the Tennessee River was dammed, and by the late 1970s, all its major tributaries were dammed. Today, the TVA operates thirty hydroelectric dams and seventeen additional dams. Barges can now travel from the Mississippi River all the way to Knoxville, Tennessee, a total of 652 river miles.[2]

The availability of affordable electricity provided by the TVA and other utilities in the mid-twentieth century revolutionized life in the Southeast. Rural residents could now use electricity around the home for heating and cooling, lighting, refrigeration, cooking, and washing. Electrification spurred industry and the growth of cities. More could

be done with less, and it could be done a lot faster. Industry arose in former farming towns along major rivers, a result of electrification and improved river navigation. Small communities enjoyed modernized hospitals and other services. Greater prosperity generated more entrepreneurship and business investment, which fueled job growth and economic diversification. City life became increasingly attractive and affordable, and the urban footprint on the landscape grew. The proliferation of hydroelectric dams was behind much of this progress.

Despite the benefits of rural electrification, there was a counternarrative to the large dam projects in the Southeast and elsewhere in the United States. Loudest among the opponents were those the reservoirs were displacing. The federal government granted the TVA the power of eminent domain. This allowed it to force landowners within the boundaries of the project—the "taking line"—to relinquish land ownership. By the end of its dam construction era, the TVA had displaced twenty thousand families. The TVA compensated landowners based on the market value of their land, and services were offered to help families relocate. Many were pleased and moved into better homes or to nearby cities.[3] However, some landowners, even those in favor of

Tellico Dam—designed solely to benefit investors—flooded two towns, displaced three hundred families, and threatened the survival of the Snail Darter. The Tennessee Valley Authority and the National Archives and Records Administration.

UNLIKELY SURVIVOR 121

the projects, protested bitterly that the TVA was assessing their land far below market value. Some accused it of offering generous payments to those who publicly supported its projects and to the landowning power brokers in the region.[4] Some resistors refused buyout offers and remained on their land until they had no further legal recourse. Among them were farmers who had worked the rich bottomland soil for decades, as had generations of their ancestors. As is usually true in cases of eminent domain, most of the displaced were poor and lacked the resources to stand up to the might of the TVA.[5]

The most controversial of the TVA's big dams was also one of its last—Tellico Dam. The reservoir was to inundate two towns, five churches, and four schools, and displace three hundred families. Ironically, some of the valley's residents had moved there after being displaced by other dam projects or the creation of the Great Smoky Mountains National Park.[6] This was par for the course, but unlike other TVA dams, the Tellico project was designed to promote economic development to benefit outside investors. The TVA imagined that the reservoir created would attract industrial, residential, and recreational development. The dam and impoundment would do little for those residents of the area who didn't need flood control and already had electricity.[7] Further infuriating the opposition was that the taking line encompassed 39,500 acres, an area over twice the size of the reservoir. The TVA was planning on taking this extra land through eminent domain eviction and selling it to investors to pay for the project.

Among the frustrated stakeholders were Native Americans. Tellico Dam would flood historical Cherokee town locations, including Chota and Tanasi, plus many burial and other significant archaeological sites, some dating back ten thousand years. The TVA planned to hire archaeologists to excavate or relocate Native American sites before inundation, but these disturbances affronted many Cherokee, whose ancestors had lived in the valley.[8] The protestations of the Cherokee and the valley's current residents were ineffective, and construction on Tellico Dam began in 1968—just as the modern American environmental movement was building momentum.

In 1970, Congress passed the National Environmental Policy Act to require any federally funded project to review potential impacts on the environment and cultural sites as part of the approval and implementation process. For projects determined to have a significant impact, planners must complete an environmental impact study and submit a statement summarizing the proposed project's environmental and cultural impact. The act also requires these environmental impact statements to include input from the public and other federal agencies.

The National Environmental Policy Act is a powerful tool for the citizenry and was the grounds for the first major legal fight over Tellico Dam. A loose coalition of landowners, historic preservationists, environmentalists, sport fishers, and proponents of fiscal conservatism sued the TVA for its failure to conduct an environmental impact study. The plaintiffs won an injunction in 1972 halting construction, but their victory was short lived. Construction resumed the next year after the TVA submitted a statement that was then approved.[9] Though no one knew at the time, however, the Tellico battle was far from over.

Heretofore, federal law had offered no protection of wildlife unless a species had significant commercial value. But after the passage of the Endangered Species Act in 1973, it was illegal to cause a species to become extinct. If the government concluded that a species was endangered—meaning that extinction was imminent—then actions hastening its extinction were illegal.[10]

Percina tanasi, the Snail Darter, is a small, nondescript fish. Measuring just a few inches long and camouflaged tan and brown, the darter dwelled in a shallow rocky shoal of the Little Tennessee River watershed. It was unknown to science until biologist David Etnier of the University of Tennessee discovered it as part of Tellico Dam's environmental impact study. The dam was destined to inundate the shoals and wipe out the only known population of the darter.

At the time, Hiram "Hank" Hill was a second-year law student at the University of Tennessee. He learned about the discovery of the darter from friends who were Etnier's students. Curious about the issue, Hill got permission from his professor Zygmunt Plater to write a research paper on the legal relevancy of the Endangered Species Act to the construction of Tellico Dam. The paper's conclusions were

The humble Snail Darter is one of the most famous fishes in the history of species conservation. Courtesy of Bernie Kuhajda, Tennessee Aquarium Conservation Institute.

clear: the fish should enjoy the protection of the federal government. Hill's argument was so convincing it inspired a new legal challenge to Tellico Dam.

Hill, Plater, and a local attorney petitioned the US Fish and Wildlife Service to list the darter as endangered under the Endangered Species Act. They were successful, but the TVA continued construction, claiming that the act didn't apply because the dam was already under construction when the act became law. In what became one of the most famous lawsuits in environmental legal history, Hill and others sued the TVA in 1976 for violating the Endangered Species Act. Hill eventually argued the case before the US Supreme Court. Much to the surprise of cynics, the court ruled in favor of Hill in 1978. Construction stopped a second time. The ruling sent shock waves through the establishment. If a small fish and a law student could stop the construction of a TVA dam, what was next?[11]

Agriculture, forestry, mining, highway building, river damming, and navigational improvements—these and similar undertakings had long been indisputable signs of progress. But during the 1960s and 1970s the public began questioning whether the benefits were worth the sacrifice of air and water quality, and other valued attributes of the natural world. Others pointed out that a privileged few usually enjoyed the dividends of natural resource use, while the least powerful in society—such as the Cherokee and rural farmers of the Little Tennessee River—were those most negatively affected. These latter observations and their implications were the seeds of the modern environmental justice movement.

With the passage of the National Environmental Policy Act, the Endangered Species Act, and similar legislation in the 1970s, powerful public and private institutions involved in the extraction and use of natural resources were to be held accountable for negative impacts of their endeavors. *Tennessee Valley Authority v. Hill* was a wake-up call for the establishment. Congress had put powerful legal tools in the hands of the citizenry.

Over in Washington, DC, legislators nervously realized that the initial (1973) version of the Endangered Species Act was more powerful than they had anticipated.[12] Within months of the Tellico ruling, Congress passed a weakened version of the act. Under the 1978 amendment, a presidential cabinet–level tribunal—the infamous "God Committee"—was convened to rule on the darter-versus-dam case.[13]

All parties were astonished when in 1979 the tribunal ruled that the TVA should not construct Tellico Dam. But the deciding issue wasn't the Snail Darter. Instead, the tribunal determined that the cost of the project far exceeded any reasonable expectation of economic benefit. The judgment affirmed accusations that Tellico Dam was wasteful "pork barrel" spending—a large expensive project without substantiated need. Nevertheless, Congress soon overrode the tribunal's ruling, construction resumed, and Tellico Dam was completed later that same year.[14] Despite the final outcome, the case demonstrated how the new federal environmental laws now empowered citizens to play a significant role in decisions about the management of natural resources.

Although Tellico Dam was ultimately constructed, the TVA no longer wields the power it once had to manipulate rivers. This became clear in 1983, when legal challenges forced it to pause construction on Columbia Dam on the Duck River, a midsize tributary to the Tennessee River. Blocking progress were two mussel species recently placed on the endangered species list: the Birdwing Pearlymussel and the Cumberland Monkeyface. At the time, neither was known to survive anywhere else but in a river reach that the new reservoir would flood. Having already sunk over $80 million into the project and poured 90 percent of the dam's concrete, the TVA sought ways to build the dam while adhering to federal law protecting the species. In 1995 the TVA concluded that the venture was a lost cause and began dismantling the partial dam. Today, with over 150 species of fish, nearly 60 species of mussels, and over 20 species of snail, the Duck River harbors the largest remaining concentration of freshwater biodiversity on the North American continent, and is among the top three rivers in the world for fish and mollusk diversity.[15]

Whatever became of the Snail Darter? As it turns out, the little fish became a big liability for the environmental movement, as Zygmunt Plater, now a professor at Boston College Law School, has explained.[16] The case was used as ammunition for those in favor of free market control of natural resources and those opposed to public scrutiny of large government and private sector projects. The Snail Darter became exhibit A that environmentalists are idealistic tree huggers willing to halt economic development for the sake of otherwise inconsequential species or ecosystems. The image of environmentalists as extreme ideologues valuing flowers and fishes over people is a propaganda tool successfully wielded by conservatives and industrialists ever since.

Plater reminds us that this slanted interpretation of the Tellico story fails to acknowledge that opponents to Tellico Dam included ordinary residents who didn't want to give up their homes and farms, Cherokee who didn't want their sacred lands flooded, and fiscal conservatives who wanted accountability for tax-funded initiatives. More importantly, distorted portrayals of the Snail Darter story fail to acknowledge that initial opponents to the dam were pro-development. They offered an alternative economic development plan for the Little Tennessee River Valley, one that allowed farmers to keep their lands and promoted tourism based on fishing and paddling as well as culturally significant sites such as the historical Cherokee towns. It was even proposed that part of the TVA purchase become a national park to complement the nearby Great Smoky Mountains National Park, within which the river originates.[17] Tellico opponents valued the livelihoods and prosperity of far more people than the few who now live in the residential developments bordering the reservoir. Though a concern for humanity's greater good is usually at the core of environmentalist arguments, their opponents conveniently ignore this.

Ironically, the one natural feature of the Little Tennessee River that survived Tellico Dam was the Snail Darter. Just prior to the completion of the dam, TVA biologists worked feverishly to capture the darters and move them to nearby creeks unaffected by the dam. And in recent years, biologists have found a few other populations elsewhere in the Tennessee watershed. The efforts paid off: the Snail Darter, the most unlikely survivor of the river's damming, is no longer considered threatened or endangered under the Endangered Species Act. By that measure alone, this was the first major victory of the new law.

16

Case Closed?

We now have the Endangered Species Act and other laws to protect species from extinction. But if we want to restore river mussel populations to help control pollution and to avoid further mussel extinctions, then we need to know what caused mussel declines in southeastern rivers. We know the culprit affected a large geographic area, killed off populations and species independent of their abundance, and did this when most mussel species or populations disappeared. We've ruled out nutrient and sediment pollution. Both are impairing river ecosystems, but as we've seen, neither can account for widespread mussel declines.

Dams may be our culprit. The construction of big dams on big rivers during the golden age of dam building meets two of our criteria: dams are ubiquitous on the American landscape, and through "death by reservoir" they would have snuffed out mussels independent of population density. But what about the last criterion—the timing? Does the history of damming match the timing of mussel extinctions and the declines of surviving species?

It's not easy to declare a species extinct. Just because a biologist cannot find a species on a field survey doesn't mean that it is not there. This is especially true for species of large rivers because they dwell in conditions where it is difficult to survey. Scientists who document extinctions are usually striving to save these species. Typically, they are looking for surviving populations of a rare species. This work often requires countless trips to museums to scour collections for specimens of the lost species. Experts must reverify specimen identifications. Collection location records of valid specimens then lead to field investigations. Biologists survey, resurvey, and re-resurvey rivers of known occurrence.

They visit other locations where the species might be surviving. These efforts are expensive and can drag out for decades. Not until all options are exhausted is a species declared extinct. And that's the easy part. Figuring out what caused the extinction often requires similarly painstaking detective work. This is why it has taken malacologists so long to piece together an explanation of what triggered the wave of twentieth-century mussel extinctions and population disappearances.

By the time that Wendell R. Haag began studying the cause of mussel disappearances, other scientists had done much of the work of documenting extinctions and population disappearances. The challenge was that this information was scattered throughout academia and across federal and state wildlife agencies. Data was in museums, unpublished reports, field notes, and scientific journal articles. There were also indirect clues in what archaeologists and historians had learned about Native American and post-Columbian history. Haag hunted down all this information and developed a framework for understanding mussel disappearances. It was Haag's synthesis that suggests pearl and shell harvesting wasn't responsible for mussel disappearances. And it's Haag's sleuthing that has cast doubt on the prevailing theory that sediment pollution is the culprit. What Haag found when he examined the evidence stacked against dams was, well . . . damning.

It was already known that dams had caused the loss of at least a few mussel species. The Leafshell and Round Combshell both disappeared simultaneously when we dammed their rivers.[1] But for others, like the Sugarspoon, Yellow Blossom, and Tennessee Riffleshell, the cause of extinction was still a matter of speculation. Haag collected all the data he could find on the occurrences of these and other species. He then mapped historical and contemporary sightings of species and related them to what was known about changes in their rivers. He didn't do this just at the level of the entire species; he examined populations within species because local disappearances can offer clues about the overall pattern.

Hagg found that mussel extinctions and precipitous population declines began in the 1930s, coinciding with the onset of the golden age of dam building. By the end of this era, in the 1980s, twenty-five mussel species were extinct, representing over 8 percent of North

CASE CLOSED? 129

American species. Haag's sleuthing revealed that half of these extinctions and the disappearances of populations of surviving species were caused directly by dams due to the ecological changes we visualized after the construction of Wilson Dam.[2]

Haag's investigations revealed other, more stealthy ways that dams caused mussel species and populations to disappear. The first species to vanish were those living only in the large rivers. The effect of dams on them was swift and direct. Mussel species with broader ecologies persisted in smaller tributaries. They survived the direct effects of damming because reservoir levels did not completely flood these tributaries. However, their populations were small and fragmented. No longer connected to a core population in the main river, they had no chance of rescue if their populations dipped. Small populations are also prone to inbreeding and low genetic diversity, threats that reduce a population's resiliency to environmental change.

These factors made small, isolated populations vulnerable to disappearance, even from threats that would not have wiped them out in the past. The story of the Acornshell's final days reveals how this can happen quickly. This small mussel was widespread in the Cumberland River and Tennessee River watersheds. In the Tennessee, its range extended from Muscle Shoals all the way into Virginia.[3] However, after both rivers were fully impounded, the Acornshell survived only in a small stretch of the Clinch River that had escaped the reach of the reservoir. Industrial chemical spills in 1967 and 1970 wiped out this last population.[4] While damming didn't pull the trigger, it created the circumstances that led to the Acornshell's extinction.

So, via direct and indirect impacts, the golden age of dam building, a period from 1930 to 1980, is to blame for most mussel extinctions and endangerments. Today, about 120 species of mussels in the United States (44 percent of living species) are threatened or endangered because they survive in small, isolated populations. A third are down to just one or two populations, and some consist of a few dozen individuals or less.[5] This is why Haag and others predict we are on the cusp of a second wave of mussel extinctions caused by dams. Haag describes this as an extinction debt for damming rivers that we haven't yet paid.

The heyday of dam building also affected another group of river mollusks, freshwater snails. In contrast with mussels, snails are smaller and more mobile, and they graze on hard substrates instead of filter feeding. Like mussels, freshwater snails are more abundant in the Southeast than anywhere else on the planet, and they perform vital ecosystem services. Acting like miniature lawnmowers, they prevent excessive algal growth on rocks and logs. Without them, streams would become carpeted with thick growths of algae, dissolved oxygen levels would drop, and fish and other wildlife would lose habitat. Unfortunately, damming caused many snail species to disappear. For example, by the time Alabama Power completed its damming of the Coosa River in the 1960s, the company had knowingly driven over two dozen snail species to extinction.[6]

Haag's synthesis also revealed that dams have a sidekick that has contributed to mussel extinctions and endangerments. The modification of river channels to promote navigation has also destroyed much river habitat. These efforts began early. The first Euro-American settlers removed trees and rock outcrops, and these efforts spread and intensified over the following centuries. Rock reefs were dynamited, and sand- and gravel bars were dug out and hauled away.[7] On rivers like the Chattahoochee, engineers built rock walls known as groins to funnel water to the center of a river. This created a faster current that scoured away sediment—and mussels—and maintained a deeper channel.[8] Some sinuous rivers were straightened. The most widespread form of channel modification has been dredging. Because currents continually move river mud, sand, and gravel, most large rivers used for commercial navigation are dredged regularly to keep channels deep enough for the passage of towboats and the barges they push. All these channel modification efforts contributed to the loss of natural habitats and the decline of mussels.

The combination of damming, dredging, river straightening, and the like during the period of 1930–80 reshaped most American rivers. Haag offered a name for this period of river history: the era of systematic habitat destruction. One of the worst and last major projects from this era in the Southeast was the completion in 1984 of a shipping route from the Tennessee River to the Tombigbee River to create the "Tenn-Tom" waterway. Engineers dug a wide canal connecting the two rivers, and the entire length of the Tombigbee was dredged, widened, and dammed to handle barge traffic. The Tenn-Tom lopped hundreds of miles off the voyage from the Lower Tennessee River to the Gulf Coast, a route that previously required a trip over to and down the Mississippi River. Project boosters promised that the project would attract industry and spur economic growth in the counties surrounding the waterway, which were among the poorest in the South. Today, the Tenn-Tom attracts only about a quarter of the cargo that planners predicted (primarily wood and metal products, chemicals, grain, and gravel). The waterway attracted a smattering of industry, but poverty in most surrounding counties is more entrenched than it was before the canal was opened.[9] Critics point to the Tenn-Tom as another example of an overpromised project to benefit a few industries at the expense (nearly $2 billion) of taxpayers. Haag marks the completion of the Tenn-Tom as the end of the era of systematic habitat destruction in US rivers. Since that time, there have been only a handful of large, federally funded river modifications in the United States.[10]

Despite the mediocre success of the Tenn-Tom, the widespread modification of large rivers to promote commerce has spurred economic growth throughout the Southeast and beyond. But from an ecological standpoint, nothing good has come of it. Channelization stripped rivers of habitats (e.g., sunken logs, gravel beds, and sandbars) and their dependent species. Canals linking rivers eased the spread of exotic nuisance species such as Zebra Mussels. The complete industrialization of the Tombigbee River destroyed one of the most biologically diverse, unpolluted, and free-flowing large rivers remaining in the Southeast. Three mussel species endemic to the Tombigbee went extinct, and a fourth endemic just barely survives.[11]

Though the era of systematic habitat destruction is over, Haag's synthesis revealed that many remaining mussel species are still in jeopardy. We've already learned about the vulnerability of small and isolated populations, but there are other, more mysterious threats. Since the 1960s, many rivers in the Midwest and South have experienced mass die-offs of their mussels. The causal agent affects all mussel species in

The Tenn-Tom Waterway helps barges avoid hundreds of miles of river navigation to and from the Gulf Coast. However, it destroyed the most biologically diverse, unpolluted, and free-flowing large river remaining in the Southeast, and has caused three mussel extinctions, thus far. The Tennessee Tombigbee Waterway Development Authority.

these streams, and at least two species have gone extinct as a result. Strangely, it's only the mussels that die; other aquatic taxa, including the most sensitive species, survive just fine. Investigators can rule out all the usual suspects—dams, dredging, mud or industrial pollution, invasive species, even poaching—for one reason or another. A pathogen may be to blame. That would explain why only mussels die. But scientists have not identified any such disease. Malacologists continue to investigate these die-offs, aptly dubbed enigmatic declines, but for now they remain a shadowy threat to mussels and efforts to save them.[12]

Construction of the Tenn-Tom Waterway in 1980. This was the last and the most damaging major project during the era of systematic river habitat destruction. US Army Corps of Engineers.

Mussel die-offs are occurring in rivers of the Midwest and South. The cause is unknown, but a disease is suspected. This mass mortality event occurred in the Clinch River in 2019. US Fish and Wildlife Service.

Another mystery involves mussel reproduction. In his review of mussel data, Haag found dozens of cases where mussel populations are no longer accumulating young. Biologists find healthy populations of adults but haven't found offspring for years, sometimes decades. Without recruitment, these populations—some the last of their kind—are doomed. Many populations of short-lived mussels—species living less than a decade—have already vanished. Individuals of other species live fifty years or more, and a few may live for centuries.[13] These long-lived species are still in rivers today, doing their thing, but young mussels are absent. Haag and other biologists may have solved this mystery, and the explanation they offer reveals some of the most bizarre interactions between freshwater species ever discovered.

17

Hitchhiking Elephantears

THE POPULATION OF ELEPHANTEARS IN the Cahaba River may be gone by midcentury, but you wouldn't know this without looking carefully. These big mussels are common in the midreaches of this Alabama river. They are blocky in shape (malacologists say rhomboidal) and covered with black periostracum, a protein layer that protects the underlying mussel shell from abrasion and the natural acids in river water, which can dissolve the shell. Wade into the cool shallows on a summer day and you'll find plenty of big Elephantears nestled between the multicolored rocks, happily feeding on the phytoplankton the current brings. Or visit in late spring and gather the many large, empty shells cast onto the sand beaches by winter freshets. Either way, a simple tally suggests the species is doing well.

But look again, this time at what is *not* there. None of the Elephantears you find are young. Most are at or near their maximum length of six inches, their periostracum is wearing away, and their shells are pitted and worn: telltale signs of age. In human terms, it's like a retirement community with no visitors. The Elephantear population is not reproducing—it is dying out.

There's nothing unusual about Elephantears to suggest why this population is having trouble. The Elephantear was never valuable as button material, and shellers considered the species a nuisance due to its abundance.[1] Nor is the species endangered or rare within its range, which encompasses much of the Southeast and the Mississippi River Basin.[2] Nevertheless, the Cahaba population is growing leaner by the year. Surveys only turn up worn and weathered Elephantears that are many decades old.[3] A 2008 study found that of eleven mussels sacrificed for analysis, counts of the annual growth layers that compose the shells revealed the youngest began its life in 1969.[4] Something is clearly wrong.

The Shoals Lily blooms in shallow, rocky reaches of several southeastern rivers each May. Pictured here are lilies in the Cahaba River, one of the most biologically intact rivers remaining in the Southeast. Photo by R. Scot Duncan.

Neither does the Cahaba River offer any strong clues as to why the Elephantear is not reproducing. A tributary to the Alabama River, the Cahaba is considered by many to be one of the most beautiful rivers in the Southeast. It is one of the only rivers in the region we never modified heavily with large dams or channelization. For this reason, the river retains much of its original aquatic fauna and currently harbors 135 species of fishes, 19 species of crayfish, 31 species of snails, and 36 species of mussels.[5]

The Cahaba does have a history of pollution. Birmingham's metallurgical industry from the late nineteenth to the early twentieth century polluted its headwaters and eradicated nine species, including mussels, snails, and fishes. The Cahaba recovered after this industry moved, but it now struggles with urban sprawl and the resulting stormwater, wastewater, and sediment pollution. These modern ailments

The Elephantear is a large mussel found in many watersheds of the Southeast. Some populations are secure, but others are fading away. Courtesy of Alan Cressler.

Elephantear mussels collected in the Cahaba River circa 2005. This population hasn't reproduced since 1970. Courtesy of Paul L. Freeman.

contribute to eleven of its species being on the endangered species list. But pollution isn't the cause of recruitment failure in the Elephantear. Most of the river's other mussels are reproducing successfully, and the Elephantear itself reproduces well in other rivers with even worse pollution. So, if it isn't the river or something about the adult mussels that's the cause of the problem, perhaps the Elephantear's process of reproduction is broken.

Most river animals—from fishes to insects—face a challenge when it comes to procreation. A river's current can sweep eggs or tiny offspring far downstream before they have a chance to mature. Because a river's ecology changes along its length, the current can carry young animals into habitats unsuitable for survival. The young of shallow headwater species, for example, can be dropped into deep, cold river sections lower in the watershed. To account for this threat, river creatures have evolved strategies for ensuring their offspring stay in the right habitat. Some fishes spawn over rocky habitats and their sticky eggs cling to hard surfaces in rock crevices. Insects crawl out of the stream to metamorphose into adults who can fly upstream to find proper habitats. Snails glue their eggs to rocks and logs. For estuarine species, outgoing tides on a full or new moon take their young larvae away, and incoming tides weeks later bring mature larvae back to shore. But freshwater mussels don't enjoy any of these abilities or circumstances. They are far less mobile than other stream animals. Once mussels settle into the streambed, they can only drag themselves a few inches a day, and at great energetic cost. For this reason, mussels have evolved one of the most elaborate systems of reproduction known to science.

Mussel sex starts out in a pretty ordinary way for aquatic organisms. Prompted by seasonal changes in water temperature, males release sperm into the current, and females capture some of these sperm as they filter water for food and oxygen. Sperm and egg meet in the female, merge, and become larvae. Instead of releasing larvae into the current, females allow them to mature for a brief time inside their shells. At this point the small larvae, called glochidia, look like miniature clams.[6]

The next step for glochidia is to hitch a ride on a fish, though some species ride on amphibians. Depending on the mussel species, a glochidium will clamp down on a fish's fins or its gill filaments. Within hours, the infected host grows tissue around the parasite as an immune response. Inside the resulting cyst, the glochidium absorbs minor amounts of nutrients from the host and then undergoes metamorphosis to become a tiny mussel. The duration of this parasitic stage varies by species and with water temperature. It can be as short as a week or as long as several months. When the young hitchhikers are ready, they break out of the cyst, drift to the stream bottom, and begin a sedentary life of filter feeding.[7] Most fish survive the parasitic infection just fine.

Mussel moms use a variety of tactics to infect a fish with their glochidia. A very few mussels simply cast their glochidia into the current in hopes that their young bump into a fish fin or a fish inhales them as it breathes. This works for species whose hosts are abundant. A few others release into the river current strings of glochidia held together by mucus to entangle a host. Both are strategies of host generalists, meaning that many fish species can be proper hosts. A proper host is one whose immune response will not kill the glochidia, and whose behavior ensures the larvae reach suitable habitats. But mussel generalists are in a minority. Most mussel mothers employ far more elaborate tactics to infect fishes.[8]

Most mussels are host specialists that can reliably parasitize only one or two fish species. To infect the right species, mussels do what any human fisher would do when trying to land a particular fish species: use a bait and fishing strategy that the target fish cannot resist. Some mussels form bundles of glochidia—called conglutinates—that resemble the shape, size, and color of foods that attract their host. Females pack hundreds of glochidia into conglutinates. Just like when we bite into a ripe cherry tomato and seeds explode in every direction, when a fish chomps down on the conglutinate, glochidia explode into their mouth and latch on to the gills. Mussel species that parasitize minnows make conglutinates that drift in the current and resemble small leeches or worms. Those parasitizing bottom-dwelling fishes such as darters release conglutinates that remain near the bottom and resemble worms or fish eggs.[9] A few conglutinates look like a larval fish (a favorite food of bigger fish), having a headlike swelling with eyespots on one end. A mussel called the Creeper releases conglutinates that wiggle like fly larvae on the stream bottom. The most elaborate conglutinate lure is manufactured by the Orange-Nacre Mucket. Its conglutinate resembles a small minnow, drifting to and fro in the current at the tip of a translucent mucus tube attached to the mother. The tube can extend up to eight feet.[10]

Other mussels draw their host fishes close for a more intimate infection. These mussels manufacture lures out of the soft tissue—known as the mantle—growing at the valve opening. Some lures are simple but colorful structures. More sophisticated mantle lures resemble small fish, complete with eyespots, fins, and camouflage coloration or body stripes. Others look like crayfishes or aquatic insects. These ruses become even more irresistible when the female pulsates the mantle lure to catch the eye of passing fishes. The goal of these deceptions is to entice the target fish to attack and, in the process, rupture modified gills turgid with glochidia. The ruptured gill releases a cloud of glochidia that infects the fish.[11]

The most aggressive method of infection is employed by a group of mussels known as the riffleshells. Like other mussels, riffleshell females grow small mantle lures resembling fish eggs or aquatic insects. But, when a fish touches the mantle tissues, the valves snap shut, trapping the fish by the head. The margins of the valve curve inward, sometimes with toothlike serrations, to hold the struggling fish. The mantle forms a seal around the fish's head, and the mussel pumps glochidia into the mouth and onto the gills of the captive. The kidnapping may last for a few seconds or up to thirty minutes.[12]

The elaborate strategies many mussels have evolved for infecting their fish hosts are unparalleled in the animal kingdom for the degree of deception the mussels employ. It is also unusual for one large organism to parasitize another large organism (some adult mussels outweigh their fish hosts). However, the specialization that most mussels have with their hosts (about 80 percent of mussel species have just one or two suitable fish hosts) is not unique among parasites. This close relationship ensures that mussel offspring receive shelter, nourishment, and transportation to the right habitat.[13] But as is true with any specialized relationships in nature, this dependency on just one or a few other species becomes a liability if the other species declines or disappears. And this brings us back around to the Elephantear....

Two fishes are known hosts of the Elephantear in the Cahaba River: the Skipjack Herring and the Alabama Shad, both members of the Clupeidae family.[14] The Clupeidae are some of the most abundant of the world's fishes. Certain among them—including herrings and sardines—are a staple for cultures around the world. The Skipjack Herring lives across the South and throughout most of the Mississippi River Basin, into the Upper Midwest. The Alabama Shad has a more restricted distribution that includes the Southeast and lower and central portions of the Mississippi River Basin.[15] Like the Gulf and Atlantic Sturgeon, the Alabama Shad is anadromous. It spends most of its time in coastal waters, then migrates up rivers to spawn over shallow sand- and gravel bars in midsize to large rivers. The Skipjack Herring is a freshwater species, but also migrates in spring to reach its spawning habitats higher in the watershed.

Migratory clupeids such as these are good hosts for young Elephantears. They migrate in large, dense schools at times of the year that are predictable for mussels based on changes in water temperature. Furthermore, they migrate to and spawn in or near the habitats where adult Elephantears thrive. It is during these migrations that the shad and herring meet drifting Elephantear conglutinates. Mussels release them in time to intercept the returning fish, who inhale the conglutinates without hesitation. All in all, it is a great system for sustaining the Elephantear in the Cahaba River, one that ran like clockwork for millions of years. But now there are no more clupeids in the Cahaba.

Even though the Cahaba has never supported a major dam, it was cut off from the Gulf of Mexico by the Claiborne Lock and Dam on the Alabama River in 1969, and then the Millers Ferry Lock and Dam in 1970.[16] Whereas both the shad and the herring were previously known from the Cahaba, the Alabama Shad hasn't been seen in the Cahaba River since 1968, and the Skipjack Herring was never common in the river basin.[17] This is why none of the Elephantears in the Cahaba are less than fifty-one years old—they've not been able to reproduce since 1969.

As Wendell R. Haag and other biologists figured out, many mussel populations no longer recruit young because dams have blocked their fish hosts from migrating. It's yet another way that the golden age of dam building broke the biology of many species and interrupted ecological processes essential to healthy rivers. The lesson is clear enough: if we want to restore and maintain mussels in our rivers, then we must also restore and maintain the fish species on which they depend. Everything in ecosystems is connected.

In addition, understanding the intimate relationship that mussels have with fishes helps explain the elusive relationship between mussel disappearances and mud pollution. Recall that mud pollution was the chief suspect for mussel disappearances until Haag reviewed the science. While most, if not all, river scientists agree that mud pollution harms mussels, sediment can be exonerated as the direct cause of mussel disappearances. However, by way of its impact on fishes, mud pollution may be an accomplice to dams for mussel declines.

Clay and silt—the chief ingredients of mud—interfere with fish reproduction in rivers. River fishes prefer to spawn in clear water habitats over stream bottoms of sand and rock. Their sticky eggs adhere to rock crevices. Clay and silt fill these crevices and adhere to eggs, blocking uptake of oxygen and causing death or stunted growth. If a mussel species relies on a fish host whose populations have declined due to mud pollution, then mud pollution is indirectly responsible for mussel decline.

As for the Elephantear's hosts, dams now block Skipjack Herring and Alabama Shad from reaching many of their former spawning grounds, including those in the Cahaba. The herring has survived better than the shad because it is a freshwater species. Some populations became trapped above dams, but their populations are strong and stable.[18] In contrast, things are getting tight for the Alabama Shad. The species has become so uncommon that there was a recent petition to the National Marine Fisheries Service to place it on the endangered species list. The petition was denied because there has not been enough research to know whether remaining spawning populations are stable or still declining.[19] Nevertheless, environmental organizations and state and federal agencies are still concerned about the Alabama Shad's fate.

The Elephantear as a species is likely to survive through this century and beyond. Skipjack Herring that still have access to spawning habitats sustain many of its populations.[20] But there's not much hope for populations like those on the Cahaba. With no fish host, Elephantears will die off one by one, until all that remains are sun-bleached shells on river beaches. But until the very end, every spring on the Cahaba the male Elephantears will send forth their sperm into the current. Fertilized females will nurture a new batch of glochidia and release the conglutinates into the current for a school of shad that will never come.

While the Elephantear's situation in the Cahaba is disheartening, at least the species will survive elsewhere. There are other southeastern mussels in much worse shape, ones dependent on fish species whose futures are far more uncertain than that of the Alabama Shad. Such mussels are on track for being our first down payments on the extinction debt we still owe, but biologists are fighting hard to save them.

18
Pearls and Caviar

IF WE STILL OWE AN extinction debt for the industrialization of our rivers with dams and dredging, the mussels of the lower Apalachicola-Chattahoochee-Flint (ACF) River Basin may be the first mussels we lose. Of the entire basin's thirty-three mussel species, two are extinct, two are extirpated (locally eradicated), and only thirteen have stable populations.[1] Seven species found in the lower section of this basin and nearby small Gulf Coast rivers are on the endangered species list. Two are threatened: the Chipola Slabshell and Purple Bankclimber. Five are endangered: the Ochlockonee Moccasinshell, Gulf Moccasinshell, Fat Threeridge, Shinyrayed Pocketbook, and Oval Pigtoe. The populations of some of the latter are down to just a few individuals.

The lower ACF is one of the region's wildest areas. The landscape has extensive forest and low human population densities. If so, then why does this area have such an endangerment burden? We can find much of the answer upstream. The ACF's headwaters originate in and near Atlanta, the ninth largest metropolitan area in the United States. Its midreaches drain a vast agricultural area that relies heavily on irrigation. And the ACF's rivers flow through three states that bicker over how much water each can extract. By the time the Apalachicola River reaches the coast, its waters have lost volume, passed through many dams, and picked up excessive sediment and nutrients. All this contributes to the decline of the lower ACF's mussels.

Conservation scientists from academia and state and federal agencies are trying to save the region's endangered mussels and other aquatic species. They are learning as much as they can, as fast as they can, in hopes they can use their discoveries to restore imperiled mussel populations. One strategy is to learn which fishes are the hosts for the rare mussels. This can make possible targeted conservation strategies, such as protections for the host species.

One way to save a rare species is to breed it in captivity and release young into the wild after they've grown beyond their vulnerable early life stages. Wildlife biologists have done this for many animals, including birds, turtles, and fishes. To do this for a mussel, they need to know which fish is the host for its larvae. Investigators have identified the fish hosts for several endangered mussels of the lower ACF, but until recently, the host for the Purple Bankclimber remained a mystery despite decades of searching.[2]

The bankclimber grows to eight inches in length and is named for the shade of iridescent purple nacre lining the inside of its shell.[3] Historically the bankclimber lived throughout the ACF, but subpopulations on the Chattahoochee and Chipola Rivers have vanished. Scattered, isolated subpopulations survive in portions of the Flint, Apalachicola, and Ochlockonee Rivers. Like with the Elephantear population on the Cahaba River, most bankclimbers are adults over fifty years old, suggesting a problem with their host species.[4] While the bankclimber isn't on the very brink of extinction like some of the region's other mussels, the proverbial writing is on the wall for it unless something is done to prevent its decline.

Recognizing the critical need to identify the hosts of endemic mussel species disappearing from the lower ACF basin, in the late 1990s Christine A. O'Brien and James D. Williams of the US Geological Survey attempted to find suitable hosts for the Gulf Moccasinshell, Fat Threeridge, Oval Pigtoe, and Purple Bankclimber. It was a huge undertaking. They screened nineteen fish species for host suitability. This entailed capturing fishes in rivers, transporting them to the lab, and keeping them healthy for months. Even more challenging was snorkeling and scuba diving in rivers to find female mussels that were incubating larvae. And after all that, there was the tedium of carefully infecting each fish species with glochidia from the four mussel species. To do this, they gently pried open each female's valves and extracted a few larvae from her modified gill, where she stored them. If the larvae were mature glochidia, they removed enough to infect a suspected host fish. They counted the glochidia, then placed them directly on the gills of a fish. If the glochidia successfully transformed into the next developmental stage, they dropped from the fish and the tank's water filter caught them. O'Brien and Williams then counted the number of successfully transformed young retrieved from the filter and calculated the percent that transformed as a measure of host suitability.

The team did this with seventy-six pairings of mussel and fish species, and many of these pairings were done in replicate.[5] As a result, they identified hosts for three of the mussels, but they only found marginal hosts for the bankclimber. The lack of a strong match despite using many potential host species suggested that the bankclimber is a host specialist, meaning that only one or two fish species can support its glochidia. What confused the team, however, is that female bankclimbers release conglutinates (packets of larvae) that disintegrate in the water when released and form a thin cloud that disperses as it drifts. This is a very nonspecific strategy for infecting a host, one used by mussels known to be host generalists, meaning that many fish species are suitable hosts. These clues about the bankclimber's host were in contradiction, and the mystery deepened.

Ten years later another team, including biologists Andrea K. Fritts, Douglas L. Peterson, and Robert B. Bringolf, took on the challenge to find the Purple Bankclimber's host fish.[6] They tested nearly thirty species of fish representing over seven families, including darters, suckers, sunfishes, and catfishes. The team also went out of their way to capture and include four varieties of sturgeon, including the Gulf and Atlantic subspecies. After three years their efforts paid off. Of all the fishes tested that were native to the bankclimber's home range, one clearly stood out as the primary host: the Gulf Sturgeon.

In hindsight, this made perfect sense. As the researchers noted, the distribution of the Purple Bankclimber falls neatly within the historical migratory and spawning range of the Gulf Sturgeon in the ACF. Both share the same river habitat, and the timing of conglutinate release by the bankclimber is precisely when sturgeon would have been migrating up rivers in schools to spawn. Just like with the shad and herring hosts for the Elephantear, the arrival of the sturgeon would have been predictable due to seasonal changes in water temperature. And because of its generous size, a lot of larval bankclimbers could piggyback on a sturgeon.

To all that I would add that the new study resolved the contradictory clues about whether the bankclimber was a host generalist or host specialist. In the precaviar era, big schools of sturgeon would have migrated up the bankclimber's rivers. Because adult Gulf Sturgeon do not feed when migrating and this is a muddy river, a lure to attract sturgeon would be ineffective.[7] However, sturgeon would likely prefer river habitats with slower currents to conserve energy, the same habitats preferred by mussels in large rivers. Furthermore, sturgeon can intake many gallons of water a minute to gather oxygen as they muscle their way upstream. Thus, the bankclimber's strategy of releasing a cloud of glochidia into the water to infect sturgeon would be very effective. And based on reports of the mussel's former abundance, it must have worked quite well.

The excitement about the discovery of the Purple Bankclimber's host has been offset by the implications for the species' misfortune to be dependent on an imperiled fish. Starting a captive breeding program for the mussel using as its host the Gulf Sturgeon, a large threatened species, will not be practical anytime soon. For now, revival of the bankclimber population depends on the Gulf Sturgeon.

However, the close relationship between these two species resolves a long-standing mystery we temporarily set aside a while back: why are the Gulf and Atlantic Sturgeon still endangered? When malacologists surveyed bankclimber subpopulations in recent decades, they learned that those closer to the coast are doing better than those that are farther inland.[8] Specifically, the populations struggling the most to recruit young are above the lowest dam on the Apalachicola River, the Jim Woodruff Lock and Dam at the Florida-Georgia border. Completed in 1957, the dam is the gatekeeper for barges, fish, and anything else heading into the Chattahoochee and Flint Rivers from the Gulf. This explains why the more inland populations of the bankclimber have so few young. Their host cannot migrate beyond the dam.

The situation also reveals an important clue to why the Gulf and Atlantic Sturgeon are still endangered. Recall that the sturgeon have had a century to rebuild their populations after the caviar fishery collapsed. They've also enjoyed many years of state and federal protections. All

The Jim Woodruff Lock and Dam is the gatekeeper for barges and migratory fishes heading into the Chattahoochee and Flint Rivers from the Gulf of Mexico. Courtesy of Alan Cressler.

this should have been enough for both subspecies to rebuild their populations. However, biologists working with the ACF's Gulf Sturgeon estimate that the population lost most of its potential spawning habitat with the construction of the Jim Woodruff. For thousands of years prior, sturgeon migrated as far as two hundred miles inland and enjoyed ample spawning habitat in Georgia and Alabama.[9] When the US Army Corps of Engineers completed the dam, the population lost 78 percent of its spawning habitat.[10] That would be more than enough to slow the sturgeon population's regrowth and prevent the recovery of the Purple Bankclimber.

For now, the Gulf Sturgeon has access to a sliver of its former spawning habitat in the Apalachicola River and its lower tributaries, and these streams have become the bankclimber's last stronghold. As for the dozen or so populations surviving in the Flint River above the Jim Woodruff Dam, they have had no contact with Gulf Sturgeon for over sixty years and are doomed unless something changes.[11] These populations recruit a few young by parasitizing suboptimal host fishes, but this cannot sustain the species indefinitely. And even below the dam, in the Apalachicola and its neighboring tributaries, the bankclimber's populations are small and isolated. Like dozens of other rare southeastern mussels, they are vulnerable to disappearing one by one until they are gone.

The Gulf and Atlantic Sturgeon, the Purple Bankclimber, and hundreds of other endangered species have endured several centuries of river industrialization. They are the lucky ones. We've eradicated dozens of others. The practices creating this extinction crisis—overharvesting for food, caviar, and pearls; deforestation; pollution; river channelization; and damming—are legacy problems that arose decades and centuries ago. If we are to save today's endangered species, we need more research, innovation, and genuine reflection and reform on how we manage southeastern rivers. The more species we can save, the better. This is not just for their sake, but also for the valuable ecosystem services that healthy river ecosystems offer us.

Regrettably, saving these species will entail more than just addressing these legacy issues. As we are about to explore, we have unleashed a new set of environmental threats in the Anthropocene. The future of life on Earth, and our very survival, now depends on how quickly and how well we address these new challenges, and those we've brought with us from the Holocene.

PART III

19
Sacrifice Zone

The Southern Appalachian Mountains are a world apart from the lowlands that make up most of the Southeast. Whereas the Coastal Plain is a landscape of lazy rivers and endless expanses of steamy fields and pine plantations, the mountains are a labyrinth of narrow valleys and ridges enveloped in thick broadleaf forest. So much is different here—the playfulness of the wind through the valleys, the birdsong melodies reverberating through the hollows, and the dialects and lifeways of the region's people.

The streams differ, too. Instead of the languid sandy creeks and clay-choked rivers of the Coastal Plain, mountain streams, confined between mountain slopes, rush downhill over the rocky terrain as if seeking the freedom of movement that only the lowlands can provide. Mountain creeks are shallow, clear, and welcoming to visitors. Shimmering through the waters are rocks of every color, texture, and shape: rough-edged cobbles of orange chert, coarse boulders of brown sandstone, jagged discs of charcoal-gray shale, and polished contortions of blue-gray limestone. Pick up one of these rocks and a crayfish is likely to scoot away, tail first and pincers raised in warning. Flip the rock over and you'll find clinging to its bottom larval insects, including stone flies, mayflies, caddis flies, and owl-flies. There are even beetles—fully aquatic species whose larvae, called water pennies, cling to the rock beneath a round, copper-brown shell. A multitiered, multispecies canopy leans over a mountain creek. These trees and those on the hillsides towering above drop leaves, fruits, and branches, whose decay supplies the energy sustaining the stream's food web.

Mountain streams are the source waters for most rivers of the southeastern landscape. They begin as ephemeral streams, slight notches in the ridgeline that carry water only after a heavy rain. Freshly revealed, jagged boulders define their streambeds. These rocks are in their first centuries of weathering into sands and clays that streams will eventually carry to the coast. Ephemeral streams slice deeper into the mountainside with declining elevation. Scramble down from the ridge and soon you'll find a cut in the bank where tiny rivulets of water seep out, gather in the streambed, then trickle downslope. That spot is the beginning of an intermittent stream, one that continually flows during the wet months.

Flip rocks in these intermittent streams and you won't see many insect larvae because the water is only seasonally present. Instead, you will find salamanders. The moist, forested slopes of the Southern Appalachians are home to more salamander species than any other region on Earth.[1] They are the most common vertebrate of the Appalachian forests, far more abundant than birds and mammals combined. Many sport bold patterns and colors. The Northern Spring Salamander is orange with black spots, the Spotted Salamander is black with yellow spots, the Yonahlossee Salamander is black with marbling of gray and a bright brick-red back. These intermittent streams are a breeding refuge for the amphibians. If they were to reproduce in the continually flowing streams below, fishes and other predators would find and eat their offspring. Other salamander species have adapted to survive without water and lay eggs in pockets of moisture they find in the surrounding forest, including in cliff faces, in cavities of rotting logs, and under thick blankets of fallen leaves. With a few hours of searching, you could find a dozen species in a Southern Appalachian forest.

These mountains are sanctuary for more than salamanders. During the 2.5 million years of the recent ice ages, they were a refuge for forest plants and animals retreating from the cold, windswept prairies and woodlands that overtook the lowlands. The mountains sheltered Native Americans of the Archaic and Woodland periods during winter, when lowland rivers were cold, swift, and dangerously unpredictable. The Southern Appalachians helped hide freedom seekers on their perilous northward trek out of slavery. In recent centuries the mountains have welcomed lowlanders seeking escape from the summer heat of the Coastal Plain. And in the past hundred years, the mountains have provided relief from darkness for millions when they flip a light switch, because deep beneath the forests of the Southern Appalachians are veins of coal.

˜

One cannot overstate the importance of coal in the modernization of the American economy. As an abundant, inexpensive, and portable source of energy, coal powered the nation through much of its Industrial Revolution. By the mid-twentieth century, coal overtook hydropower as the primary source for generating electricity. One of coal's key assets was that unlike hydropower, coal could reliably produce electricity during droughts. Coal continued to be the predominant source of electricity until 2015, especially in the East, South, and Midwest. These days, coal is best known for its significant contribution to the accumulation of carbon dioxide in our atmosphere and subsequent climate change.[2]

What is less well known is the legacy of coal in the Southeast. Its extraction in the Appalachians has left behind broken bodies, impoverished towns, poisoned streams, and mountains shredded beyond recognition. Some say the Appalachian coal fields have become a national sacrifice zone, where the landscape has been forever ruined for the benefit of outsiders.[3] This zone begins in the headwaters of the Mobile River Basin of Alabama, and it stretches across the upper Tennessee and Cumberland River Basins of Tennessee, Kentucky, and Virginia. The sacrifice zone includes the mountains of West Virginia, Pennsylvania, Ohio, and Kentucky—the states Americans most associate with coal mining. The mining and use of coal from these regions are part of the origin story of the southeastern river crisis.

Coal mining has been around for centuries. It began in North America in the 1600s in the James River watershed and will likely continue in some form for decades into the future.[4] Throughout this long history, the extent and form of coal mining have been shaped by maximizing profit, minimizing costs, and, in recent decades, adhering to regulations. Many factors play into this equation, including coal's market value, available technologies, and the costs of technology and labor. One of the greatest physical and financial challenges to mining Appalachian coal is the amount of rock, known as overburden, that miners must remove to access the coal seam. The more overburden removed, the costlier the operation. The removal and disposal of overburden is also the most environmentally challenging and controversial aspect of coal mining in Appalachia.[5]

Appalachian coal mining in the twentieth century came in several forms, each of which handled the overburden differently. When coal seams are wide and deep below ground, underground mining is usually the most economical approach. This was the predominant form of mining in the Appalachians until late in the twentieth century. Tunneling down to the seam leaves much of the overburden in place, which reduces the costs and environmental burdens of its extraction and disposal. Once workers reach the seam, mining takes place from within the seam. Miners extract the rock removed when tunneling, as well as unusable rock near the seam. This rock, known as spoil, is used as filler by workers when closing the mine or left in piles at the surface. Due to the long history of underground mining, countless spoil piles litter the valleys of Appalachia.

If spoil piles were inert heaps of rock, they would be little more than an eyesore. Instead, they are massive slow-motion, toxic chemical reactions. The reasons for this begin with the origins of these rocks. Like the rock layers surrounding it, coal is a sedimentary rock over three hundred million years old. It formed over a twenty-million-year period from the accumulation of plant material in tropical swamps next to the young Appalachian Mountains. Floods or changes in sea level periodically deposited sand, mud, or reef materials over these layers, which compressed and protected these deposits. Millions of years of burial and slow chemical reactions led to the transformation of these swamp plants into coal. Coal is a useful fuel because most of it is carbon.

Coal is not a clean fuel. It holds metals and other contaminants that are toxic to most life forms in high concentrations, including mercury, arsenic, selenium, cadmium, and lead. Because bits of coal are frequently intermixed with the sedimentary rocks above and below coal seams, these contaminants are present in much of the overburden and spoil.[6] One of the most common contaminants in coal and surrounding rocks is pyrite, a sulfur- and iron-rich mineral also known as fool's gold. When exposed to water and oxygen in the atmosphere, pyrite slowly transforms into a strongly acidic mixture. Bacteria that obtain all their energy by stripping electrons from the iron atoms in pyrite catalyze the chemical reaction.[7] The acids produced from this process chemically liberate the metals embedded in the coal fragments. Environmental problems begin as this acidic water, laden with heavy metals, seeps into the groundwater and nearby streams. The acidity and toxicity of this pollution kill or sicken stream animals and degrade water quality for communities downstream. Though the spoil piles created by underground mining are a continuing threat to streams across Appalachia, this issue pales in comparison to the problems created by surface mining.

By the end of the twentieth century, thick coal seams were becoming tapped out and underground mining slowed. There was still plenty of coal in the Appalachians, but it was in thin seams, varying in thickness from a few inches to a few feet, thus too narrow to support underground mining.[8] In addition, many seams were so close to the surface that there was not enough overlying rock to support tunneling, but too much rock to make it affordable to mine it from the surface. Miners might have therefore left the remaining coal untapped, but the value of the coal in these seams changed overnight when the 1990 Clean Air Act Amendment became law.

We can trace the inspiration for the amendment back to the 1980s, when it became clear that the burning of coal to produce electricity was creating severe air pollution problems. The most well known among these at the time was acid rain, produced when the exhaust from burning coal—especially sulfur dioxide—reacts with other molecules in the atmosphere.[9] Acid rain was identified as the reason that forests and wetlands were dying across the United States, especially in the Northeast. In 1990, Congress passed and President George H. W. Bush signed into law a revised Clean Air Act to address acid rain and other air pollution problems. The act required that coal-fired power plants emit less sulfur in their exhaust. Congress left industry to figure out how to do this, and the most economical method was to burn coal having a lower sulfur content. Because coal in the Central

Appalachians is low in sulfur, the thin black seams tucked away in the mountains were now golden.[10]

To mine this coal, Appalachian miners adopted technologies and methods used for surface mining in Wyoming. The Wyoming mines use huge earth-moving machines known as draglines, which scoop up tons of rock at a time. Just a few miners can run draglines, and they are efficient at removing the overburden above seams. The combination of high efficiency and minimized costs gave birth to a new form of surface mining called mountaintop mining.[11]

Mountaintop mining disassembles a mountain. First, machines scrape the forest and soil away. Then the top of the mountain—the overburden—is blasted apart. This creates a tremendous amount of spoil that miners move out of the way by dumping it in an adjacent valley, a process called valley fill.[12] After miners extract the first coal seam, they remove the next layer of overburden to access the next coal seam. Miners dump the newer spoil on top of the older spoil in the valley. This continues downward until miners extract the last retrievable seam, sometimes 650 feet or more below the height of the mountain's original ridge, which is now long gone.

Mountaintop mining has reshaped Earth's surface in Appalachia. From valley highways this destruction is hard to see, but a view from the air reveals an ugly patchwork of barren, rocky islands in a sea of forest green. The Alliance for Appalachia, an environmental group, reported that by 2008 mountaintop mining had partially or completely gutted at least five hundred mountains.[13] According to the US Environmental Protection Agency (EPA), by 2012 surface mining had consumed about 1,260 square miles of forested land in Central Appalachia, a region spanning northern Tennessee to West Virginia. This is the equivalent of 80 percent of the area of Rhode Island.[14] Matthew R. V. Ross and colleagues quantified the impact of this in southern West Virginia, where mountaintop mining has been most intense.[15] Using satellite data, Ross's team found that the region is now significantly flatter. Prior to mining, the most common slope measurement within the mined area was twenty-eight degrees. After mining, it was nearly flat at two degrees. Their findings illustrate the damage done by coal mining to watersheds throughout Appalachia, including those portions in the Southeast.

Mountaintop mining for coal has permanently reshaped the Appalachian Mountains. This form of surface mining has disfigured over five hundred mountains and destroyed over twelve hundred square miles of forest. Thousands of miles of streams are filled with rubble or are polluted with acidic waste and toxic heavy metals. Courtesy of iLoveMountains.org, flight provided by SouthWings.

Mountaintop mining also obliterates creeks and rivers. Ross's team tallied 1,544 headwater valleys that mining had filled with spoil. These valley fills contain 1.5 cubic miles of spoil, enough to cover all of South Carolina with three inches of rubble. And that's just from southern West Virginia. The EPA estimated that by 2012, about 2,416 miles of headwater streams were filled, or were projected to be filled, as a result of mining in Central Appalachia.[16] These valley fills are immense—the largest measured by Ross's team was over 600 feet in depth, although it only takes a few inches of rubble to crush a stream or a salamander.

Considering this devastation, one might ask what laws apply to these activities and whether they can help minimize or prevent the destruction. Recall that for most of the twentieth century, the federal government left states to regulate their environment. Most states kept weak laws to encourage industry investment and job growth, but it was at the expense of miner safety and the environment. In 1977, the rising body counts of dead miners and conspicuous negative environmental impacts led Congress to craft, and President Jimmy Carter to sign into law, the Surface Mining Control and Reclamation Act (SMCRA, often pronounced "smack-ra").[17] Reclamation requires eliminating hazards such as highwalls (man-made cliffs) and restoring ecological function to the landscape. This usually entails contouring the landscape to minimize erosion and planting vegetation.

SMCRA helps ensure the landscape isn't left in complete ruins when surface mines close, but it doesn't set the bar high in terms of ecological restoration. SMCRA doesn't require mining companies to reforest the landscape. Some will plant trees, but usually mine owners plant the terrain with non-native species of grasses and weeds. Planted trees struggle to grow roots in the rocky, compacted soils that shed rainwater and are usually dry. The rate of natural tree establishment and growth on reclaimed mines is slow, suggesting to scientists that centuries will pass before a mature, species-diverse forest returns—if it even can.[18] As for the mountain, it is lost forever, replaced by a rubble-capped plateau.

The Clean Water Act applies to the streams buried beneath valley fills. If stream burial cannot be avoided, then mining companies must complete a process of compensatory mitigation, whereby the ecological function lost by burial is offset by restoring the ecological function of previously damaged off-site streams or creating new, ecologically functioning streams on the mine site.[19] The extent to which mining companies restore or create streams depends on whether ephemeral and intermittent streams fall under the reach of the Clean Water Act (this is the WOTUS issue, "waters of the United States"). At the time of this writing, it is unclear how the 2023 US Supreme Court ruling in *Sackett v. EPA* will affect this.

Regardless, compensatory mitigation in the form of creating new stream ecosystems on mine sites is failing to meet Clean Water Act standards set by federal law, according to scientists who have studied stream reclamation. Mine companies create intermittent streams on the graded surfaces of the leveled mountain. These sometimes have a structure approximating a mountain stream, but their ecological function is negligible.[20]

Mining companies have tried to solve the problem of acids and metals seeping from mine sites and valley fills, but their solutions have created other problems. Alkaline chemicals can neutralize acidic chemicals. In much of Appalachia, overburden layers include rocks that are alkaline (limestone and dolostone). To limit the acidic discharge, miners will mix alkaline spoil with the pyrite-rich spoil that creates acidic drainage. This neutralizes the acids, but the resulting seepage into groundwater and streams becomes too alkaline and carries concentrations of selenium, sulfate, and other contaminants. Selenium is toxic to humans and wildlife, and sulfate can be found downstream at levels thirty to forty times above normal.[21] These contaminants increase the electrical conductance of drainage waters, and this interferes with essential biological functions of stream creatures, such as the uptake of oxygen through gills.

By changing the chemistry of mountain streams, mountaintop mining and valley fills are causing entire components of the stream ecosystem to disappear. Biologists studying creeks and rivers downstream from mountaintop mining and valley fills have compared them with nearby similar streams unaffected by mining. While the streams below the mines are clear and surrounded by forest, surveys of aquatic animals reveal disturbing problems. Mussel species diversity and abundance are much lower in streams below the mines.[22] Mayfly larvae are particularly sensitive to changes in electrical conductance. Normally, they are up to half the aquatic invertebrates in Appalachian streams, but in streams below surface mines, they are scarce.[23] In one of the most damning studies yet, biologists found that fish abundance, species diversity, collective mass, and range of ecological roles were significantly lower below mines than in similar streams with no mining in the same watershed.[24]

Mountain ridges, lush forests, ephemeral and intermittent streams, mussels, mayflies, fishes, and unthinkable numbers of salamanders —these are among the sacrifices made for millions of us to have affordable, reliable electricity. Environmentalists have complained, protested, and even sued, but the warnings and resistance haven't been enough to halt mountaintop mining. Thus far, vanishing wildlife and mountains haven't swayed enough public opinion. As disheartening as this may be, the destruction of mountains and streams in Appalachia is only a fraction of the debt we are paying for Appalachian coal. The people at ground zero in our national sacrifice zone are making the greatest sacrifices.

20
Coal's Curse

BEHIND THE TALLIES OF MINED mountaintops, tons of extracted coal, and gigawatts of energy produced, a public health disaster has unfolded in Appalachia. Mining accidents, lung damage, respiratory disease, and cancer have torn holes in the social fabric of the communities providing us with coal. We can trace all these problems back to the mines. Some of the links are direct, such as high workplace injury rates. But others are elusive—clearly related to the mining, but via mechanisms that science hasn't yet discovered. Miners are not the only ones affected. Wives, children, and others who have never set foot in a mine are suffering for reasons not understood.

One piece of this puzzle is that the entrenched poverty in the region—and its entourage of poor health care, obesity, and smoking—multiplies the public health threats linked to coal. It's as if coal has cursed the region. Coal keeps promising to bring prosperity, but instead the communities of Appalachia stay mired in poverty and poor health. To understand the relationship between coal mining and the current public health crisis of Appalachia, we must explore the origins of the region's poverty and its perpetuation.

The first white settlers in the region were poor English, Scottish, and Irish immigrants, who valued independence and wanted to be left alone by the outside world. The rugged terrain offered the isolation they wanted. What emerged was a culture of self-sufficiency, kinship, and religious faith.[1] However, the region's physical isolation, sparse population, and lack of extractible resources other than coal prevented the Central and Southern Appalachians from enjoying the economic prosperity the Industrial Revolution brought elsewhere in the Southeast. Several twentieth-century federal initiatives designed to bring prosperity to the region had limited success, but poverty is still entrenched, exacerbated by the steady growth of unemployment during the twentieth century as mining companies adopted technologies that reduced the need for laborers.[2]

Though working in the underground mines offered income, mining perpetuated the region's poverty. During the late 1800s and early 1900s, miners and their families lived in towns built and owned by the mining company, complete with houses, schools, churches, and the company store. Companies paid miners with company scrip, a currency only valid at the company store, where prices for goods were inflated. Families would outspend their paychecks and become indebted to their employer and trapped in the system. Making matters worse, mine owners often neglected maintenance in company towns, and living conditions in some were deplorable.

Frequent injury also perpetuated poverty. An injury could result in weeks or months of recovery with no pay. Fatalities were common in the mines, and the death of a miner was both emotionally and financially difficult for the next of kin. There were many ways for an underground mine to injure or kill a miner. Falling rocks could hit miners, or machinery could crush them. Drilling sometimes tapped into underground streams, which would flood mines within minutes, drowning trapped miners in the dark. Methane, another of coal's impurities, was a constant threat. Also known as natural gas, methane is dangerous because it is odorless. If miners hit a methane pocket and the gas was undetected, they would succumb to asphyxiation before they knew what hit them. Miners often took a small, caged bird into the mine as an early warning system. With its high respiration rate, the bird would die quickly when a methane cloud arrived. A dead bird was a warning to drop everything and run. This trick saved lives and is the origin of the proverbial "canary in a coal mine." Methane's other life-threatening property is its flammability. Explosions happened when a spark—perhaps from an electrical short or metal striking metal or hard rock—ignited the gas. The accumulation of coal dust generated through drilling caused similar explosions. The dust would hang in the air in poorly ventilated mines until ignited by a spark.

Accidents such as these were common for decades. Between 1905 and 1930, the period of peak employment in underground coal mines, an average of 2,402 miners died per year, averaging 3.3 deaths for every thousand miners. The worst accident was a 1907 explosion in Monongah, West Virginia, that killed 362 men and boys and shook buildings eight miles away.[3] Poverty contributed to high mortality of miners. Because jobs were scarce in Appalachia, the surplus of men willing to work in the mines despite the dangers allowed mine owners to keep wages low and minimize investment in mine safety.

In the late 1800s and early 1900s, miners began unionizing to lobby for higher wages and better living and working conditions. Mine owners met these efforts with forceful opposition that fomented violence. From 1890 to 1930, a period known as the Coal Wars, skirmishes were frequent between armed miners and the armed security forces hired by mine owners to enforce their rules and suppress union activity. Escalating tensions erupted in the Battle of Blair Mountain in Logan County, West Virginia, in 1921, when ten thousand armed miners confronted three thousand law enforcement and private security forces. By the end of the battle, the two sides had fired over a million rounds, planes had dropped homemade bombs on miners, dozens were dead, and many more were wounded. The battle ended when the federal government declared martial law and sent in troops to restore calm. It was the largest armed insurrection in the United States since the Civil War. Despite continued efforts to unionize, conditions for miners didn't improve until Congress intervened in 1935 by passing the National Labor Relations Act. Signed into law by President Franklin Delano Roosevelt, the legislation protected workers' rights to unionize, collectively bargain with companies, and strike.[4]

Underground and surface mining are much safer today. There were 207 deaths in mine accidents from 2008 to 2017, a rate of 0.2 deaths per thousand miners. These improvements are due to federal mine safety requirements plus a precipitous drop in the number of coal miners employed due to mechanization. But the work is still risky. Coal miners are six times more likely to die on the job than the average American worker.[5] And, a culture of cutting corners to maximize profits continues in some companies, as was seen in 2010, when an explosion killed twenty-nine miners in Massey Energy's Upper Big Branch Mine in West Virginia.[6] Federal investigators determined that the ultimate cause of the explosion was deliberate safety violations by Massey Energy managers, intended to reduce operational costs.

If a coal mine doesn't kill or injure a miner quickly via falling rock, explosion, drowning, or asphyxiation, it can do so slowly via lung damage. Coal dust is ever present, and its accumulation in the lung tissues causes health problems. The most well known is black lung disease, named for the blackened condition of the lungs caused by the accumulation of coal and silica dust. It and similar ailments are known more properly as coal workers' pneumoconiosis, or CWP. Advanced CWP results in severe shortness of breath, coughing spasms, and related problems. These lead to a significant decline in quality of life. In severe cases former miners can't walk far and must continuously breathe from an oxygen tank.[7] Prevention is the only cure, and CWP often leads to an early death.

For decades the coal industry systematically denied the existence of CWP to avoid having to compensate affected workers. However, in the face of mounting evidence and pressure from mine unions, Congress passed the Coal Mine Health and Safety Act in 1969.[8] The act improved health and safety practices, and initiated programs such as free screenings to detect the first stages of CWP. Still, coal miners today endure more nonfatal injuries (mostly CWP) than do workers in any other private industry, including oil and gas extraction, construction, and manufacturing.

The data gathered on CWP after the 1969 act testifies to the sacrifice coal miners have made to hold a steady job and to bring us cheap electricity. Between 1968 and 2014, over seventy-eight thousand deaths were attributable to CWP.[9] Mortality dropped steadily over that time, but CWP is still a problem. The screenings help, but many miners today decline them because early symptoms are easy to ignore, and miners fear what the screenings will reveal. Some are concerned that a positive finding will eventually result in their termination by managers wanting to avoid future disability payments for workers diagnosed with CWP. For all these reasons, many miners put off screening until retirement, but by then the disease has done its damage. For severe cases the only remedy is a new pair of lungs, though lung transplant recipients rarely live more than a few extra years.[10]

Recently, there has been an alarming uptick in the worst form of CWP, a condition called progressive massive fibrosis. Investigators have attributed the surge to lax enforcement of and noncompliance with federal regulations. A decline in unionized labor also contributes to the problem. Without the unions, miners work longer hours, which increases their exposure to dust, and they get fewer breaks to breathe clean air to help them shed the particles. Unions also provided a way for miners to report safety violations without risking retaliation.[11] But, the major cause of the uptick is the mining of thin, low-sulfur coal seams that began in the mid-1990s because it requires more blasting and grinding of silica-rich overburden than in the past. Silica dust is far more damaging to lung tissue than coal dust. In the past, CWP patients were mostly retirees. Today, clinics are treating CWP in young and midcareer miners. In 2014, the Obama administration issued new rules to protect miners from dust exposure.[12] But, in November 2020, the US Department of Labor's Office of Inspector General concluded that the Mine Safety and Health Administration was still not doing enough to protect miners from silica dust. Hopefully this scrutiny will lead to improved protections for miners.[13] Regardless, such measures focus only on the miners and do nothing to address the fact that coal mining has killed or debilitated many who never set foot in a mine.

On the sleepy Saturday morning of February 26, 1972, a thirty-foot-tall tsunami of water, coal particles, and industrial chemicals burst from a coal-processing retention dam and surged down Buffalo Creek, West Virginia. In its path were sixteen small towns where coal industry workers and their families lived. The flood swept up everything in its path, including trees, trucks, and people. The roaring waters lifted entire houses, with families inside, off their foundations. Terrified occupants screamed from the windows as the wooden houses careened downstream and were torn apart when they smashed into one another, trees, and bridges. Survivors in the first towns hit by the wave raced in their cars down the valley's road honking their horns and yelling warnings to others downstream in the path of flood. Some heeded the warnings and ran to high ground; those who didn't were soon fighting for their lives.[14]

Alerted by the honking cars, Sylvia Albright grabbed her nine-month-old boy, Kerry, and ran outside her home accompanied by her teenage son. Not far up the creek, a wall of water and debris tumbled toward them. They raced for the hillside at the yard's edge, but the sludgy waters overtook them. Struggling against the mud and strong current, they couldn't reach the arms of their neighbors on the higher ground. With the mud above her waist and rising fast, Sylvia, in one final act of desperation, threw baby Kerry toward the hillside. But it was not enough. Kerry landed in the sludge, and the current swept him away. Sylvia and her older son soon lost their footing and slipped under the dark waters.

Immediately after the flood subsided, those who had escaped searched the wreckage for survivors. They soon found the bodies of Sylvia and her older son among the debris downstream. They also found baby Kerry, face down in a pile of debris. Miraculously, he was alive. Coal oil covered him, grit filled his mouth, and a gash on his thigh exposed the bone. Rescuers treated baby Kerry, and he eventually recovered. His father raised him, and Kerry now lives in Brooklyn, New York. He was fortunate to have survived. The Buffalo Creek flood killed 125 people that day and injured 1,100 in the span of a few terrifying minutes. The surge also destroyed 546 homes, damaged another 943, and left 4,000 people homeless.[15]

Pittston Coal Company, the owner of the dam that burst, declared the Buffalo Creek flood to be an act of God, pointing to the heavy rainfall in the area during the previous days. But investigators concluded that the company had not designed its dams properly, probably to minimize expenditures. The failed dam had withheld wastewater from cleaning coal. After mining, processors crush coal and bathe it in chemicals to improve its combustibility and remove impurities. This creates a slurry of water, coal particles, impurities, and residual processing chemicals. Processors store the slurry in ponds built in valleys near coal-processing facilities. They create ponds by building a dam across the valley out of mine spoil and backfilling the upstream area with the slurry. Sediments settle to the bottom of the pond, while water rises to the surface and evaporates or is reused. When they need more capacity, processors build the dam taller or build new dams and retention ponds higher in the valley. To close an impoundment, workers cap it with mine spoil or dirt, grade the surface and embankments, and plant it with vegetation. It is another form of valley fill. Hundreds of these impoundments in Appalachia store billions of tons of sludge.[16]

The Buffalo Creek tragedy inspired Congress to institute a dam safety inspection program overseen by the US Army Corps of Engineers. While this has helped, slurry spills still happen. The largest in recent history was near Inez, Kentucky, in 2000. Slurry from an impoundment broke through the rock of the valley wall and poured into mine tunnels near the surface. After surging through several miles of adjacent mines, 281 million gallons of slurry gushed into two separate watersheds. Fortunately, the flood did not injure anyone, but it caused the shutdown of several public water supply plants and a power plant, and it annihilated the ecology of the creeks that were flooded.[17]

Slurry spills are dramatic threats to Appalachian coal communities, but other threats are more mysterious. Scientists have detected an association between mountaintop mining and high rates of death and illness in adjacent communities.[18] Cardiopulmonary disease, birth defects, and overall mortality rates are higher in communities near mountaintop mining than in those with no mining. And cancer rates are rising in the region while declining across most of the United States. This latter pattern persists even when confounding variables such as Appalachia's high smoking and obesity rates are factored out.[19] An analysis of county-level data by an advocacy group, Appalachian Voices, revealed a strong association between levels of mountaintop mining and cancer in Central Appalachia.[20] Michael Hendryx, a University of Indiana scientist, estimated that in areas where mountaintop mining occurs in Central Appalachia, an additional 1,200 people die annually from respiratory disease, respiratory cancer, and other ailments than would die otherwise.[21]

These results could seem to be a slam dunk: mountaintop mining is bad for community health. Slow down, say critics. The above studies analyzed health data at the level of the community, not individuals. Such studies have limited ability to nail down causal linkages. Correlation does not prove causation, as the old scientific adage warns. In 2017, scientists with the National Institutes of Health and outside colleagues published a careful review of ninety-five studies with a focus on mountaintop mining and valley fill operations and community health.[22] The authors found that harmful materials are released from mountaintop mining, but that the evidence linking the mining to human disease was inconclusive. The problem is a lack of research linking an individual's disease to exposure to contaminants. Such precision is necessary to identify the exact mechanism causing disease so that industry can develop preventative measures.

Scientists are working to find the causal link between mountaintop mining and disease, but the mechanism remains elusive. One possibility is drinking water contamination. In rural Appalachia, many households are beyond the reach of public water systems, so residents use well water for drinking, cooking, bathing, and washing.[23] A 2006 study by the US Geological Survey compared wells near mountaintop mines to wells farther away but with similar underlying geology. Compared to well water away from the mines, water samples close to mountaintop mines had higher levels of dangerous gasses such as hydrogen sulfide and methane, as well as elevated levels of contaminants leaching from mine spoil. While this suggests that well contamination may play a role, residents with uncontaminated wells are still getting sick. Thus, some scientists believe something else is causing the broader pattern of disease and mortality across the community.[24]

Fugitive dust from mountaintop mining may be to blame. The blasting, drilling, crushing, and hauling of coal and mine waste generate clouds of fine particulate matter. Winds carry this dust away from the mines and into nearby settlements, where people inhale it. According to the EPA, particulate matter (defined as particles less than or equal to one one-hundredth of a millimeter in diameter) causes or worsens cardiovascular and lung diseases and respiratory disorders. These particles enter the body through the lungs and the gastrointestinal track. They are so tiny that they interact with living cells and may enter the bloodstream.[25]

Wind carries fugitive dust away from mountaintop mines, and it accumulates in the sheltered air of nearby valleys. Rock-hauling trucks rumbling down rural highways also stir it up. EPA scientists found that particulate matter sampled in communities near mountaintop mines has the same composition as particulate matter from coal and spoil. Furthermore, particulate matter levels in the air of these towns exceeded EPA health standards.[26] Though this fugitive dust is a plausible cause of the health problems in communities near mountaintop mines, much more research is needed to determine the extent of the exposure and whether and to what degree it is responsible for human disease.

Based in part on these new findings, in 2016 the Office of Surface Mining Reclamation and Enforcement commissioned the National Academies of Science to conduct a new review of existing research on the issue and make recommendations as to whether and what type of research needs to be done. However, the Trump administration halted the study in summer 2017, while it was in full swing. The White House offered no explanation, but critics widely interpreted this as meddling in science to protect the coal industry.[27] What happens next will depend on whether leaders in the federal government decide to allocate funding to enable scientists to solve this mystery.

Appalachia has given too much in its role as a national sacrifice zone. Forested ridgelines are gone, and creeks are swimming with contaminants instead of fishes. And despite the decades of employment coal mining has created, Appalachia is one of the poorest regions of the United States. It is easy to blame the coal company owners and their allies. After all, they have a centuries-long record of putting profits before people. But those of us who use electricity produced by burning Appalachian coal are also to blame—our complicity perpetuates the injustice. And now, we are finding that coal's curse extends far beyond Appalachia. Communities across the Southeast are experiencing hazards and public health risks due to our use of coal.

21

Toxic Chemistry

THE NINE HUNDRED PLUS WORKERS conducting the seven-year cleanup of the largest environmental disaster in US history are among the latest human test subjects in one of the most expansive chemistry experiments humanity has ever performed on the environment, and itself. The disaster was the spilling of coal ash slurry from an immense retention pond at the Tennessee Valley Authority's (TVA) Kingston coal-fired power plant in 2008. Over two hundred of the cleanup workers and their family members are now suing Jacobs Engineering, the company that hired the people to remove the waste. Though there were no deaths or injuries on the day of the spill, hundreds of cleanup workers have fallen ill. Over forty of them have died of brain cancer, lung cancer, blood cancer, and other illnesses the plaintiffs attribute to exposure to the coal ash. Cancers and other diseases riddle the bodies of several hundred surviving workers. Many will not live much longer. To understand what happened to these people, it helps to think of coal use as a gigantic, poorly managed toxic chemical experiment.[1]

Coal is a chemical reaction frozen in time. The formation of the rock halted the chemical processes of decay that were slowly happening in nearby swamp forests when the Appalachian Mountains were young. But recall that coal formation captured far more than the carbon in plants; it also trapped and concentrated trace elements within plants that at high concentrations are toxic to life forms, including us.

The instant we pull coal from the ground and expose it to the atmosphere, we reactivate the chemical reactions trapped inside. It's like opening a Pandora's box of toxicity. We've already seen the collateral damage from this in the mountains and valleys where we mine and process coal. The spoil piles and valley fills leftover from Appalachian mining are massive unplanned chemical reactions from which ooze a stream of acids and heavy metals, or bases and salts, that contaminate well water and poison streams. And, for reasons not yet understood, this coal's toxicity is causing disease in the people living near mountaintop mines. As tragic as they are, these unforeseen chemical reactions are just the beginning.

Burning coal releases the worst of its contaminants. In places where we burn a lot of coal, stagnant weather conditions can cause the formation of coal smogs—thick clouds of coal smoke, often in combination with water vapor in the form of fog. This is the most dangerous of the inadvertent chemical reactions we've created by burning coal. London has a long history of coal smogs, some of which would settle on the city for days at a time, causing mass mortalities and illness, especially among those whose health was already weak. Even as recently as 1952, over twelve thousand Londoners died during a coal smog. In 1948, a coal smog settled on the Rust Belt town of Donora, Pennsylvania, killing at least twenty and sickening thousands. These two events encouraged the development of modern clean air policies in both the United Kingdom and the United States.[2] In recent decades, coal smogs have plagued China due to a heavy reliance on the fuel for modernization, in combination with weak pollution controls.

Another toxic chemical reaction we've unleashed by using coal is mercury contamination of ecosystems. Mercury is a heavy metal that damages organs in the vertebrate body, especially the nervous system. Exposure to minuscule amounts can inhibit neurological development in a human fetus or child. When we burn coal, the metal is set adrift, settles on the landscape, and washes into rivers, reservoirs, and the ocean. After microbes absorb the mercury, it begins ascending the food chain. Because animals do not excrete it, mercury becomes increasingly concentrated with every link in the food chain. Levels in fish and shellfish can reach such high concentrations that state governments in the Southeast issue annual consumption advisories. Due to the burning of coal at power plants, thousands of river miles in the region are under such advisories. For example, in North Carolina women of childbearing age and children under fifteen are advised to never eat Largemouth Bass, and all others should limit their intake to no more than one meal a week.[3] Because mercury is an element, it cannot be destroyed. Nor is there any known way of removing it from ecosystems—it is a problem we will contend with for the foreseeable future.

To reduce levels of mercury emissions into the environment from burning coal, power plant operators have added yet another chemical reaction into the mix, one with nasty unintended consequences that are an emerging threat to the safety of drinking water for communities throughout the United States. Bromide is a coal impurity found in low levels. Bromide itself is not known to be toxic in low doses to humans or wildlife. Power plants have begun adding bromide to the coal they burn because it can bond with mercury after combustion and prevent it from entering the atmosphere. This has helped plant utilities comply with regulations on mercury emissions. But as a result, elevated levels of bromide are in the wastewater produced from burning coal. Rivers are usually adjacent to coal-fired power plants, and they offer utilities a convenient way to discharge this wastewater.[4]

Bromide creates a public health threat downstream from power plants, where water systems uptake river water and produce drinking water. Many public water systems use chlorine to disinfect source water. Unintended chemical reactions occur when bromide mingles with chlorine and other molecules commonly in river water. This produces a family of dangerous compounds known as trihalomethanes (THMs). Even in low dosages, THMs can cause cancer and damage the kidneys, liver, and central nervous system. And, they can persist in water for months.[5]

This chemistry is causing trouble for residents in North Carolina. The first problems arose downstream from Duke Energy's Belews Creek facility, one of the company's largest coal-burning plants. Bromide from the facility was contaminating the nearby Dan River, the source of drinking water for many downstream communities. THMs spiked to levels flagged as unsafe by the US Environmental Protection Agency (EPA) in the drinking water of Madison and Eden. Both communities were forced to install expensive upgrades to their water treatment systems. Duke Energy agreed to pay for the upgrades, but only after a court ordered it to do so as part of a 2015 criminal plea agreement for a different pollution issue.[6] As word spread, other communities in the area began testing their water for THMs. At least six North Carolina communities found that bromide from Duke Energy's coal-fired plants was causing THMs to appear in their drinking water. All total, the THM problem was affecting 12 percent of the state's residents, including the one million residents of Charlotte.[7]

It is a contentious issue. The reduction in mercury emissions from power plants is saving lives and protecting ecosystems. But THMs have periodically exceeded safe limits as set by the EPA in affected communities. Duke Energy is in a tough spot. Adopting a different technology to remove mercury from the superheated chemical soup created by burning coal would be expensive. These costs would either cut into Duke's profits or would need to be passed on to customers. Duke Energy doesn't want to pay for more expensive upgrades to water treatment plants for the same reasons. Meanwhile, most communities affected by the pollution, especially smaller ones, cannot easily afford upgrades to their water treatment systems to remove THMs, and alternative water sources are either not available or are expensive to develop.

Another issue is that THMs are dangerous at any level, yet the EPA allows low levels in drinking water. This is the agency's attempt to avoid forcing upgrades at water treatment plants and avoid requiring utilities to install more expensive technology to remove mercury. The EPA often makes similar compromises between protecting public health and protecting industry when setting permittable pollution limits. In this case, it's a compromise that would be unnecessary if utilities used cleaner sources of energy.[8]

The disaster at the Tennessee Valley Authority's Kingston coal-fired power plant is another unintended and dangerous problem created by coal's toxic chemistry. The burning of coal in power plants creates by-products of combustion in solid form (fly ash, bottom ash, boiler slag, and flue gas desulfurization material), technically known as coal combustion residuals, but commonly referred to as coal ash. The bulk of coal ash is not hazardous, but there are traces of over two dozen toxins including arsenic, barium, cadmium, chromium, lead, mercury, selenium, and radioactive isotopes.[9] According to the American Coal Ash Association, the United States produced 102 million tons of coal ash in 2018. That's a lot, but this represents a 23 percent decline since the peak of coal ash production, in 2008.[10]

The modern-day coal ash crisis that led to the Kinston spill began in 1990, with the passage of the Clean Air Act Amendment. The act required that industry emit less sulfur and other pollutants when burning coal. To meet the new standards, power companies began buying low-sulfur coal (this is what led to mountaintop mining and valley fill operations in the Appalachians) and installed technology that pulled contaminants out of the gasses released from burning coal. The result has been improved air quality across the United States, less acid rain, less smog, and less mercury fallout.[11] These are good outcomes.

Unfortunately, the new processes used to clean the exhaust from burning coal have created far more coal ash than ever before. Much of it has been repurposed in construction materials. Since the 1940s, industry has used coal ash as an ingredient of cement and drywall, and as fill for road construction projects. Industry scientists say the toxins stay locked up in these materials and are not a threat. But after the 1990 amendment, utilities produced so much coal ash that the construction industry could not absorb it all. In 2018, the construction industry used only 55 percent of the coal ash produced.[12]

To deal with the surplus coal ash, utilities began mixing it with water pulled from the adjacent river and piping it to a nearby impoundment. Coal ash impoundments are earthen dams surrounding a large pond that receives the slurry. Included in the mix is the wastewater with the bromide. Like what happens in coal slurry ponds in the mountains, solids settle in the pond and become compacted over time. The water is discharged into the river or evaporates.

Coal ash impoundments cover dozens of acres, and when they reach capacity, the earthen dam is built higher so more ash can be added. Most tower several stories over the surrounding landscape and hold massive amounts of ash. Power utilities had planned to keep these impoundments in place for the foreseeable future, but the future is a long time, and coal ash and its contaminants have had more than enough time to find ways to escape.

Until the past decade, most coal ash ponds created were unlined, meaning they lacked a barrier preventing leakage from the impoundment into the groundwater or surface waters. About 1,400 unlined ponds are found across the United States, and many are now leaking.[13] The heavy metals and other contaminants in the leakage and the discharged water have worried environmentalists and nearby communities for years; the problems with bromides and THMs are just the latest concern. But the slow seepage of toxins from the ash ponds is a minor issue compared to what happens if an impoundment fails.

During the night of Sunday, December 21, 2008, a fifty-seven-foot-tall dam holding back 5.4 million cubic yards of coal ash slurry failed at the Kingston Steam Plant in Tennessee. The slurry poured over the landscape like wet cement, burying three hundred acres of a residential area under several feet of gray sludge. The spill destroyed or damaged dozens of homes and ruptured a major gas line. Most of the waste poured into the adjacent Emory and Clinch Rivers. This was the largest industrial waste spill by volume in US history at the time of this writing, even larger than the Deepwater Horizon oil spill in 2010.[14]

The small army of people hired to clean up the spill over the next few years became involuntary test subjects in our ongoing, poorly planned chemical experiment with coal. The workers drove trucks or operated machinery, and most worked ten to twelve hours a day. Dry days were a problem because fugitive dust could blow off the site and into the surrounding neighborhoods. Water trucks sprayed to keep the dust down, but the work by heavy machinery corralling the ash and the continual comings and goings of the trucks hauling it away unavoidably created clouds of dust.

It was already well known within the power utility industry that coal ash contained toxins, yet Jacobs Engineering, the firm hired by the TVA to clean up the spill, didn't issue the workers dust masks or other personal protective equipment (PPE). One former worker alleges that Jacobs managers fired him for asking for a mask, and others were told by managers they'd face the same fate if they did the same. The failure to protect employees has been the subject of several lawsuits against Jacobs Engineering brought by cleanup workers and their families. In one of the suits, the plaintiffs claim that the lack of PPE led to over forty deaths and four hundred illnesses (thus far) among former cleanup workers. The plaintiffs and reporters investigating the incident say that the TVA and Jacobs Engineering didn't want the public to see the workers wearing PPE because that would advertise the toxicity of coal ash. They allege this was a publicity strategy designed to limit the risk of litigation, one that backfired and caused great loss and suffering among the workers. In 2018 a federal jury ruled that Jacobs Engineering had failed to protect its workers, and that lawsuits against the company for damages could proceed. The settlement phase of the trial and other lawsuits against Jacobs Engineering and the TVA are still ongoing.[15]

The Kingston spill triggered new scrutiny of coal ash and the ways power companies handle it. After an extensive review, in 2015 the EPA under the Obama administration issued new regulations for coal ash impoundments. Rules now require liners in new impoundments to prevent leakage. If unlined impoundments contaminate groundwater through leakage, owners must close or upgrade them.[16] Power companies must secure closed impoundments or move the ash to a lined landfill. Because utilities have complained loudly about the cost of implementing the new regulations, the EPA under the Trump administration issued new rules giving utilities more time and flexibility in meeting the requirements of the earlier ruling.[17]

Meanwhile, several southeastern states are acting on their own. In 2019, North Carolina's Department of Environmental Quality ordered Duke Energy to excavate all its unlined coal ash ponds adjacent to water bodies and move the waste to lined landfills.[18] In Virginia, the state legislature required Dominion Energy to do the same for three leaking coal ash ponds, including one on the James River. In South Carolina the state's three utilities agreed to excavate all unlined coal ash ponds, and in Tennessee a coal ash pond on the Cumberland River is being removed.[19] These outcomes were the result of public pressure and legal action from the Southern Environmental Law Center, a nonprofit legal watchdog organization. Though these actions will cost these utilities tens of millions of dollars, these changes will save lives and safeguard rivers.

Dozens of these massive coal ash impoundments loom over nearby rivers across the Southeast. In a region subject to heavy rainfall and flooding, it should come as no surprise that there have been other coal ash spills since Kingston. There was a spill in February 2014 at Duke Energy's Dan River Steam Station in North Carolina and another spill at a Duke Energy plant on the Cape Fear River in 2018 due to flooding from Hurricane Florence.[20] More spills are inevitable until utility companies move coal ash into lined landfills and away from rivers. But even then, the plague of coal ash will continue. Coal ash doesn't disappear. We are stuck with it forever. That said, some communities are bearing this burden more than others.

The Black Belt region of Alabama was once some of the most valuable real estate in America. The dark prairie soils inspiring the region's name grew cotton better than anywhere else. Exploitative labor practices through the use of enslaved people and, later, sharecropping were key to making cotton plantations profitable. However, when soil fertility declined, the Black Belt's agricultural economy crashed and never recovered. Today, the region is one of the poorest in the nation. In recent decades, the abundance of cheap, rural land has attracted opportunists, especially companies that build and operate landfills. Black Belt landfills are profitable because many distant states, counties, and cities have no room for their waste or would rather pay someone else to handle it.

In affluent and populated areas of the United States, residents often mount successful campaigns to prevent landfills in their communities. They don't want the permeating stench, continuous rumble of garbage-hauling trucks and trains, and declines in property value. They also raise concerns about landfill toxins in the groundwater and air. And because landfills never go away, landfill problems are intergenerational problems. So, there are many reasons for residents to be concerned about a landfill in their community.

In contrast, it's been easy to build landfills in Alabama's Black Belt. There's little resistance from residents, owing to the rural location, low population density, and extreme poverty. Moreover, local politicians have welcomed landfills because of the tax revenues they generate. As a result, Alabama has thirty-five landfills for every million residents. In stark contrast, affluent states such as California and New York have just three to five landfills per million residents. And even though most residents in Black Belt communities are African American, landfill owners and their clientele—who unreservedly send them trainloads of sewage, garbage, and toxic waste—brush aside accusations of environmental racism.[21]

Uniontown is a small, rural Black Belt city. Most of its 2,300 residents are African American, and poverty rates hover around 40 percent. Unlike elsewhere in the Black Belt, in 2003 when residents learned of plans to build a landfill just beyond the city limits, they organized and protested to stop its construction. The Perry County commissioners who struck the deal with Arrowhead Environmental Partners to build the massive landfill dismissed their concerns, and Arrowhead Landfill opened in 2007. The residents didn't give up. They formed a group—the Black Belt Citizens Fighting for Health and Justice—to lobby for safeguards against landfill pollution and other environmental problems in their communities.[22]

Within months of the 2008 Kingston disaster, the TVA asked the Perry County Commission to allow Arrowhead Landfill to accept some of the coal ash. Enticed by the revenue it would generate, the commission didn't hesitate. One month after the commissioners' visit to Kingston (and less than a week after the commission sought feedback in a public meeting) trainloads of ash began arriving in Uniontown. For six months the landfill accepted coal ash. As the ash mountain grew, residents complained of fugitive dust settling on their houses and cars.[23]

Though the facility hasn't accepted coal ash since late 2010, Uniontown residents are still worried. Like the Kinston cleanup workers, they've become the latest involuntary test subjects in the grand chemical experiment we've set in motion by extracting and burning coal. The landfill looms high above the treetops as an ever-present threat that will be there for centuries. The city has received media attention as one of the latest sobering lessons on the interaction of poverty, race, and exploitation. Uniontown has more allies now, including environmental lawyers who help ensure that landfill owners follow state and federal regulations. What will happen in Uniontown is unclear. Will there be more fugitive dust? Will coal ash toxins leak through the landfill's liner and into the water supply? Will any of this affect public health? Let's hope not. But no one knows for sure—we've never done this experiment before. The only silver lining is that many Americans in recent years have turned against coal because of its toxic legacy. This opposition, combined with other influences on energy markets, means that King Coal is being dethroned.

22

Quitting Coal

THE LONG SAGA OF OUR relationship with coal is troubling. The exploitation of miners. The cancer rates in Appalachian communities. Poisoned wells. Disfigured mountains. Forests and streams crushed by valley fills. Yes, much of this destruction was done before we knew about the toxic chemical reactions we unleased by extracting and using coal. And many of the problems can be blamed on simple greed and convenience, such as cutting corners on mine or impoundment safety, not providing protective dust masks to the spill workers at Kingston, or shipping coal ash to landfills in poor communities. But now, with the power and responsibility of hindsight, we can no longer plead ignorance or tolerate wrongheaded intentions. The only reasonable choice is to quit coal.

In light of the social and environmental collateral damage it has caused, many citizens, environmental groups, scientists, and politicians now call for the end of coal usage. Mountaintop mining, for its environmental and social impacts, has received much of the scorn. There have been petitions and protests. Even former miners have turned against it. Activists have called out state agencies for lax inspection of mines and poor enforcement of regulations. Watchdog groups have filed lawsuits against companies accused of being noncompliant with existing regulations. Scientists studying the impacts of coal extraction and use repeatedly come to the same conclusion: coal harms us and our environment. And finally, abandoning the use of coal for power is possible because we have other ways of producing energy—ones that are dependable and affordable, and that cause far fewer negative social and environmental problems.

The rising concerns about coal extraction and use led to new regulations during the Obama administration. Under the president's direction, the US Environmental Protection Agency (EPA), Office of

Surface Mining, Mine Safety and Health Administration, and other federal agencies took measures to strengthen regulation of mining activities to protect miners, adjacent communities, and the environment.[1] It became harder to get permits for valley fills. The government commissioned studies to examine the impacts of Appalachian surface mining on human health and streams. Stiffer rules were issued on the regulation of valley fills and dust in mines.[2] Science and public feedback shaped policy. However, the Trump administration halted these initiatives. At the direction of the new president and his agency appointees, there was a massive rollback of a broad range of environmental regulations, but especially those for coal.

The continued support for coal is no surprise. Coal has loyal and powerful supporters, from the corporate boardroom to miners and their neighbors.[3] These boosters argue that coal is widely available across the United States, and we have enough to use it for decades, possibly centuries. The reliance on domestic coal helps maintain America's security in the global economy. Unlike with electricity generated from wind and sunlight, power companies can generate electricity on demand by burning coal. Coal has been a reliably inexpensive source of energy, and we already have the infrastructure to produce and use it. Switching to new energy sources could inflate the cost of electricity. There are counterpoints to each of these assertions, but coal's platform is strong enough to sustain its many supporters.

A deeper explanation for coal's popularity is that there's an entire economic ecosystem that has evolved around its use and has enabled company owners, shareholders, and executives to accumulate wealth. These include thousands of companies that mine, process, and ship coal; the electric utilities that burn coal; and the businesses manufacturing the technology these companies use. Supporters include the industries using coal as a fuel to produce materials such as aluminum, steel, and cast iron, plus all the manufacturers who depend on these materials. A long supply chain of corporate stakeholders and their employees can thank coal for their economic security and wealth. And this coalition does everything possible to protect the economic ecosystem supporting them.

These oppositional views on the future of coal—a natural resource that can keep the United States strong or a dirty fuel that is taking lives and destroying the environment—have been at the center of political debates in recent years. Some people (typically Republicans) support the continued reliance on coal, and others (typically Democrats) advocate for a transition to cleaner technologies. Both sides use coal miners as a political football. Pro-coal voices argue that we need to revive the coal economy and put miners back to work. Anti-coal voices argue that we need to end our reliance on coal and offer job training programs for laid-off miners. What both sides often overlook is that the future of coal and coal miners has less to do with who shapes the policies, and everything to do with energy innovation and the free market.

Coal mining jobs have been on the decline for a century. The 1920s were the peak era for employment in the coal industry. Demand for coal was high, and coal-fired power plants rapidly gained ground on hydroelectric dams in the race to provide the nation's electricity. Mine owners continually adopted technologies that improved production efficiency, but as the mechanization of coal mining advanced, the number of coal mining jobs declined. In 1923, the year of peak employment, there were 862,536 coal miners in the United States. By 2017, this number had dropped to 82,843.[4]

The loss of mining jobs in Appalachia has compounded the region's economic insecurity. Counties where coal mining has historically been the major source of employment are among the poorest in the Southeast, with poverty levels as high as 30 to 45 percent of all residents as of 2016.[5] Laid-off miners can't easily find new jobs near their families because there are few other large employers in the region. As unemployment spread, rates of poverty, illegal drug use, and domestic violence also increased.

Some scholars describe these problems as symptoms of the "resource curse." This is a nickname for an economic syndrome befalling countries or regions for whom a single, nonrenewable resource dominates the economy. When the market value of the commodity rises, there is prosperity and job growth. But when commodity values drop, or the resource is depleted, there is an economic slump. Perhaps coal could never have brought lasting prosperity to Appalachia.[6]

The recent history of coal mining in Appalachia validates the resource curse. First, owners began closing many underground mines with depleted coal seams. Then, with its thick deposits of low-sulfur coal, Wyoming became the lead coal-producing state in the late twentieth century. Beginning in 2000, exports of coal to Asia declined significantly as the Chinese expanded the mining of their own coal reserves. Collectively, these factors led to the layoff of over forty thousand workers—34 percent of US coal mining jobs—between 2006 and 2016. In 2015 alone, 22 percent of the US mine workforce lost their jobs, with the most losses coming from Appalachia. Thus, while Obama-era regulations caused a decline in coal mining, the impact of these policies was just a small part of much more influential economic trends of the resource curse.[7]

But while these trends have taken their toll, the greatest cause for the loss of Appalachian coal jobs in recent decades has been the rise of hydraulic fracturing, or fracking.[8] Fracking begins with pumping a high-pressure mixture of water, sand, and chemicals through a well to fracture rocks far below the surface. The target is shale, a carbon-rich sedimentary rock formed from ancient muds. Shale fracking releases trapped methane that then rises from the well.[9] Methane, also known as natural gas, is a highly flammable and efficient fuel. The industry emerged in the 1980s, when fracking was combined with technology allowing rig operators to drill laterally through horizontal shale deposits. Fracking is extensive in Alabama, Kentucky, Tennessee, and portions of western Virginia.

The vast quantities of methane produced through fracking in recent years are rapidly changing the energy sector, and coal is not faring well.[10] Natural gas is an attractive fuel for power utilities due its low cost, abundance, and favorable production forecasts. When utilities switch to natural gas, they also avoid producing coal ash and the dangerous toxins that have become increasingly expensive to handle under new regulations. Due to utilities switching to methane, there was a 12 percent reduction in coal-fired plant capacity between 2002 and 2016. In 2015 natural gas surpassed coal as a power source for generating electricity in the United States.[11] This was a landmark event, because coal had dominated the electricity markets since World War II. Natural gas as a fuel is so attractive that a growing number of coal-fired power plants are now slated for retirement by 2030, perhaps up to 18 percent of those operating in 2017.[12]

In combination with the factors noted earlier, fracking has caused significant reduction in coal mining. Production from mountaintop mining declined 62 percent between 2008 and 2014, and total US coal production declined 15 percent over the same period. Just in 2015, 17 percent of coal mines closed across the United States.[13] Experts predict these trends will continue for the foreseeable future.

There are, however, dreadful consequences of fracking. For one, the process uses enormous amounts of water that people in the region could use for other purposes. Some of it returns to the surface, and some stays underground, but all of it is contaminated. Fracking has also polluted groundwater reserves for nearby communities. Shockingly, so much methane comes out of the faucet in some homes that residents can ignite it. The litany of other problems caused by fracking is lengthy and includes dangerous air pollution, noise pollution, and even earthquakes. And although the burning of methane releases half the amount of carbon dioxide as coal for the same amount of heat, methane is a potent greenhouse gas, and colossal amounts of it escape during extraction and delivery.[14] It's too early to know for sure, but we may find that fracking is just as dangerous as coal for people and the environment.

As coal's influence on American society wanes, it leaves us with a complicated legacy to grapple with. Coal deserves much of the credit for the modernization of the American economy and the country's rise as a superpower. But coal also leaves behind a ruinous legacy. Foremost are the sacrifices made by generations of miners. Sure, they worked for a wage, but considering the history of exploitation, injury, illness, and death, miners and their families sacrificed more than what was fair to ask. Our nation is indebted to them, and we should find a way to repay mining communities with more than just gratitude.

Some of coal's legacy, like trihalomethanes, coal workers' pneumoconiosis, coal smog, and acid rain, will fade away as we stop using it. But other features of coal's legacy will endure for generations, including mercury pollution, toxic seepage from mines and spoil piles, coal ash and coal slurry disposal sites, and defaced mountains. These problems should be more than enough incentive for us to leave all remaining coal in the ground. But the most important reason we should quit coal—the reason why it will be notorious for millennia to come—is that by burning it, we are destabilizing Earth's climate.

23

Hothouse Earth

LET THERE BE NO DOUBT: the scientific evidence is overwhelming in scope, depth, and clarity: Earth's climate is changing rapidly, and we are to blame. Temperatures are rising rapidly due to the accumulation of too much carbon dioxide and other greenhouse gasses in our atmosphere from the burning of coal, methane, and petroleum products, and the expansion of agriculture.

When I began this quest to better comprehend the Southeast's current environmental crisis, I already understood the fundamentals of climate change and its impacts. But I was blissfully ignorant of how dire our situation is right now. The problem begins with carbon dioxide. It's a natural constituent of our atmosphere, and as a greenhouse (heat-trapping) gas it ensures that Earth is warm enough to sustain life. Over Earth's history, carbon dioxide levels have varied. When they rose, Earth's average atmospheric temperature and sea level rose. When levels fell, Earth's temperature and sea level fell. Atmospheric temperatures affect sea level because they control how much ice is on land in the form of polar ice sheets and mountain glaciers. When this ice melts, water spills into the oceans; when glaciers and ice sheets grow, sea level drops.

For example, during the 2.5 million years of the Pleistocene—the epoch that we often call the ice ages and that ended 11,600 years ago—the concentration of carbon dioxide in the atmosphere averaged 200 parts per million (ppm).[1] This means that out of every million gas molecules in the atmosphere, 200 were carbon dioxide. During the Pleistocene's ice ages, sea level dropped hundreds of feet below today's ocean level. When the ice ages ended and Earth entered the warmer Holocene, the most recent epoch, carbon dioxide in the atmosphere rose to about 280 ppm and ocean levels reached their current height. Unlike the Pleistocene, which was a series of ice ages and intervening warm periods, the Holocene was relatively stable, and this allowed humanity to master agriculture and build civilizations.[2]

With the onset of the Industrial Revolution, we initiated an era of widespread deforestation and burning coal and other fossil fuels. These practices have loaded our atmosphere with greenhouse gasses. At the time of this writing, in 2023, carbon dioxide levels are at 421 ppm and still rising.[3] Global change scientists warn us that if we continue on this path, we will trigger runaway climate change, when Earth's systems begin causing global warming and other planetary changes on their own, regardless of what humanity does or does not do.[4] For example, with enough warming, the arctic permafrost will melt and release massive volumes of methane currently trapped in the frozen soil. Methane is another greenhouse gas, one that traps twenty-five times more heat than carbon dioxide. Its release to the atmosphere would trigger rapid atmospheric warming that we could not stop. Other examples of such events, known as tipping points, include the loss of the Greenland and Antarctic ice sheets, changes to oceanic currents, and the loss of tropical forest in the Amazon. Any one of these events would push us closer to Earth having an extremely warm climate by today's standards. Earth has seen such conditions many times in the past: scientists call it the Hothouse Earth climate scenario, and it would last for millennia.[5]

So how close are we to the tipping points that could trigger a Hothouse Earth? We are close—too close. Scientists predict that as early as 2030 we will have added enough greenhouse gasses to the atmosphere to begin hitting tipping points. The warming threshold that climate watchers fear is a rise of 3.6°F (2°C) above preindustrial levels. We are already at 2°F above preindustrial levels, and rising fast. To illustrate what is at stake: the last time atmospheric carbon dioxide levels were above 400 ppm, Earth was 4–7°F warmer than it is today. This was hot enough to raise sea level to between seventy-two and eighty-two feet higher than it is now.[6] This would be more than enough to inundate the entire southeastern coastline for dozens of miles inland and drown many of the world's major cities. Furthermore, the last time Earth was as warm as it is now, sea level was twenty to thirty feet higher than today.[7] We are not experiencing these high temperatures and sea level yet because there is a delay between increases in greenhouse gas levels and rising temperatures and ocean levels. The delay gives us a window to prepare for and adapt to global change but doesn't allow us to fix the problem. With our current technology, we can only prevent climate change from getting worse.

The magnitude and severity of the changes we are forcing is why some of the world's top scientists concluded that of the nine planetary systems on which humanity depends for survival, climate stability is one of four for which we have exceeded the safe operating limits. Put another way, our climate manipulations have caused Earth to enter a new climate phase. Instead of the predictable, mild conditions of the Holocene epoch, we are in for a wild and difficult ride in the new geological era, the Anthropocene.

The delay in warming, ice melting, and other effects of climate change isn't long enough to spare us from substantial changes—they are already here. In the 1980s climate scientists estimated that noticeable impacts of climate change wouldn't occur until midcentury, but already polar ice is undergoing substantial loss, sea level is rising rapidly, oceans are warming and becoming acidic, plants and animals are shifting their distributions or dying out, coral reefs are vanishing, and wildfires are becoming more frequent and dangerous. This is just a partial list of what the world is witnessing, and much, much more change is coming. Even if we were to halt greenhouse gas emissions tomorrow, it will be centuries, possibly millennia, before these gasses are captured and stored by Earth's biogeochemical systems.[8] Carbon capture technologies may one day help, but they are in an infantile stage of development, and are decades away from having an impact, if any.

We do have control over how bad it gets. It all depends on how soon we can stop pumping greenhouse gasses into the atmosphere. But even in the best-case scenario, as modeled by the world's top climatologists working through the United Nations' Intergovernmental Panel on Climate Change, if we applied stringent measures to reduce greenhouse gas emissions, we would still reach carbon dioxide levels of 430–480 ppm by the year 2100. If we don't restrict emissions, then we will exceed 1,000 ppm by 2100.[9] Clearly, we must cut greenhouse gas emissions quickly and deeply, as soon as possible. This should be every nation's top priority right now.

Action to curb emissions here in the United States has been agonizingly slow. The main problem is the skepticism and denial of climate change by right-wing politicians and their supporters. Investigative reporting in recent years has revealed the origins of this resistance. It began with those in the private sector who have become wealthy and powerful due to fossil fuel use and extraction. They are justifiably concerned that if we phase out fossil fuels, they will lose their economic stronghold. Beginning in the 1980s and using the vast resources at their disposal, they mounted a political and counterfactual campaign to encourage skepticism of the scientific evidence that Earth's climate is changing, and we are to blame.[10]

This publicity campaign has been successful. According to Yale University researchers, in 2019, only 67 percent of Americans polled agreed that climate change is happening, and only 53 percent of Americans believe that humans are to blame for climate change. Thus, when Americans see reporting on worsening wildfires in California, accelerating permafrost melting in Alaska, and flooding and heat waves in the Midwest, only half understand that humans have intensified these "natural" disasters. Even more discouraging is that even though 100 percent of Americans will be affected by climate change, only 42 percent of those polled believe that climate change will harm them personally. These low numbers, ultimately the result of the disinformation campaigns, explain the weak public demand for actions addressing climate change. If people don't think it's a problem, they are not going to demand a solution. Among southeastern states, the average number of those who believe climate change will affect them personally ranges from 34 percent in Kentucky to 43 percent in Florida and Virginia.[11] And yet, the Southeast is predicted to suffer economically from climate change more than any other region in the United States due to storms, flooding, drought, extreme heat, and related problems.[12]

One of the looming questions for the Southeast is whether the region will have enough water in the coming years. This affects not only the region's economy and environmental security, but also the future of its aquatic biodiversity. Rainfall plays a lead role in affecting water availability in the Southeast, so how are rainfall patterns expected to change?

Global warming has already, over the past few decades, changed rainfall patterns in the Southeast. To understand why, it helps to be familiar with the climate forces influencing the region's rainfall. If not for the warm, moist air blown in from the Gulf of Mexico and the Atlantic Ocean, the region would have an arid climate like Central California or the Mediterranean. Instead, the Southeast receives more rain than any other region in the United States. The coasts and high elevations in the Appalachians receive the most rainfall, but tropical depressions, tropical storms, and hurricanes periodically bring high volumes of rain into the interior. Every few years the El Niño/Southern Oscillation disrupts the average rainfall patterns of the Southeast. When oceanic waters off the Peruvian coast are abnormally warm, a condition known as El Niño, the Southeast is unusually wet and cool during the winter and spring. A La Niña climate phase often follows an El Niño. During a La Niña event, abnormally cool waters off the Peruvian coast cause the Southeast to have a dry, warm winter and spring.[13]

Until recently the above factors were nearly all that climatologists understood about rainfall patterns in the Southeast. But in the era of climate change science, researchers began studying weather patterns over the long term to investigate other forces shaping the southeastern climate. There are several methods for detecting climate trends. The most direct approach is to examine past weather records for patterns. In doing this for the Southeast, scientists found no strong trends in annual rainfall from 1895 to 2019. However, there has been a recent and sharp rise in the yearly variation in total rainfall. For example, the frequency of exceptionally wet or dry summers more than doubled from 1990 to 2019.[14] The major reason for this was a shift in the behavior of the Bermuda High.

The Bermuda High is one of the most important forces on southeastern rainfall patterns.[15] It is a large region—hundreds of miles across—of high pressure created by dry air descending on the Atlantic Ocean from high altitudes. The high has a dual influence on rainfall. Beneath it, there is little to no rain because the dry, falling air prevents storm formation. Around it, clockwise winds steer the movement of storms and moisture that has evaporated off the ocean.

The Bermuda High migrates across the Atlantic during the year. In winter it is closer to Europe and Africa. This allows winter cold fronts and warm fronts to be the main drivers of southeastern weather. These fronts bring shifts between warm-wet and cold-dry conditions. In the summer, the Bermuda High migrates toward the United States. During a normal summer it hovers over Bermuda, and its clockwise winds push moisture evaporating off the Gulf and Atlantic into the Southeast. This water vapor feeds the region's summer rains.[16]

Climate change has altered the Bermuda High's normal behavior. Since the early 1980s, the high has become stronger and stretches farther west in the summer than it did in the past.[17] When the extension bends to the northwest, the high covers the Southeast and causes a dry summer. When the extension bends to the southwest, the high reaches into the Caribbean and Gulf of Mexico and its circulating winds steer moisture into the Southeast, causing a wet summer. Recent research by Wenhong Li and colleagues suggests that some of this change in the Bermuda High's behavior is attributable to unusually warm sea surface temperatures in the Atlantic Ocean, caused by rising atmospheric temperatures.[18] Climatologists expect that these trends will continue and the Southeast will see increasingly variable amounts of summer rainfall. This is a prediction based on the past, but when climate scientists look to the future, even more changes affecting rainfall appear to be coming.

Predicting the future isn't easy for anyone, but climate scientists have a powerful tool at their disposal—the global circulation model, or GCM. The GCM is the crystal ball of climate science. But instead of relying on sorcery, GCMs use complex mathematics run by supercomputers to predict the future. Because Earth's climate is one of the most complex phenomena studied—and our future depends on getting it right—the GCMs scientists have built are among the most complex computer models ever constructed. Nearly everything that scientists say about the future climate is derived from GCMs, so understanding the basics of what GCMs can do, and what their limitations are, empowers us with clarity about climate change predictions.

GCMs simulate how the atmosphere, oceans, and land surface interact to shape Earth's climate. They model the flow of heat, greenhouse gasses, and water using contemporary observations and historical weather records. The models also factor in historical changes to Earth's systems that influence the climate, such as deforestation, rising levels of greenhouse gasses, and changes in the Sun's radiative output.

When climatologists build or alter GCMs, they refine them by hindcasting. This involves setting model conditions to match Earth's climate at some point in history (e.g., before the Industrial Revolution) and then running the model to see how well its outcomes match the climate patterns that actually happened. The closer the match, the better the model is at simulating Earth's climate. After running hindcasting simulations, climatologists improve GCMs by adding new computations and data until they accurately hindcast the past climate. The more data modelers use, and the better the results of hindcasting, the most confidence modelers have that their GCMs accurately predict the future climate.

However, there are limitations to what scientists can do with GCMs. Accurate weather observations from around the world don't exist prior to the 1880s, and this limits the capacity of hindcasting to refine models. For modeling Earth's longer-term history, scientists use measurements from natural features such as tree rings, coral growth rings, and layers of glacial ice to reconstruct earlier climates. Known as proxy data, these indirect sources of information are less accurate than direct observations. To compensate, modelers use multiple sources of proxy data, and this produces a clearer picture of the past climate.

Another limitation is computing power. A GCM divides the planet's atmosphere, oceans, and lands into a three-dimensional grid. As the model runs, it computes changes at the scale of each cell in the grid. In each forward step through time, the model calculates interactions between each cell and its neighbors. The more cells a model has, the more computing power the model demands. Early GCMs used large cells measuring over three hundred miles across. As computer power improved, climatologists developed GCMs with ever-smaller cells to offer more precise predictions. The most recent models use cells about forty miles across. Cell size will decrease as computer technology advances.

The GCM results that climate scientists share usually come with statements about a level of confidence (e.g., high, moderate, or low) with outcomes. This is necessary because predictions come with a degree of uncertainty. The quality of the data used to build the model, evaluations of the model's accuracy, and results from other studies shape these levels of confidence. Another influence on their confidence is variation in model outcomes. GCMs are so complex that each run of the model produces a different outcome whether hindcasting or forecasting. This is typical of any complex system. Even if we could magically conjure a thousand exact copies of Earth circa 1900 and then watch how they changed over a century, the climate on each Earth at the end of our experiment would be different. To minimize the risk of overemphasizing an extreme outcome that is less likely, scientists run models thousands of times for each novel set of predictions. The more often a particular outcome is produced, the more confident scientists are that it represents the future.

Another way that climate modelers guard against spurious model outputs and therefore build confidence in their predictions is to combine the results from independently derived models. Academic and governmental organizations around the world have built GCMs, and scientists at these institutions share their findings to help improve the fleet of models. Some of the latest models integrate the forecasts from an ensemble of models to better capture the range of possible future climates. Modelers presume this increases the accuracy of modeling efforts.

The greatest uncertainty of all when modeling the future climate is human behavior. No one knows how long and to what extent humanity will continue to load the atmosphere with greenhouse gasses. To contend with this, beginning in 2014 scientists developed models—known as representative concentration pathways (RCPs)—for different future scenarios of greenhouse gas concentrations in the atmosphere. The optimistic RCP 2.6, or early peak emissions scenario, assumes that greenhouse gas emissions peaked in 2020, carbon dioxide concentrations rise to 490 ppm at midcentury, and warming is only 2.7°F (1.5°C) above pre-industrial levels. While emissions did not peak in 2020 (despite the COVID-19 pandemic) many nations and organizations are working hard to meet the other goals of this scenario.[19]

RCP 4.5, the midcentury-peak emissions scenario, represents the next-best outcome. This model assumes that emissions peak in 2040, atmospheric carbon dioxide concentrations rise to 650 ppm, and global warming reaches 4.3°F (2.4°C). This much warming would disrupt life as we know it.[20]

Meanwhile, in the RCP 8.5 business-as-usual emissions scenario the world fails to limit emissions, carbon dioxide emissions rise to 1,370 ppm, and global warming reaches 8.8°F (4.9°C) above pre-industrial levels. The consequences of this would be catastrophic to all nations, and some argue the disruption would threaten humanity's survival. Unfortunately, the business-as-usual scenario represents our current trajectory (at the time of this writing) for rising global greenhouse gas emissions.[21]

Given that we now have sophisticated computer models for predicting possible futures based on the best available climate data, what do these models say about rainfall in the Southeast? GCMs suggest that total rainfall in the Southeast will increase moderately if we continue emitting greenhouse gasses at the current rate. Under the business-as-usual scenario, by the end of the century most of the region will receive more rainfall than it did in the period between 1986 and 2015.[22] Springs will be marginally wetter (up to 10 percent) for most of the region. Summer rainfall will increase slightly (up to 5 percent), while autumn and winter rainfall will increase by as much as 10 to 15 percent. Though an increase in predicted annual rainfall is encouraging, scientists caution that they are only moderately confident in these predictions.[23]

In contrast, climatologists are highly confident that the Southeast will see a sharp uptick in the frequency and severity of intense rainfall events that can cause dangerous flooding. Since the mid-twentieth century the region has already experienced a steady rise in the number of days with more than three inches of rainfall. Models predict that this trend will increase under both the midcentury-peak emissions scenario and the business-as-usual emissions scenario.[24] Relative to 1980–2000, most of the region will see an increase of at least 10 percent in the number of days with an inch or more of rain. Northeastern Florida, southern Georgia, and the Southern Appalachians will see up to 20 percent more of these days.[25] And under the business-as-usual emissions scenario, by the end of the century the Southeast will see a doubling of intense rainfall events persisting for at least two days.[26] So, though much of the Southeast will be seeing an overall increase in precipitation, this rain will be delivered in more intense bursts that will cause flooding problems.

Flooding issues aside, a prediction of more rainfall in the Southeast, even one offered with moderate confidence, sounds like good news for the region. With an increase in rain, it seems we might have enough water to meet human demand and sustain rivers and aquatic biodiversity, right? Unfortunately, water supply experts say that if we continue using water like we do now we will still face water shortages. The reason for this involves a basic property of water we all know quite well.

24

A Thirsty Future

Each spring I plant basil in containers on my porch in Birmingham. When I water my basil in April, the soil stays moist for days. But when I water in July, the soil is dry the next day. Why the difference? It's because the evaporation rate has changed. Evaporation happens when heat excites water molecules, and they transition from a liquid to a gas state. Evaporation speeds up as temperatures rise. This simple property of water will cause the Southeast a lot of trouble in the coming years. A warmer climate will cause greater loss of water to evaporation.

But evaporation is only part of the problem. Like nearly all plants, the basil I grow uses soil water to conduct photosynthesis, the process of converting sunlight into energy stored in molecular form. At the end of photosynthesis, leaves release water into the atmosphere, a process called transpiration. Plants release more water into the atmosphere in summer than in cooler months because of the stronger sunlight and higher temperatures. As average temperatures warm in the Southeast and around the world, vegetation will be growing more and sending more soil water into the sky via transpiration.

Because the Southeast has so much vegetation, transpiration will join with evaporation (together known as evapotranspiration) to reduce water availability as the climate warms. Historically, about 70 percent of rainfall in the Southeast has returned to the atmosphere via evapotranspiration.[1] This is already a huge fraction of the region's water budget, and temperatures are rising quickly.

Since the beginning of the twentieth century, average annual temperatures in the contiguous United States have risen 1.8°F.[2] The warming hasn't been consistent over time or geography. Temperatures in the United States peaked in the Dust Bowl era of the 1930s and then declined until the 1970s. Average annual temperatures rose quickly thereafter. Based on records from tree rings and other proxy data, the most recent decades have been warmer than any other time in the past 1500 years. Curiously, the Central South, particularly Mississippi and Alabama, didn't warm as much as the rest of the United States during the twentieth century. Climatologists call this region a "warming hole" and are still investigating potential causes.[3] Regardless, the Southeast has since joined the rest of the United States and is in a phase of rapid warming consistent with predictions of the global circulation models.[4]

Climatologists say that if we begin a moderately concerted effort to reduce greenhouse gas emissions (midcentury-peak emissions scenario), then average temperatures in the Southeast will rise 2–4°F by midcentury (relative to 1976–2005) and an additional 2°F by the year 2100. That's a lot of extra heat to cope with. However, if greenhouse gas emissions continue to rise at their present rate (business-as-usual emissions scenario), by midcentury the bulk of the region will warm by 4–6°F, and another 2°F by 2100.[5] A temperature rise of this much will have substantial impacts on southeastern culture, economy, and ecology.

In both scenarios, rising temperatures will eat away at water availability. Water yield is the term scientists use for precipitation not lost through evapotranspiration. Historically, the water yield for the Southeast is just 30 percent. That's all that's available to meet our water needs and fill aquifers and streams. Increases in rainfall will spare a few portions of the Southeast from water shortages, including the Florida peninsula, and portions of eastern Georgia and the Carolinas. But for everywhere else, hydrologists expect that water yields will decline by up to 5 percent by midcentury. So, although annual rainfall in the Southeast is on the rise, water availability for most of the region will decline.[6]

While water yield declines for much of the region, the demand for water will increase simply because the Southeast's population is rising by about 8 percent per decade. This will place more stress on water supplies.[7] Water scarcity will be more intense in some areas of the Southeast than others. For example, major metropolitan areas in the Piedmont are prone to water shortages due to large populations and limited availability of surface and underground water. The Piedmont cities of Charlotte and Atlanta regularly experience water supply problems already. Periods of water scarcity in the Southeast will also be episodic, such as during summers when the Bermuda High sits over the region and prevents rainfall. These dry summers are hotter than usual because clouds and humidity do not reflect incoming sunlight. In these situations, evaporation rates spike and water availability in our soils, rivers, and reservoirs declines rapidly. Climatologists are very confident these dry summers will be more frequent in the coming years.[8] Furthermore, these conditions can trigger droughts, prolonged periods of water scarcity and little precipitation.

The Southeast is not prepared for an era of water scarcity. Over the past two centuries, the region has developed systems of water use that depend on a surplus of water. Water conservation has not been a priority, and this will worsen water shortage impacts during dry periods. For example, a declining water supply puts the Southeast's power production at risk. At present, coal, natural gas, and nuclear plants produce most of the region's power. These plants generate heat through nuclear fission or the burning of fossil fuels. The heat creates steam for spinning turbines that generate electricity. The steam is cooled back into water before reuse, and the coolant is usually surface water taken from an adjacent river. By far, this is the Southeast's greatest use of river water.[9] However, during dry spells or droughts, surface waters become hotter than normal. Already some power plants in the Southeast must shut down or reduce production during dry summers or droughts because available surface water is too hot to be effective as a coolant. Sometimes plants shut down because they are unable to chill coolant water enough to discharge it legally into streams. These challenges will intensify in the coming decades.[10]

Agriculture will also struggle with a declining water supply. About 16 percent of the Southeast is cropland, and only 17 percent of this is irrigated because farmers have historically been able to rely on rainfall. Farmers will need to irrigate more with river water and groundwater as temperatures and soil water evaporation rates rise.[11] In addition, more of the nation's agricultural production will shift to the Southeast as the multidecadal drought continues in California and Arizona. All these factors will lead to increasing competition for water between agriculturalists, power utilities, municipalities, and those tasked with protecting wildlife populations.

These and other water scarcity problems will be devasting during droughts. Droughts are a natural phenomenon in the Southeast, but global warming is causing them to establish faster and be more intense than in the past, both in the Southeast and elsewhere. Furthermore, the Southeast is prone to megadroughts, droughts persisting for at least two decades. Scientists learned this by studying tree rings. Because trees grow more during periods of high rainfall and slow their growth during times of drought, tree rings are reliable indicators of past rainfall. This research has led to the discovery that a megadrought contributed to the failed colony at Roanoke and the near failing of Jamestown.[12] A megadrought today triggered by higher temperatures would cripple the southeastern economy.

What does all this mean for the region's rivers and their biodiversity? The amount of water a stream carries is a function of rainfall. If a storm delivers enough rain, some of it flows across the ground to the nearest stream, but much of it sinks into the ground. Depending on local geology, groundwater can join an aquifer or seep downhill through the soil to feed rivers and wetlands for days to months after a major rainfall event.

Local conditions affect these processes. For example, areas with forest retain more rainwater as groundwater than urban or agricultural areas do. Thus, each river's future will unfold differently as the Southeast warms further. Those with a forested watershed may see minor changes in streamflow. Streams within agricultural and urban watersheds will see great declines in stream volume due to higher rates of evapotranspiration and water extraction. In general, as water supply declines and demand increases, it may be inevitable that aquatic wildlife loses out to prioritization of water for power generation, irrigation, and drinking water.

The worst times for southeastern streams will be during the low flows of summer and fall. Already, rainfall during these seasons has become less predictable, and above-average temperatures are forcing rapid water loss through evapotranspiration. These conditions cause water quality to decline. Warm water holds little dissolved oxygen and encourages algal blooms. And, because of greater evapotranspiration, small creeks high in the watershed run dry and flowing streams carry a higher concentration of pollution. Unfortunately, climate change will make the warm season even longer and hotter, and this will impose increasing stress on aquatic biodiversity.[13]

Concerned about how much stream conditions will deteriorate in the Southeast, Daniel A. Marion with the US Forest Service and colleagues modeled streamflow for the region's streams at midcentury. They estimated that on average the South will see a 6 percent decline per decade in warm season stream volume until midcentury. The cumulative decline for southeastern Alabama and the Florida Panhandle will be as much as 25 percent. The rest of Alabama, western Georgia, South Carolina, and central Tennessee will see declines up to 10 percent.[14]

A related study by Ram P. Neupane and others examined how climate change under the midcentury-peak emissions and business-as-usual emissions scenarios would affect flows in three Gulf Coast rivers. In the business-as-usual pathway, they predicted that annual streamflow by 2080 will decline by 9 percent, 3 percent, and 13 percent for the Mobile, Apalachicola, and Suwannee River Basins, respectively. Minimum summer flows will decline by 16 percent, 9 percent, and 25 percent, respectively. Streamflow declines were less severe in the midcentury-peak emissions scenario, but still enough to impair stream ecosystems.[15] These declines in stream volume will further imperil southeastern aquatic biodiversity.

All told, the Southeast is entering an era of water insecurity that will affect every use of fresh water in the region. Though climate change is forcing rainfall to increase, population growth and rising evapotranspiration rates will commandeer most of this extra water. Some areas within the region will have it worse than others, but dry summers and droughts will cause water shortages in communities across the region. The impact of declining water availability on biodiversity will be severe, especially in watersheds whose ecosystems are already stressed by loss of forest cover.

Declining freshwater availability is just one of several major challenges that climate change is bringing to the Southeast. Another serious threat to the region's communities and ecosystems also involves water, and this threat is coming from the ocean.

25
Rising Waters

When I return to the homestead in Gulf Breeze, the first thing I do after swapping hugs with Mom and Dad is greet the bay. I walk down the homestead's old brick sidewalk and stand below the canopy of oaks at the edge of the fourteen-foot bluff overlooking the water. I scan the familiar vista: our simple dock, the seawalls protecting the shore, the demarcation where the sandy shallows meet the deeper water, and the Pensacola skyline. But these days, I also look for evidence of sea level rise: wear and tear on the dock, weak spots in our seawall, and shoreline modifications around the bay. The estuary's waters have already risen and are rising faster each year. This coastline will be different by the end of the century.

Already I've seen a lot of change. When I was a child, there was gigantic tree on our shoreline. It was a sprawling Sand Live Oak, whose curvy limbs, draped with Spanish Moss, reached out over the water. Dad and I built a platform in the tree, where I would read or play for hours on end. On the other side of the seawall, low tide revealed a narrow sandy beach. Today, that beach and the oak are gone. Coastal erosion stripped away the sand. The oak went down in 1985, when the storm surge from Hurricane Elena breached the old seawall and washed away the soil beneath it. We still miss that tree thirty-five years later.

My parents have seen even more transformation at the homestead. When my father, Bob, was a child, he could play on the wide beach for a half-mile in either direction. By the time he and my mother, Lucy, married in the 1960s, erosion had already swept away most of the beach and neighbors were beginning to protect the remaining shoreline with seawalls. Since then, the beach has vanished. In its place are thick wooden bulkheads, metal barriers, and boulders quarried hundreds of miles away in the Appalachians. The shoreline—dangerous and unwelcoming—now resembles a battlefield fortification.

This is a lot of change for just sixty years, but it pales in comparison to what the bay has seen in the past twenty thousand years. If I were standing on the bluff at the homestead eighteen thousand years ago, at the peak of the last ice age, there would be no bay. I would be gazing over a three-mile-wide river floodplain of cypress and tupelo swamp through which the Escambia River flowed. The Gulf's beaches that today are just two and a half miles away were sixty miles due south. Sea level was lower because there was so much ice on the land.

The Pleistocene-Holocene transition occurred over several thousand years, allowing time for coastal ecosystems to adjust. During the thaw, glacial meltwater filled the oceans, sea level rose, and marine waters crept inland. The freshwater swamps of cypress and tupelo that had filled the basin where Pensacola Bay is now migrated upstream. Salt marshes replaced them, but as seas rose higher, open water replaced the marshes. Similar transitions played out across 150,000 square miles of flooded southeastern coastline—an area larger than Alabama, Georgia, and South Carolina combined. Earth's temperatures stabilized about 11,600 years ago, but sea level rose until 4000 years ago because of the lag between changes in atmospheric heat, ice melt, and sea level rise. After that, sea level remained constant until recently.[1] Sea level is rising again, this time because humanity has altered the chemistry of our atmosphere. Over the next few centuries and millennia, the southeastern coastline will undergo changes like those it endured as the most recent ice age ended. As for the remaining decades of this century, we will see changes to our coast that are far more profound than those my father and I have already witnessed.

We often think about sea level as if the oceans and seas of the world rise and fall in unison everywhere. They do not. Periods of high oceanic rainfall can cause a temporary rise in local sea level. Intense winds or high atmospheric pressure can push water against a distant shore. Ocean currents vary in strength over time and can cause water to pile up against a coast. Changes in the heat and salinity of water also affect volume.[2] For these and other reasons, sea level rise caused by climate change is affecting regions of the world differently. Scientists modeling sea level rise account for this variation. They must also account for the rise and fall of land independent of sea level. New Orleans, for example, is sinking.

South Louisiana and many other coastal lands around the world's coasts are sinking because they are composed of soft sediments deposited over millennia by rivers. As these sediments settle, the land surface drops. Rivers can rescue these lands by depositing new sediments during floods. However, when residents constructed levees along the margins of the Mississippi River to prevent flooding of New Orleans and adjacent areas, this starved the land of sediment, and it began to sink. Much of the city is now below sea level and vulnerable to flooding.[3]

The Crescent City is not the only metropolis that's sagging. From Georgia to New England, the US Atlantic Coast is also sinking. It's a gradual slumping that began when the most recent ice age was waning. The ice sheet covering the northern portions of the continent was two miles deep at its maximum, and its weight caused Earth's crust beneath it to sag. This caused adjacent areas not buried by the ice to rise in elevation, including most of the East Coast. With the ice block gone, the East Coast is now sinking, a process called glacial isostatic adjustment. Humans are also causing portions of the southeastern Atlantic Coast to sink. Many cities are pumping fresh water out of coastal aquifers faster than new water can replace it. No longer buoyed by as much subterranean water, the land is slumping.[4]

Because land elevation and ocean volumes can change simultaneously, and because the scale of these changes varies around the world, scientists calculate relative sea level change for positions along a coast. For example, one inch of subsidence plus one inch of actual sea level rise would yield a relative sea level rise of two inches.

Climate change is causing sea level to rise in several ways, not all of which are obvious. About a third of the present rise is from losses of the Greenland and Antarctica ice sheets (thick layers of ice covering large areas of land). Greenland has contributed most of this water, owing to its large ice sheets and the rapid warming of the Arctic. Another third is contributed by the rapid melting of the world's montane glaciers (rivers of ice in valleys at high elevations).[5] What is less intuitive is that sea level is also rising due to oceanic warming. The oceans have absorbed much of the heat initially captured by greenhouse gasses in our atmosphere. This has slowed atmospheric warming, but as water warms its volume increases. This process, known as thermal expansion, accounts for the last third of present-day sea level rise.

Climate change is subtly tinkering with sea level in other ways. Oceanographers recently discovered that the shape of our ocean basins is changing. Earth's crust beneath the seafloor is much thinner than continental crust, and it is sagging under the growing weight of the oceans. This has been good for us because the sagging has increased the ocean's capacity to hold water. Sea level rise thus far is about 8 percent less than it would be otherwise.[6]

Another effect of climate change on sea level involves gravity. Coastal ice sheets and glaciers are so large and dense that they exert a gravitational pull on ocean waters, creating slight bulges in the ocean around them. As they melt, the gravitational tug relaxes and the released water slowly sloshes against shorelines on opposite sides of ocean basins, causing faster rates of relative sea level rise on those coasts.[7]

After scientists factor in meltwater, thermal expansion, coastal subsidence, and all the other variables affecting sea level, what can they say about sea level rise? Based on historical data from a global network of tidal gauges, average sea level around the world rose 7–8 inches between 1900 and 2000. Since 1993, satellites have helped us measure changes in sea level. The combined data shows global sea level rise has accelerated since the end of the twentieth century. It is rising 1.1–1.3 inches per decade, totaling about 2.75 inches from 1990 to 2015.[8]

An inch per decade might not seem like much. But the story on sea level rise isn't so simple. Some places in the world, such as cities or nations on Pacific atolls or river deltas, are just a few inches above sea level, and many are already experiencing flooding. And due to relative sea level rise, some coastal communities here in the Southeast are also dealing with flooding, or soon will be. Relative sea level rise along the southeastern Atlantic Coast is over twice the global rate, at 2.6 inches per decade. Relative sea level rise in the eastern Gulf of Mexico matches global increases, but in the western Gulf average relative sea level rise is about 2.3 inches per decade, with South Louisiana dealing with a rise of 3.5 inches per decade.[9] Unlike the rest of America's coastline, most of the southeastern coast is at a very low elevation. Thus, just a few inches of sea level rise can affect places far inland.

The future of sea level rise is much more alarming than the changes we are currently experiencing in the Southeast. Federal agencies recently developed baseline models to forecast sea level rise under future greenhouse gas emissions scenarios. Because sea level rise is accelerating, the incremental changes we have been measuring in inches in the past few decades contrast sharply with predictions measured in feet. In the midcentury-peak emissions scenario, sea level will rise 1.2–3.1 feet by the year 2100, and if greenhouse gas emissions continue at the present rate, sea level will rise 1.6–4.3 feet by 2100. Even the low end of these estimates is enough to threaten many coastal communities in the Southeast.[10] However, these are rosy scenarios compared with the latest estimates. More recent studies suggest that our situation is even more dire.

Lurking at the bottom of the planet is Antarctica, a continental-sized water bomb. This is Earth's deep freeze, holding enough ice that if all of it were to melt, the oceans would rise hundreds of feet. With

Scenarios for sea level rise for this century. Based on greenhouse gas emissions at the time of this writing in the early 2020s, we will see six to eight feet of sea level rise by the year 2100. Note that in all scenarios, sea level will continue to rise after 2100. Graph reprinted from Hayhoe et al., "Our Changing Climate."

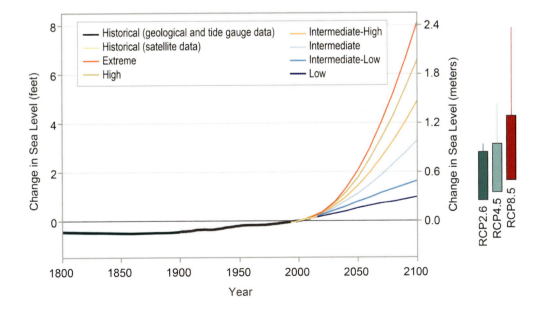

full melting, at least a third of the Southeast would vanish, including most of Florida. Hundreds of coastal cities around the world, even entire countries, would disappear. A lot is at stake, so scientists have been warily eyeing Antarctica ever since global warming came onto the scene. Until recently, models of sea level rise have assumed a gradual melting of the Antarctic ice sheets. These sheets—averaging 1.3 miles thick—seemed too massive to catastrophically melt in the foreseeable future.

Unfortunately, there's mounting evidence that they are far more unstable than scientists had assumed. If you looked at a map of Antarctica without its ice, the continent would resemble an archipelago with a large, irregularly shaped island at its center. This means that much of the continent's ice is resting on seafloor. Currently, sea ice and coastal glaciers act as dams, preventing the continent's ice sheets from spilling into the ocean. Unfortunately, the Southern Ocean is warming rapidly, and its seawater is melting these dams. We are much closer to losing the Antarctic ice than scientists had long assumed.[11]

Models factoring in these new discoveries suggest that in the midcentury-peak emissions scenario, Antarctic ice will contribute another foot of sea level rise by the year 2100. In the business-as-usual scenario, Antarctic melting will add an additional 2.5 feet of sea level rise by 2100. To these values we must add the contribution of the Greenland ice sheets, which are also melting faster than originally assumed. Altogether, we are looking at sea level rise of about 8 feet by 2100 if humanity does not curtail greenhouse gas emissions.[12]

This much change is already difficult to grasp, but I am afraid we need to look beyond 2100 to understand what's at stake. Though our oceans are soaking up much of the heat from anthropogenic warming, they will cling to this heat and prevent the formation of new ice in the Antarctic archipelago for thousands of years to come. Under the business-as-usual emissions scenario, Antarctic ice will contribute a total of 16.4 feet of sea level rise by 2200. And by the year 3000 the oceans will be 66 feet higher than today. The rosier scenario of peak emissions at midcentury has Antarctica contributing 20 feet of sea level rise by 3000.[13] That is 2000 years in the future and may seem irrelevant for us today. Though most of us struggle with long-term thinking, consider that some of the major cities of Europe, the Middle East, and Asia are well over 2000 years old. This much sea level rise will swallow many of them. Even more sobering is that we can't engineer ourselves out of this predicament. Because water stubbornly holds on to its heat, future technologies that may be developed to draw carbon dioxide out of the atmosphere will have only a minor influence on reducing sea level rise.[14] Simple physics will ensure that we pay a debt, plus plenty of interest, for the fossil fuels we are so eagerly burning.

I'll admit that when I'm back at the Duncan homestead, standing on the bluff and looking out over Pensacola Bay, it's hard for me to worry about sea level rise several thousand years from now. Usually my anxiety is about the coming century. What losses will I see due to sea level rise over the remaining decades of my life? How much will it cost to protect this place? What changes will my daughters and my niece witness after I'm gone? Will the homestead even survive to the end of the century? I didn't worry much about sea level rise until I did this research. I had assumed that there'd not be much impact until next century. A few inches per decade—the current rate of sea level rise—didn't seem like enough to cause much harm. What I now understand is that areas throughout the world are already struggling with sea level rise, and this includes dozens of communities here in the Southeast.

26

Salty Floods

Tidal floods are the most visible signs that sea level rise is gnawing away at the southeastern coast. Locals call them sunny day floods because the streets can be underwater even though the sky is clear. Oceanographers label them high tide floods or nuisance tides. Regardless, for many residents along the southeastern Atlantic coastline, these floods are disrupting their lives. And, they are a clear warning that the ocean is moving inland.

High tide floods naturally occur on low coastlines, but they are increasingly frequent along the southeastern Atlantic Coast because sea level is rising so quickly there. In 1960, Savannah, Georgia, saw an average of three nuisance floods per year. By 2014, this had increased to seventeen. Over the same interval, Charleston, South Carolina, had a rise in nuisance flooding from three to twenty-six per year, or about two per month. Wilmington, North Carolina, has seen some of the most dramatic increases in nuisance flooding along the East Coast. In the early 1960s, high tide floods came about once a year. Now floods inundate low-lying neighborhoods forty-one days a year on average. That's one out of every nine days. More broadly, high tide floods are increasing in frequency across the region due to sea level rise. In 2015, nuisance floods were two to three times more frequent in the region than they were in 1995.[1]

In communities where floods are frequent, residents carefully watch the lunar cycle and the weather to plan their day. They know a nuisance flood is more likely when the lunar phase is full or new. During these times the gravitational pull of the moon and sun on the ocean is at a maximum. However, the resulting tides are not enough to cause a nuisance flood on their own. Flooding occurs when a strong wind, a storm, or a shifting ocean current pushes extra water against the coast during a spring tide. Sometimes rainwater coming down the river

collides with an incoming tide to produce the flooding. On occasion, these phenomena can move enough water to cause a nuisance flood without the help of the moon.

Nuisance flooding isn't just a problem for neighborhoods close to the shoreline. Tidal creeks and stormwater systems funnel flood waters deep into the interior of coastal communities. People who cannot see the waterfront from their home awaken to find that floodwaters have covered the yard and swamped their cars. City streets become impassable, and commuters must find alternative routes to drive, bike, or walk to work. Schools close or delay their start because busses cannot reach neighborhoods. Businesses and hospitals are short staffed because employees are late, or never arrive. Floodwaters trap the elderly and disabled in their homes. Simple tasks, like walking a dog or checking the mail, are more complicated. They also bring risk. Floodwaters carry pollutants and harmful bacteria floating up from sewers. Injuries are more frequent because people slip and fall on wet walkways.[2]

High tide flooding is so frequent and disruptive in some communities that it is causing the decline of property values. In Florida, coastal

Flooding caused by sea level rise is already afflicting many coastal US cities. For now, most of this flooding occurs during particularly high tides or storm events. Today's flood-prone areas, including downtown Annapolis, Maryland (*pictured*), will be permanently flooded in the coming years unless preventative measures are taken. Courtesy of the Chesapeake Bay Program.

SALTY FLOODS 211

property values have declined by over $5.4 billion this century due to nuisance tidal flooding. Collectively, the other southeastern states have seen declines of over $2.4 billion.[3] Local governments suffer, too. Repeated exposure to salinity and saturation damages infrastructure. And flood preparation and recovery efforts strain the budgets of maintenance and emergency departments.

Today's nuisance flooding is a wake-up call. It is just the beginning of a difficult future for the southeastern coast. For now, a high tide needs help from a wind, current, storm, or swollen river to cause a nuisance flood. But as sea level rise accelerates, a high tide flood will be the new normal tide by midcentury in many communities. A recent study by the Union of Concerned Scientists estimated that by 2035 and under a moderate sea-rise scenario, about 170 coastal communities in the United States will see a nuisance tide at least every two weeks, and these tides will flood 10 percent or more of the community. This number jumps to 270 communities by 2060, and 490 communities by 2100. By the end of the century, nuisance tides will flood 25 percent of the community for a majority of American coastal cities. In the high-emissions future, the entire southeastern coastline will experience a nuisance flood every day by 2100.[4] And in both these scenarios, high tide floods will cause excessive damage when winds, rain, or currents amplify them.

High tide flooding is also threatening sources of drinking water for coastal communities. Through a process known as saltwater wedging, tides send plumes of salty water up local rivers and creeks. This brings seawater far inland, even while more buoyant fresh water flows downstream above the wedge. Saltwater wedging is a natural event in rivers of the coastal lowlands, but sea level rise is forcing salt water farther upstream than in the past, and this can contaminate the freshwater supply for communities.

Newport News, Virginia, near the mouth of the James River, is one of these communities. The city is in one of the Atlantic Coast areas where sea level is rising fastest. One of its freshwater sources is a reservoir behind Walkers Dam on the Chickahominy River, a tributary to the James River. Saltwater wedges already bring dilute seawater to the dam, despite it being twenty-one miles from Chesapeake Bay. Under the business-as-usual climate scenario, by the end of this century average salinity below the dam will be six times higher than the recommended limits set by the US Environmental Protection Agency (EPA) for drinking water during a typical year, and ten times higher during years when rainfall is scarce. Similar levels of saltwater intrusion will reach positions over thirty miles up the James River.[5]

Shallow aquifers near the coast are also vulnerable to saltwater contamination. Many coastal cities have wells that tap into aquifers for fresh water. Deep aquifers are not vulnerable to salt contamination from sea level rise if the amount of water withdrawn is no more than the amount of incoming fresh water.[6] However, some coastal cities and coastal farmers rely on shallow aquifers. These freshwater sources are highly vulnerable to saltwater contamination brought by sea level rise and high tide flooding, especially when users extract water faster than surface water can replace it.[7]

Beyond the urban landscape, the appearance of so-called ghost forests is an ominous sign that the rising ocean is slowly annexing land by way of high tide flooding. Ghost forests are stands of coastal trees that have died due to salt exposure. Despite their proximity to the ocean, coastal ecosystems fringing southeastern bays and estuaries and their rivers cannot tolerate high salinities for very long. However, tidal floods are exposing them to more of these deadly salinities as the waters rise. This is causing the dieback of freshwater and brackish-water plants and the ecosystems they create. Salt-tolerant plants are gradually replacing them. The net effect is a slow-motion migration of ecosystems. Open water is replacing salt marshes, salt marshes are replacing brackish marshes, brackish marshes are replacing freshwater marshes, and freshwater marshes are replacing swamps. Because river deltas become narrower farther inland, there will be less area for all these ecosystems than is available today. Populations of some animals specializing in these habitats, such as the Saltmarsh Sparrow, are now in severe decline. Signs of ecosystem migration are obvious in the ghost forests, where the sun-bleached trunks of dead trees tower over the brackish marsh that has overtaken them.[8]

Southeastern coasts are already haunted by ghost forests, stands of trees killed by salt water pushed inland by sea level rise. Ghost forests are replaced by marsh, and eventually open water. Pictured here are dead pines along the Alabama coast. Photo by R. Scot Duncan.

The loss of these ecosystems threatens more than biodiversity; it also jeopardizes the economic security of coastal communities. For example, coastal salt marshes are nurseries for young shrimp, snapper, crabs, and other species that sustain lucrative sport and commercial fisheries. The economic value of these services to the Southeast isn't known, but ecosystem services in the Mississippi River Delta have been calculated to be between $12 and $47 billion a year.[9] If we scale this up to the entire Southeast, the region's coastal wetlands may yield services exceeding $100 billion a year.

The losses of coastal wetlands due to sea level rise could be extensive this century. One team of scientists modeled the future of salt marshes along the Georgia coast.[10] They predict that by 2100, the overall coverage of tidal marshes could decline by 20 to 45 percent of their current extent, depending on how high seas rise. Not only will this harm fisheries, but the loss of these ecosystems will contribute to climate change as swamps filled with carbon-storing trees vanish and dying marshes release carbon dioxide and methane into the atmosphere.[11]

Sea level rise is also reshaping the coast by making hurricanes more destructive. Hurricanes are a triple threat. Fierce winds can far exceed one hundred miles an hour. Sustained downpours cause flooding hundreds of miles inland. And the storm surge—the bulge of ocean water that hurricanes sling against the shore—tears apart coastal towns and damages ecosystems. Sea level rise enhances destructive power of storm surges. As the oceans increase in volume, surges are higher, cause more damage, and spill further inland.

Hurricanes are the most destructive natural phenomenon to threaten the Southeast. Their return interval ranges from five to thirteen years for any point along the coastline.[12] Hurricanes are costly because the region's coasts have a low elevation and a high density of people and infrastructure. Since 1980, the Southeast has incurred more billion-dollar weather disasters than any other US region. In 2018 Hurricanes Florence and Michael caused $24 and $25 billion in damages to the Southeast, respectively. The year before, Harvey and Irma caused $125 and $50 billion in damages, respectively. Eighteen additional hurricanes striking the Southeast have each caused over $1 billion in damages since 2000.[13] Hurricanes are also deadly—the four storms mentioned above caused 306 deaths.

North Atlantic hurricanes have become more frequent since the 1970s. Climate scientists suspect that there is a link between rising storm frequency and the warmer ocean temperatures in the Atlantic caused by global warming, but they are still investigating. Climatologists also don't know whether hurricane frequency will continue to rise. Some climate models predict more hurricanes, while others predict fewer.[14]

However, scientists know for sure that hurricanes are striking the Southeast with greater force than ever before due to climate change. Not only are storm surges more powerful due to sea level rise caused by global warming, but today's hurricanes last longer and bring stronger winds and more rain than in the past.[15] As a result, category 4 and 5 hurricanes have increased in frequency since the 1980s. There is broad agreement among climatologists that natural climate variation and human-induced warming of the Atlantic Ocean have caused these changes. Hurricanes are also moving more slowly than ever before due to the warmer atmosphere. Slower tropical cyclones cause more damage with winds and flooding than those that move quickly over a region.[16] We saw this when Hurricane Dorian lingered over the

Bahamas for over twenty-four hours as a category 4 and 5 storm in 2019. Because the oceans are warming and hurricanes grow in strength by absorbing heat from the sea surface, most climate models predict we will see an increasing number—as much as a doubling—of category 4 and 5 storms this century.[17]

Sea level rise will affect the Southeast more than any other region in the United States. Other coasts have shorelines fortified with cliff and rock, but the southeastern coastline is one of soft sediments that move with every wave, tide, and storm. The early European and American settlers to the region learned this, sometimes the hard way. Shipwrecks were common because the shifting of shoals, channels, and barrier islands made the latest navigation maps obsolete. Hurricanes flooded settlements on land that was too low. But with time, settlers found places and ways to build to minimize flood risk. Centuries later, we have invested trillions of dollars in urban infrastructure on the coast.

Now the pace of change along the coast has quickened. After being stable for several thousand years, sea level is rising at a rate that is accelerating. As a result, the coast is changing before our very eyes, with high tide flooding in cities, the inland migration of coastal ecosystems, salt contamination of rivers and aquifers, and more deadly storm surges delivered by the most powerful storms ever recorded.

What happens next to the southeastern coast? Much depends on whether and how soon we curtail greenhouse emissions. If we act quickly, we will avoid the worst of climate change and sea level rise. But this is a global decision made by governments, businesses, and citizens, and the Southeast plays only a minor role in this endeavor. And regardless of that outcome, sea level will rise for the foreseeable future—well into next century. It's just a matter of how long, how fast, and how much. Will we migrate inland as coastal ecosystems do? Or will we try to defend our ground? Either way, tough times and tough decisions lie ahead.

PART IV

27
Ecological Free Fall

Freshwater supplies are declining, pollution seems intractable, flooding and droughts are on the rise, the coast is disappearing, and the endangered species list keeps growing. These are some of the daunting problems we've uncovered while trying to understand the southeastern river crisis. Three questions have guided us while exploring the causes and consequences of this crisis. The first is *how did we get here?*

The short answer is that we industrialized the southeastern landscape, including its rivers. It began with the Native American genocide. We then converted the region's forests and prairies into agriculture and managed forests, mostly through the blood, tears, and sweat of enslaved Africans. All the while, we hunted and harvested wild species, including sturgeon and freshwater mussels, for commercial markets. Industrialization has brought comfort and prosperity to many, but the costs include the ongoing oppression of people from two continents, species extinctions, and the pollution and reengineering of southeastern rivers to the point where aquatic ecosystems are failing.

Multiplied the world over, industrialization has forced Earth out of the stable and mild Holocene climate that over the course of ten thousand years nurtured the spread of human populations and the development of complex cultures. We have now initiated a new geological age, the Anthropocene, one in which *our* hands are on the throttles controlling the planetary systems sustaining us. One system we have pushed too far is climate stability. Climate change is now reshaping our culture and our planet in ways that we've only begun to understand. Consequently, the Southeast is facing a gradual decline in water availability over the next decades, punctuated by increasingly frequent and intense droughts.

The climate crisis prompts our second question: *will the Southeast have enough water in the future?* Thus far, the answer is no. Despite climate models suggesting that the Southeast will be receiving more rain, we will not have enough water if we continue using it as we do now. Continued population growth will cause some of the shortage, and evapotranspiration rates are rising as the climate warms, removing water from the region's soils, rivers, aquifers, and reservoirs. Shortages will be most acute during hot, dry summers and droughts. Because the Southeast has enjoyed an abundance of water in the past, the region is ill prepared to handle water shortages, and this threatens drinking water supplies, power generation, irrigation, river transportation, and recreation.

The prediction of water scarcity, in combination with other threats, leads us to our third question: *can southeastern river biodiversity survive the Anthropocene?* The region's aquatic biodiversity is unrivaled in the temperate zones of Earth, and the region is the epicenter of continental, sometimes global, species diversity for several life forms, including fishes, mussels, and snails. But many of these species are gone, and hundreds more face extinction. This is the ecological collateral damage from our industrialization of the landscape. Will the remaining species survive? It's hard to be optimistic based on our past performance.

Considering what we've learned thus far, the future for rivers seems bleak: lower flows, higher water temperatures, algal blooms, and disappearing river deltas. These and other climate change problems will worsen the legacy problems of biodiversity loss, pollution, and other downsides of river industrialization. Other problems created by climate change in the coming decades will draw attention away from river conservation. Longer and hotter summers, forcing changes to how we live and work. More heat-related illness and mortality. Increasingly powerful storms and more flooding. Agriculture struggling to adapt. Rising wildfire frequency and intensity.

One of the most troubling features of climate change is how fast it is happening. Until late last century, most climate scientists said we had plenty of time to prevent or prepare for climate change. I didn't expect to see any impacts until I was older. Then two things happened.

I got older, and the climate scientists were proven wrong. Now climate change disasters are constantly in the headlines. Our younger generations are seeing more environmental change in their first decades of life than was collectively witnessed by generations before them.

What alarms me most about climate change is that if we keep emitting greenhouse gasses at the present rate, some analysts foresee a future of increased political instability. Coastal land loss and failing agriculture will force mass migrations of people. As this is happening, the world population will approach or exceed 11 billion people during this century (we are at 7.8 billion in 2020).[1] This combination will cause chain reactions of natural resource scarcity, tensions between rival groups, internal conflicts spilling across international borders, and regional destabilization. Governments reeling from climate disasters and legacy problems will be ill equipped to handle these challenges. Wars may be inevitable. Due to the possibility of futures like this, some analysts view climate change as an existential threat to humanity.[2]

I am not worried about the survival of civilization in some form. We will endure climate change because we are innovative, adaptable, and prolific. Cooperation, not conflict, is the norm. But I am deeply concerned about the outlook for our quality of life and the future of biodiversity. If we continue to postpone a significant reduction in carbon dioxide emissions, by midcentury we will be in an ecological free fall, where climate change and other unintended side effects of planetwide industrialization will be eroding the livability of the planet and humanity's upward cultural trajectory.

I've always hoped that we are on a path of progress. For me, this would mean building an inclusive future, where everyone enjoys a life of security and opportunity. Milestones along the way would include an end to economic insecurity and the hate and conflict it fuels. We would cure diseases. Technology and culture would prosper. Extinctions would end. Attaining these goals will be far more difficult with the disruption climate change is bringing.

When I think about the future, I now selfishly worry about my daughters and my niece, and the world they are inheriting. I want them to live in a hopeful time for humanity, not an age of calamity and despair. I also cannot avoid other selfish worries, especially the prospect of watching the places I love along the coast wash away. Each harbors memories of visits with family and friends, and encounters with the wildlife that share these spaces. But even if greenhouse gas emissions peak around 2040, then by the time I am in my eighties (in the 2050s), sea level could be two feet higher than it was in the year 2000. The Duncan homestead will still have several feet of freeboard, but major hurricanes will bring storm surges ever higher into the yard. Waves will erode The Point down to a nub. High tides will regularly flood my friend Ann's house. Hurricanes will shatter the barrier island protecting Pensacola Bay into an archipelago of islets and shoals. If we continue spewing greenhouse gasses into the atmosphere at the current rate, much of the southeastern coastline will be unrecognizable by midcentury.

But these are the concerns of economic privilege. My family and I are among those least likely to suffer greatly from climate change. It is cruel irony that those who are and will be enduring the greatest hardships due to climate change are those who have benefited the least from the industrialization that brought on the Anthropocene. The poor suffer first and suffer most with environmental degradation, because they have the fewest options for adapting to a changing climate. If the rice crop fails in South Asia, tens of millions go hungry, but I'll hardly notice the higher price at the grocery store. When a heat wave strikes, the poorest of my neighbors, without air conditioning, are prone to heat-related illness and death. Those that do have air conditioning will have to rework their budgets to pay the higher utility bills. As for me, I may not even notice the higher charge.

Truthfully, until I began this investigation into the future of the southeastern environment, I didn't know the outlook for humanity's future was so grim. At the beginning I was optimistic. I thought we were on the cusp of overcoming some long-enduring problems. I hoped the conservation movement was rebooting to finish the river revival begun in the 1970s. Although we've found a few hopeful trends, the complexity and magnitude of the coming challenges overshadows our progress. Should I recalibrate my expectations and simply hope that we can minimize the damage as we stumble through the rest of this century and beyond?

Maybe my concerns about biodiversity loss are frivolous. In the coming decades we may need to focus on self-preservation. Thousands of species will disappear during the Climate Change Extinctions because our own survival consumes all our attention. The Age of Loneliness that Edward O. Wilson sagely warned us about will have begun.[3] Only the common species will survive. Weeds will be the wilderness of the future. Ecosystems will comprise a handful of generalist species that can survive in our ditches, hedgerows, and alleyways. Today's wild places will become seafloor, urban sprawl, climate-blighted landscapes, or cropland to feed the new billions. What will remain of today's biodiversity will be inadequate facsimiles in our museums, libraries, and memories. It will be a future of massive, costly infrastructure to replace the ecosystem services lost in the ecological free fall of the Anthropocene.

Are we really going to let this happen?

28

Ivan's Wisdom

On September 16, 2004, Hurricane Ivan made landfall near Gulf Shores, Alabama, as a category 3 storm and offered a warning of what the future may bring in the era of climate change. Sustained winds at 120 miles per hour made Ivan the strongest storm to hit the area in over a century. The storm surge and winds pummeled Gulf Shores and surrounding communities. Cities along the coast received five to ten inches of rain. Sustained winds up to 70 miles per hour buffeted areas far inland. Toppled trees crushed buildings and blocked roads. Flooding and power outages were widespread for days. Ivan caused $19 billion in damages.

Pensacola, thirty miles east of where the eye made landfall, received some of Ivan's worst fury. There were sustained winds near 100 miles per hour for several hours. Over fifteen inches of rain fell. A ten-foot storm surge along the city's waterfront flooded homes and businesses. Small boats ripped from the harbor floated through downtown streets. A twelve-foot surge at the top of the bay tore apart an interstate bridge.[1]

Three days after Ivan struck, I drove down to Gulf Breeze to help my parents and their friends. I began to see damage long before I crossed the border into Florida. Fields flooded. Streams surging. Hundreds of thousands of pines leaning to the west. When I entered Pensacola, it was obvious the city was in bad shape. Power lines down. Storefronts tattered and darkened. Roofs torn. Debris hanging from trees. Streets crowded with work crews. Fallen trees on yards, cars, and houses. Long lines at gas stations. Police and the National Guard directing traffic through intersections.

The Duncan homestead weathered the storm well. By settling on a bluff on the north-facing shore, the first Duncans in Gulf Breeze chose a position protecting the original house from the worst surge and winds a hurricane can bring to Pensacola. At an elevation of sixteen feet, my parents' house had a comfortable margin of safety from the surge, which crested at seven and a half feet along their shoreline. The live oaks that cover the yard were in tatters, but true to their reputation they were still standing. My brother, Will, also came to help, and we spent most of our time clearing the tangle of heavy limbs piled waist deep. The worst damage was to the dock. It was banged up badly by the surge and the drifting debris its waves tossed about.

My friend Ann's property sustained considerable damage. The house was intact. She and her husband, Dan, had designed it to rest on pilings thirteen feet above the ground. But the storm surge gutted the carport and woodworking shop beneath the house and scattered the contents throughout the neighborhood and adjacent bayous. The water had been so high during the storm that waves smacked against the underside of the house. When my family came over to help clean up, there was little we could do. Piles of debris were too heavy for us to move, and our chain saws were too small for the pines that were down. Months of tedious and backbreaking repairs were ahead for Ann, Dan, and their neighbors.

Pensacola Bay was a mess, too. Angry, brown river water was still rushing through the bay. There were no signs of life. And there was an odor. Long ago, Pensacola had built its main sewage treatment plant in the city's lowlands, a block from the shore. With the plant overrun by the storm surge, raw sewage poured into the bay. The stench of human waste mingled with the oily smell of diesel leaking from a fifty-foot-long workboat that was listing heavily on a shoal near my parents' house.

After two days of chainsawing and hauling debris, my brother and I walked to The Point. We needed a break from the work and were curious about what had washed up or washed away. The last of The Point's beaches at its tip were gone. The splintered remains of docks and boat houses littered the shore. Accenting the debris piles were colorful artifacts from houses and apartments blasted apart by the violent storm surge that had swept over the barrier island. Houses on the south-facing bluffs were fine, but the surge had overwhelmed homes in the adjacent lowlands and battered them with drifting rafts of destroyed boat docks.

The destruction by Ivan, and the many other hurricanes that have battered the Southeast since then, offers a glimpse of our future with climate change, especially if we do not significantly reduce greenhouse emissions within the next few years. And of course, it is not just hurricanes. River flooding, coastal land loss, droughts, wildfires, and heat waves—climatologists predict that the Southeast will see more than its fair share of disruption, suffering, and loss in the Anthropocene.

Ivan also lashed the ecosystems along the Alabama and Northwest Florida coast. Over the next year, I revisited many of my favorite natural areas along these coasts. Some places were unrecognizable. Trees I remembered from my youth were gone. The dunes had vanished. Swamp trees had been delimbed or snapped in half, and the waters beneath filled with appliances, building fragments, and every variety of household belonging that can float. Salt water carried inland by the storm surge had pooled in low areas and killed acres of forest. It was devastating to see.

But just as coastal residents rebuilt their communities, these natural areas recovered in the months and years that followed. Ivan wasn't anything new for the coastal ecosystems in its path. They'd weathered dozens of storms just as intense over the past several thousand years. On the barrier islands, where the damage was worst, nature wasted no time. Plants and animals recolonized, tides and breezes brought sand back to the island, and winds began rebuilding the dunes. The scars from the storm became part of the new landscape. Slash pines killed by salt water became nesting trees for Ospreys. Channels cut through the barrier islands by the storm surge became lagoons and marshes. Some eventually filled with sand, while others persist two decades later. At Gulf Islands National Seashore, south of Gulf Breeze, the surge had swept away the dunes, leaving behind flat expanses of sand littered with chunks of the destroyed road. These areas were a boon to birds that breed along the shoreline. Terns, skimmers, and plovers—species that nest on the bare ground—bred here by the thousands. They liked the ability to see approaching predators and the absence of human interlopers because the roads had washed out. Over the next few years, native beach plants colonized these overwashed areas, and blowing sand collecting against them became the first new dunes.

In ecology, succession is the name we use for the process of ecological recovery after a disturbance. Species colonize a damaged area and enjoy an abundance of resources, and an ever more complex and diverse ecosystem arises. Ecological succession occurs wherever there is life and periodic disruption. Even where the most violent phenomena have bludgeoned the landscape—lava flows, landslides, and wildfires—plants and animals will soon colonize. All nature needs are time, opportunity, and resources. To clarify, this is not a coordinated, intentional response. The emergence of ecosystems after a disturbance is the cumulative result of organisms seeking out places to grow and reproduce. Nor are the rebuilt ecosystems replicas of the ones they replaced. Often the environment is different, and there are new combinations of species. Nevertheless, natural landscapes heal. Nature comes back. It's been this way for hundreds of millions of years.

Ivan and the ecological changes it set in motion offer a couple lessons for us as we face climate change and the other environmental challenges of the Anthropocene. The first is that just like the plants and animals that endured Ivan, the species alive today are survivors. If we give them an opportunity—and sometimes a little help—most of them should be able to adjust to the new era. It's in their DNA to never give up. Even the rarest of the endangered species alive today have endured tremendous amounts of environmental change. They survived the transition from the globally tropical temperatures of the Pliocene (5.5–2.6 million years ago) into the deep chill of the Pleistocene (2.6 million to 11,600 years ago). Throughout the Pleistocene the climate swung, sometimes rapidly, between ice ages and periods as warm as today. Then, during the Pleistocene-Holocene transition, sea level rose nearly four hundred feet over the span of 14,000 years. This caused great disruption, yet all the species on Earth today survived. More recently, today's species have endured several centuries—millennia in some regions—of landscape industrialization. Having withstood all this, the species still with us deserve a chance. And if they do survive, we'll be better off because they will help sustain the ecosystem services we need and enjoy. The key is giving them a chance to fight for their survival.

While we should do everything possible to minimize or halt climate change, ecosystem loss, and the extinction crisis, some change will be unavoidable. Thus, like the species that prospered after Ivan by shifting how and where they lived, we need to adjust our way of thinking and living if we are to thrive—not just survive—in the Anthropocene. This is the second lesson Ivan offers. The challenges we are facing are occurring early in our history as a scientifically aware, globalized civilization. Instincts such as tribalism and greed that served us well in our past now threaten our future. The cultural systems we developed for stability and prosperity—agriculture, immobile infrastructure, and geographic-based self-governance—worked well during the past 11,600 years, when Earth's climate was stable. But now, because of us, the world is changing quickly. We are confronting our first climate endurance test, and we are not ready.

Anthropogenic climate change is a problem of our own design. It is the fallout from industrializing most of the easily inhabitable landscapes of the planet and, in the process, filling the sky with heat-trapping pollutants. Environmental change resulting from our actions is nothing new for humanity. It has happened regionally for ages. When it has been done to excess, cultures adapted, moved to new locations, or collapsed. But now, as a first for humanity, we are degrading our environment at the planetary scale. There is nowhere else we can go, and the future of thousands of cultures and millions of species around the world is now at risk.

Though the challenges are daunting, self-inflicted extinction or cultural decay is not inevitable. We have a choice. Our science has caught up with our technological innovation and newly globalized economy. We now understand the consequences of our actions; we know that our prosperity and security depend on the very Earth systems that we've been blindly altering. Because of this self-awareness we have a choice in how we respond to the climate endurance test.

We will fail the test if we choose reactive adaptation. Instead of heeding the warnings of scientists and planners, we preserve the status quo until forced to do otherwise. We avoid the work and expense of preparing for change. There are no contentious debates about rezoning our communities, developing contingency plans, or paying for infrastructure upgrades. Everything seems fine until disaster strikes. Only then do we realize that we should begin planning for the problems that climate change is causing. However, the recovery phase after a disaster is the worst time for long-term planning and investment. More immediate concerns overwhelm our government. Responding to the disaster strains our talents and resources, and most of us are too traumatized and distracted to take part in any public planning process. It's a struggle to restore any form of normalcy, and the easiest path for us is to rebuild in the way it was before the disaster. Then the next catastrophe strikes, and the next. We become so preoccupied by responding to the most recent calamity that we lose our capacity to prepare for the next. Last year was a hurricane. This year it's river flooding. Next year will be drought and a water shortage. Recovery takes too long because there has been no investment in community resiliency, the ability to bounce back after a disruption. Businesses relocate to communities or regions that are more prepared and secure. Tax revenue and jobs disappear. Tourism declines. We shift public funding to disaster recovery instead of education, job training, public health, and infrastructure maintenance. Our community's future becomes a tortuous downward spiral.

There is another way. We can pass the climate endurance test through proactive adaptation to climate change. It begins with believing the science and abandoning the expectation that the status quo should and can continue. Our top priority is to minimize greenhouse gas emissions as much and as swiftly as possible. The sooner we act on this, the more disruption and misery we prevent for us and future generations.

We must also be proactive and prepare for the new climate of the Anthropocene. Regardless of how soon we curtail greenhouse gas emissions, a considerable amount of climate change is unavoidable. Fortunately, our climate models inform us of the changes we can expect. Heat waves, flooding, sea level rise, wildfire, water shortages—we have ways to prepare for these and other climate-induced problems. Climate adaptation is the name for such actions, and the field of adaptation professionals is broad and growing. They include building and landscape architects, civil engineers, urban planners, and scientists from disciplines ranging from coastal geomorphology to sociology. Proactive adaptation is the wise and graceful way to pass the climate endurance test. We take charge of our future. We reengineer our way of living to minimize the crisis and prepare for change. The public is engaged, and we plan for a diversity of concerns and vulnerabilities. Principles of community resilience guide the way. None of this will be easy, but it avoids the epic fail of a reactive response.

Climate change is forcing us to alter our way of living. But we have a choice between a reactive or proactive response. Both require tremendous effort. One minimizes destruction and suffering for us and future generations. The other is a descent into chaos. The wise choice is clear.

What will a future of proactive climate change adaptation be like? How should we manage our rivers differently? Can we prepare for water supply shortages, heat waves, and droughts? Is it possible to adapt to climate change and also build a fairer and more inclusive society? What can we do about sea level rise? And, as we address all these and related challenges, can we make room for biodiversity?

We must find answers to these questions. Fortunately, we can learn from people and communities around the world, and even in the Southeast, that are proactively adapting to climate change. Our next step is to visit a southeastern city sinking into the Atlantic Ocean and learn how it is preparing for sea level rise.

PART V

29
Armor, Adapt, Retreat

NORFOLK IS SINKING, AND THE waters are rising. Due to subsidence, the rate of relative sea level rise in this coastal Virginia city of 242,000 is three to four times faster than the global average. Norfolk sits at the mouth of Hampton Roads, the estuary receiving the James River and connecting it with Chesapeake Bay. It is the third busiest port in the United States and is famous for hosting the world's largest naval base. Lately, the city has become infamous for its frequent floods, which shut down roads and neighborhoods and damage infrastructure. The flooding comes in three flavors—rainfall flooding, nuisance tides, and storm surges—and each is increasing due to climate change. Even with just a moderate amount of sea level rise, up to half the city will be in a high-risk flood zone by the year 2100.[1]

Norfolk is not alone. Hundreds of other communities in the Southeast are facing a soggy future thanks to sea level rise. A 2018 report by the Union of Concerned Scientists found that by 2045 under a scenario of moderately high sea level rise, properties currently valued over $18.4 billion will be at risk of frequent or continual inundation in Alabama, Georgia, the Carolinas, and Virginia. This includes 51,655 homes occupied by 79,603 residents. Florida is especially vulnerable, considering the length of its coastline, high population density, and low elevation. The scientists predicted that the Sunshine State will see property worth $26.3 billion at risk of inundation by midcentury, affecting 64,039 homes and 101,428 residents.[2]

Sea level rise in the Southeast and elsewhere threatens more than homes. We've built a lot of important infrastructure within the grasp of projected sea level rise. The list includes businesses, airports, rail lines, shipyards, military bases, shipping docks, marinas, parks, hospitals, schools, churches, libraries, museums, fire stations, and city halls.

When we built all this no one was thinking about climate change and sea level rise.

Some of the most important and threatened infrastructure is out of sight. As sea level rises, the growing pressure of seawater forces water inland through the soil beneath the land. This sneak attack forces a rise in elevation of the water table (the water you would find if you dug a deep hole). This worsens surface flooding because the ground has less capacity to absorb rainwater during heavy storms. The rising water table also threatens critical infrastructure, including building foundations, basements, and thousands of miles of conduits shuttling electricity, communications, gas, water, stormwater, and sewage.

Norfolk is one of the few southeastern cities that has been taking a proactive adaptation approach to climate change. By the early 2010s, the city was experiencing up to fifteen nuisance tides a year, and city planners were raising concerns about the specter of sea level rise.[3] Then, in 2012, Hurricane Sandy lashed thousands of communities from North Carolina to Maine with strong winds, rains, and storm surges. It was a wake-up call for Norfolk and other cities.

Norfolk began planning for a future with sea level rise. In 2013 the Rockefeller Foundation invited the city to join its 100 Resilient Cities program. The foundation funded the hiring of a chief resiliency officer in the city's government to lead planning for maximizing innovation and minimizing both the damage caused by catastrophes and the city's recovery time. The foundation also provided city leaders and planners with information, technology, and connections to dozens of other cities planning for resiliency.[4]

At the center of Norfolk's strategy was the process of community planning. Several well-tested models for community planning are in use, and they all include a process of guided dialogue and cooperation among multiple stakeholders. The planning process brings together people from different socioeconomic, racial, ethnic, and professional backgrounds to share concerns and plan for their collective future. Also invited are experts from governmental agencies and nonprofit organizations with relevant expertise. Through guided, data-informed discussions, the community identifies its assets and vulnerabilities, then determines what adaptations are appropriate to reduce risk. By inviting all stakeholders, the planning process creates opportunity for synergy and creativity. It also builds trust between stakeholder groups, which lays a foundation for further cooperation.

The participants in Norfolk's planning process quickly realized they faced a daunting task. The city has 144 miles of coastline to defend and numerous low-lying neighborhoods. In addition, the city has a growing population, several underserved neighborhoods, and an aging infrastructure. A profusion of questions arose early in the planning process. How much will sea level rise? Would it be necessary to abandon low-lying neighborhoods already experiencing flooding? How will adjusting to sea level rise affect existing social inequities? But before the participants could answer these and other questions, they needed to learn about coastal flooding adaptation strategies.

Adapting to sea level rise may be the most technically, economically, and socially complex endeavor humanity has ever tackled. Sea level rise quite literally shifts the line on the map between where we can and cannot live. We've learned to cope with storms, droughts, fires, heat waves, and river flooding. But how do we prepare for a rising ocean?

The practice of adapting to sea level rise has been creeping into the Southeast for several decades, though most residents don't recognize these changes as climate adaptation. They simply see upgrades to stormwater systems, new homes and businesses built with ground floors above the hundred-year flood elevation, or new bridges built taller than the ones they replaced. These modifications are helping people live smarter and safer than before in the face of rising tides and other climate change threats. Those planning these changes know they are preparing communities for sea level rise. But because climate change has been a divisive political issue in the Southeast since the 1990s, most planners, engineers, and architects, seeking to avoid stirring up controversy, simply talk about their work as preparation for future flooding.

Meanwhile, academics around the world have been studying climate change and developing strategies to prepare for it. Major universities offer college courses on the topic. Design and material technologies have improved. Urban planners and civil engineers now have detailed elevational maps showing areas most prone to flooding. Building architects design structures to withstand rising temperatures, stronger winds, and higher floods. Landscape architects can design urban and suburban settings to be cooler and more resilient to flooding.

As for sea level rise, three adaptation strategies exist: keep the water out, live with the water, or get out of the way. The first of these—known as coastal armoring—may initially be the most attractive. If we can repel the water, then we can go about our business as we always have. Cultures around the world have been doing this for millennia, using levees. These are broad walls of dirt, sometimes reinforced with rock or steel, engineered to repel floodwaters.

A more common coastal armoring approach is to harden shorelines with bulkheads, seawalls, groins, or walls of rock or rubble. These structures protect the land from floods and waves. Nearby jetties and breakwaters reach into the open water, reducing wave energy before waves reach the shore. Armoring is relatively simple and already in wide use throughout the Southeast. In the continental United States, we have already hardened 14 percent of the coastline with these structures, and this will be the primary strategy of southeastern coastal communities this century.[5]

Some of the world's largest coastal cities protect their shorelines with surge barriers, also known as flood barriers. The most complex form of shoreline hardening, these are massive structures that function as gates between the open water and harbors. Most surge barriers are kept open when sea level is within the normal tidal range. This allows for the ebb and flow of tides, the passage by boats, and the movements of wildlife. However, when storm surges or unusually high tides are expected, harbor managers close the barrier's gates to protect infrastructure within the harbor. Surge barriers protect some of the world's most vulnerable and busy ports, including New Orleans and London.

While shoreline hardening as a strategy can be effective, it has well-known disadvantages. Large surge barriers are tremendously expensive, sometimes costing billions of dollars to build and millions more annually to maintain (small surge barriers are less expensive). And while surge barriers can repel high waters, they do nothing to prevent flooding from rainfall and long-term sea level rise.

Levees also have many significant drawbacks. Though dirt is cheap, levees are not. They are expensive to build, require monitoring and upkeep, and are not readily adjustable if designers underestimated flood levels. Levees also disrupt local ecosystems and block the viewscapes that attract people to the coast. More troubling is that levees do nothing to prevent seawater from seeping through the sediments beneath them and causing the water table to rise. To compensate for this, cities with levee protection often build networks of canals and pipes to channel soil water and stormwater to pumping stations that send the water back over the levee. Finally, levee failure can be catastrophic, as we saw in 2005 with Hurricane Katrina, when over 1,300 people— mostly poor African Americans—in the Lower Ninth Ward of New Orleans died.[6]

Even hardened shorelines have significant drawbacks. Walls of rock, wood, or metal are expensive and need periodic maintenance or replacement. These structures replace natural shorelines such as beaches and marshes that offer valuable ecosystem services. Like levees, seawalls can block the views of the open water that make coastal living attractive. Hardened shorelines also reflect wave energy, which worsens erosion in adjacent areas and encourages more armoring. On shores with heavy wave action, hard structures interrupt the flow of sand, causing nearby beaches to disappear.[7]

Considering the costs and limitations of coastal armoring, many communities will need to live with flooding. Fortunately, flooding adaptation strategies are available. One is to elevate structures above the reach of floodwaters. Owners can have existing buildings lifted and placed on new foundations. New structures can perch above floodwaters on pilings. Buildings can sit on mounds built with soil taken from the adjacent area and the borrow pits used to lower the water table. Other approaches include designing infrastructure to withstand inundation. For example, a condominium may have a parking garage on the ground floor that absorbs floodwater with minimal damage. Or a community park doubles as a flood basin during storms.

A less conventional approach is to build floating infrastructure. While the concept may seem fanciful, several cities in the world have neighborhoods of house boats or floating houses that rise and fall with the tides or floods. Floating neighborhoods can increase housing density but only where there is little wave action. In the coming decades we will see similarly innovative approaches to creating flood-friendly communities.

There will, however, be circumstances when armoring or flooding adaptations are either ineffective, unpalatable, or prohibitively expensive. In these cases, retreat will be the only choice. Retreat will be necessary in some communities after just a few more inches of sea level rise. Unplanned retreat will occur when residents abandon properties because nuisance flooding becomes too troublesome. It will also happen when the recovery after repeated flood events becomes too expensive or wearisome.

Because unplanned retreat will be disorderly, if not chaotic, a few communities in the United States are experimenting with what's known as managed retreat. Through this process the government helps homeowners and businesses move and takes ownership of the property. The Federal Emergency Management Agency (FEMA) offered a buyout program after Hurricane Sandy struck New Jersey and New York in 2012. FEMA has bought six hundred homes at prestorm values from willing sellers in heavily damaged neighborhoods likely to flood again. It demolished the homes, revegetated the lots, and changed the deed so that no one could build on the land again.[8]

Armor, adapt, or retreat. These were the strategies available to Norfolk residents when they gathered to discuss their future. Over three years they developed a plan titled *Vision 2100*, which outlines a strategy for adapting to sea level rise this century. Completed in 2016, the plan divides the city into four color-coded zones based on elevation and economic and social functions. Red zones are the economic engines the city will defend with armoring and flood control infrastructure. These areas include the navy base, docks and shipyards, hospitals, universities, and the waterfront and downtown business districts. High-density, mixed-use (residential plus business) developments are a priority in red zones.[9]

Yellow zones are Norfolk's low-lying, flood-prone residential neighborhoods. Because the city cannot afford to defend all these neighborhoods, residents must adapt to flooding on their own. The city will support essential infrastructure such as roads and utilities, but residents and businesses will need to make their neighborhoods and homes more resistant and resilient to flooding. This can include filling in basements, placing homes on taller foundations or pilings, and implementing stormwater mitigation strategies such as retention ponds, rain gardens, rain barrels, and green roofs. The city will also encourage a gradual decline in housing density in yellow zones.

Green zones in Norfolk are areas away from the coast at higher elevations that are currently underutilized. Roads are wide, and buildings and parking lots are large, and largely empty. With flooding on the rise along the coast, the green zones offer room for establishing new centers of commerce and high-density, mixed-use housing.

Finally, the purple zones of Norfolk are stable residential neighborhoods at relatively high elevations. Well-maintained older homes are common, as are neighborhood parks, libraries, groceries, and restaurants. *Vision 2100* acknowledges the need for improvements in the purple zones, but planners deemed these neighborhoods unsuitable for large-scale transformation.

One complexity of planning like this is addressing social inequities. As coastal communities respond to increased flooding, they should be mindful and deliberate about addressing the needs of all their residents. Compared with low-income neighborhoods, wealthy communities are better organized, are able to hire more attorneys and lobbyists, and have more influence in state and local governments. This is one reason why cities perpetually neglect poor neighborhoods—which are disproportionately communities of color—for decade after decade while other neighborhoods prosper. Funding for coastal armoring and flood adaptation will disproportionately flow to wealthy neighborhoods unless a community deliberately addresses social inequities.

A related problem to avoid will be climate gentrification.[10] This arises when climate change flooding causes properties in higher elevation, low-income neighborhoods to become more valuable. Climate gentrification can play out in several ways. In one, rising rents and property taxes force poor, working-class residents to move. Then developers buy up these properties and build housing and businesses that attract wealthier residents. Gentrification is already a problem in most US cities, and it is one that coastal flooding can aggravate as it has already in parts of Miami.

Another version of climate gentrification begins with cities condemning low-income houses and apartments at risk of flooding. Cities then allow developers to buy the property and build upscale housing that is resilient or resistant to floods but is too expensive for former residents. City governments will do this because of the higher property tax revenues generated by the new development. Unfortunately, cities throughout the United States have a long and shameful history of selling out low-income communities in this way to enrich developers and, sometimes, corrupt officials.

Norfolk is already facing climate gentrification pressure but is committed to finding a win-win solution for all its residents. Near downtown and within the zone that the city will protect from flooding is a struggling public housing community known as St. Paul's. Most residents are low-income African Americans. The city built St. Paul's in the 1950s on low-lying land that now floods often. The mid-twentieth-century experiment to concentrate the poor in public housing projects like St. Paul's did not help many of the families taking part in the program. Instead of being a stepping stone to escape poverty, public housing projects led to fewer economic opportunities, lower educational attainment, and increased participation in criminal activities.[11]

Because of flooding and entrenched poverty, Norfolk is considering a plan to tear down St. Paul's and several other public housing developments and allow investors to build mixed-income housing. But instead of this becoming a textbook example of climate gentrification, the city plans to control the rent on some of the apartments and make them available only to low-income tenants. Other St. Paul's residents would have to move, but they'd be eligible to apply for housing vouchers to help them pay market-based rents in neighborhoods of their choice. Mixed-income housing and rental vouchers are tools in the latest social experiment to alleviate urban poverty. Because it doesn't work to simply dilute poverty, the City of Norfolk also plans to offer new social services such as job training to current and future residents of St. Paul's to help them attain greater economic security. Plans for all this are still in development and have drawn criticism from those wary of a typical gentrification outcome. But if Norfolk can reduce poverty and other social inequities while also preparing the city for sea level rise, it will have a lot to teach the rest of the country.[12]

In the years since Norfolk published the *Vision 2100* plan, the city has been busy. It has developed new zoning codes to guide development in the four zones. The city boasts that these codes are some of the most advanced for addressing coastal flooding in the United States. Norfolk has also obtained a $115 million grant from the US Department of Housing and Urban Development for developing flood resiliency in a low-income neighborhood and on the campus of a historically Black college.[13]

Norfolk may eventually get help with its flooding from the US Army Corps of Engineers. Following Hurricane Sandy, Congress directed the corps to study ways to defend areas from Virginia to Maine that are most vulnerable to coastal flooding. Due to Norfolk's vulnerability, size, and economic and strategic importance, the corps selected it as one of nine locations to be part of the study. In 2018, the corps released its preliminary plan. It includes storm surge barriers, eight miles of floodwall, a one-mile levee, eleven tide gates, and seven pump and power stations. If developed, the plan would offer significant protection for Norfolk's most vulnerable areas. The project will cost $1.4 billion, with expenses shared by the corps and the city. While the plan is not cheap, the corps estimates that it will provide the city $122 million per year in net benefits by reducing flood damage and improving resilience.[14] The project is now in a more advanced phase of planning, but construction can only begin if funding becomes available. Because the corps has a backlog of projects totaling near $100 billion to complete, whether and when the Norfolk project will begin is anybody's guess.[15]

Cities like Norfolk are the laboratories for climate change adaptation, and the nuances of geography, culture, and economy make each city a different experiment. With time, we'll see the emergence of new engineering solutions and strategies for promoting economic development while simultaneously fighting social inequities. Because it is being proactive, Norfolk is likely to fare better than many other coastal communities in this first century of sea level rise.

In contrast to mainland communities like Norfolk, barrier island communities are extremely vulnerable to sea level rise. In their natural state, these low-elevation landscapes are little more than narrow strips of loose sand capped with a thin layer of vegetation. On one side they face calm estuarine waters. But on the other side they face an open ocean and the full fury of its storms. Wind, waves, currents, and storm surges continually reshape barrier islands. Their instability and vulnerability were known to early Europeans and Americans, who opted to build their settlements farther inland.[16]

We began settling on barrier islands when industrialization and economic development made wealth and leisure time more abundant. We built roads, installed utilities, platted the land, and sold lots. Hurricanes occasionally blew through but were never frequent enough to cause a full retreat. Over time, barrier communities that began as a dusting of fishing shacks and beach houses morphed into crowded jungles of concrete, asphalt, steel, and plastic.

But these days sea level is rising, and hurricanes are growing ever more destructive. Though these are warnings that we should pull back, a retreat is unlikely until survival on the barrier island is no longer affordable. These communities infuse far too much money into local economies and government coffers for local governments to abandon them, even though many are already struggling with the few inches of sea level rise we've experienced thus far.

Tybee Island, Georgia, is one of those struggling barrier island communities. It is a small beach town on a barrier island twenty minutes southeast of Savannah. It has seen ten inches of sea level rise since 1935, and the number of nuisance tides has tripled since the mid-1980s, to over nine per year. Floods often shut down the only highway connecting the island to the mainland, and tidewater infiltrating through the stormwater system regularly floods streets and yards. In 2011 city officials reached out to regional experts at the Georgia Sea

Grant and the University of Georgia about their flooding concerns. They recruited more partners from other academic institutions and county, state, and federal governments.[17]

The outcome was an analysis of Tybee Island's vulnerabilities and options for reducing flood risk. The city adopted several proposed actions: protecting freshwater infrastructure, improving the stormwater system, requiring that new buildings be built above expected flood levels, and elevating the highway to the mainland. A few other cities along the southeastern Atlantic Coast have followed Tybee Island's lead, including St. Marys, Jekyll Island, and Brunswick, all in Georgia.[18] It's encouraging to see these coastal communities proactively adapting to sea level rise. But given their exposed position to storms and low elevations, it's possible these communities will not survive the two to eight feet of sea level rise that's expected this century.

Tough decisions on whether and when to defend, adapt, or retreat are inevitable for communities like Norfolk and Tybee Island. Even under a moderate sea level rise scenario, by midcentury there will be hundreds of coastal communities dealing with monthly or weekly nuisance tides and periodic flooding catastrophes. Funding will be more of a limitation than available engineering technology. If the estimated cost of protecting Norfolk from a moderate degree of sea level rise is $1.4 billion, it will cost hundreds of billions—perhaps trillions—to protect all the vulnerable coastal communities of the Southeast. Many communities simply won't be able to afford these protections, even if the federal government pays a generous proportion. Clearly, coastal communities should right now be finding the most effective and affordable solutions to prepare for rising waters. Fortunately, one of the best solutions available isn't made of concrete, rock, or metal: it comes in the form of natural ecosystems.

30

Working with Nature

THE SOUTHEASTERN COASTLINE MAY RESEMBLE a landscape under siege by the end of this century. In the coming years we will fortify thousands of miles of our estuarine and barrier island beaches with high levees and rock walls. Surge barriers will stand guard at the mouth of bays and harbors. Breakwaters and jetties will repulse waves.

The security these structures can bring comes at a cost. Coastal communities will lose some of their allure as they wall themselves in. It will be challenging to find beaches unbroken by concrete and rock for long walks, or to find unbroken horizons for watching a sunrise or sunset. Such changes can affect the desirability of coastal living and tourism. This can be harmful to the economy of coastal communities that have already spent a fortune on hardening their shorelines.

There is an alternative to armoring the coast. Flooding adaptation experts have begun focusing on a different form of infrastructure—something as ancient as the land but relatively new as a strategy. These systems are inexpensive to install, self-repairing, and able to adjust to sea level rise as needed and without our help. And the best part is that these systems support native biodiversity and lots of it. In fact, they have the potential to rebuild depleted fisheries for overfished coastal species. What is this novel approach? Collectively these systems are known as green infrastructure. But you already know them by other names: salt marshes, oyster reefs, coral reefs, and barrier islands.

Green infrastructure is a new term for ecosystem services that can provide what we try to achieve with traditional engineering, or gray infrastructure. For example, salt marshes absorb floodwaters and the energy of incoming waves, and the grasses hold sediments in place and prevent coastal erosion. Oyster reefs and coral reefs absorb wave energy and dampen wave height. These living shorelines also offer benefits that gray infrastructure cannot. Oyster reefs and salt marshes can grow in height or migrate inland as sea level rises. And when storms damage these ecosystems, they can self-heal. For these and other reasons, flooding adaptation experts have begun promoting the restoration of marshes and reefs along our coastlines.[1]

Flood scientists also consider barrier islands and beaches as green infrastructure because they protect estuaries and inland settlements from oceanic waves and storm surges. In recent years, some coastal communities along the Gulf and Atlantic Coasts have begun restoring eroded barrier islands by pumping in sand from offshore locations to increase their width and height. Beach nourishment is similar but done along small sections of the coast to rebuild public beaches where storms have swept sands away. Dune restoration is often part of both efforts. To do this, restorationists install sand fences and plant vegetation to trap blowing sand. Once established, dunes then sustain themselves and rebuild when damaged by storms.

The two big downsides of beach nourishment are that it is expensive and only persists for a few years. Because of this, the Netherlands and England are experimenting with pumping in massive amounts of sand from offshore to create peninsulas over a half mile wide. Engineers hope these so-called sand engines will supply sand to adjacent beaches for several miles along the coast for many decades. If successful, sand engines will preclude frequent, costly beach nourishment projects. As an added benefit, sand engines offer space for beachgoers and habitat for wildlife.[2] If the sand engines are successful, they may be a useful tool for protecting outer beaches and barrier islands in the Southeast.

Beaches, salt marshes, and oyster reefs are the primary natural shorelines of the region, but by the end of this century another type of living shoreline will protect the southeastern coast. In the southern portions of Texas, Louisiana, and Florida, where the climate is warmer, small trees known as mangroves grow in the intertidal zone of estuaries (*mangrove* can refer to the trees of these ecosystems or the ecosystem itself). Whether they grow as patches or as entire swamps, mangroves offer the same flood and erosion protections as do salt marshes.

Though mangroves are tropical and subtropical plants, climate change is allowing their northward expansion.[3] Seeds germinate on the parent plant and develop into propagules that drop into the water and can float on currents to distant shores. They start as individual trees in the intertidal zone, often in oyster reefs and salt marshes. Individual trees grow outward as well as upward. With time they recruit more trees and form small clusters resembling islands rising above the marshes or estuarine shallows. In the past, periodic freezes killed mangroves that had expanded north of their historical range. But the warming climate is allowing them to migrate north and stay. Biologists estimate that even under a scenario of moderate climate change, by 2100 mangroves will be common and taking over salt marshes in the northern Gulf of Mexico and on the Georgia coast.

From a biodiversity perspective mangrove expansion is both good and bad. Mangroves will radically change the character of the southeastern coast. We will lose—and many of us will miss—the sweeping vistas the marshes provide. Wildlife species found only in salt marshes will become scarce. And gone will be the exquisite contrast of an expansive sky against browns, grays, and subdued greens of a marsh extending to the horizon. On the other hand, mangrove estuaries shelter many wildlife species and offer emerald swamp labyrinths to explore. And, mangrove swamps may be just as good, if not better, than salt marshes at absorbing floodwaters.

The value of restoring living shorelines to reduce flooding risk has received considerable attention from coastal flood experts lately because these living shorelines can do things that gray infrastructure cannot. These ecosystems can match the pace and height of sea level rise and self-heal after incurring storm damage. They pull carbon dioxide from the atmosphere and clarify waters by trapping sediments. Marshes and reefs are essential nurseries for commercially valuable marine species, and by absorbing nutrients they reduce occurrences of harmful algal blooms. These ecosystems attract tourists and residents for fishing and wildlife observation. The minerals in dead oyster shells neutralize the rising acidity of marine waters (caused by absorbing carbon dioxide). And nearshore oysters produce offspring that can settle farther offshore and sustain commercial oyster fisheries.[4]

Coastal ecologists have been promoting the value of restoring these

ecosystems for decades. But as a response to sea level rise it can be a tough sell. A wall or levee is an obvious and measurable defense against floodwaters. One can see the vertical feet of protection it provides, and there's psychological comfort in having a firm structure built by engineers to repel waves and floods. Marshes and beaches, on the other hand, are soft barriers. Their vertical gradients are difficult to notice without survey equipment because they extend over dozens to hundreds of feet.[5]

In recent years some biologists have used economics to assess the value of green infrastructure, and their findings offer strong support for these nature-based flooding adaptations. A study published in 2008 revealed that when hurricanes arrive, each acre of wetland prevents an average of $2,000 of damage to adjacent properties.[6] In 2018, Borja G. Reguero and colleagues published a study on the economic benefits of restored natural shorelines as an adaptation strategy to mitigate flooding risk and damage on the Gulf Coast.[7] They estimated that if sea level continues to rise at the present rate, by 2030 the US

Construction on a salt marsh restoration project led by the Nature Conservancy of Alabama. Completed in 2020, the forty-acre project restored coastline lost to erosion in Bayou La Batre, Alabama. Compared with alternative strategies to protect the coast, marsh and reef restoration is economical and revitalizes coastal ecosystems. Courtesy of CrowderGulf for the Nature Conservancy in Alabama.

WORKING WITH NATURE 247

Gulf Coast will sustain at least $134 billion in damages from coastal flooding each year. That's more than the combined economic damage from Hurricanes Florence, Michael, and Irma. Natural defense systems such as marshes, oyster reefs, beaches, and barrier islands can prevent 37 percent of these damages, saving $50 billion annually. And, they found that these nature-based adaptations are inexpensive to install and maintain compared with gray infrastructure. Building a levee costs $53 million per mile of protected shoreline, but restoring a salt marsh costs only $25 million per mile. Oyster reef restoration costs only $1.5 million per mile. Compared with levees, a restored barrier island offers five times more benefit per dollar spent. Oyster reefs provide seven, and wetlands in flood-prone areas provide nine times more value than levees. Thus, the economic argument alone should be enough to motivate coastal communities to embrace green infrastructure.

Though green infrastructure has many advantages, there are situations where it is not a suitable solution to sea level rise and stronger hurricanes. Along some shorelines oyster reefs and marshes are incompatible with human activities such as swimming, or infrastructure such as shipyards, docks, and marinas. Nor can reefs or marshes work where dredging or hardened shorelines have created artificially deep waters near the shore. Even where living shorelines are possible, oyster reefs and marshes cannot grow above the high tide line. They also need time to grow or heal after storms. And like all varieties of coastal armoring, green infrastructure doesn't prevent sea level rise from forcing the water table higher in protected areas.

Some limitations of green infrastructure arise because the species forming the ecosystem has a limited range of physical stress it can tolerate. This is the Goldilocks principle: too much or too little, and a species does poorly or dies. But a species can thrive in its zone of optimality. Thus, if we are going to use green infrastructure to prepare for sea level rise—and we should—then we'll need to restore conditions that allow coastal ecosystems to thrive.

Freshwater availability supplied by rivers is one of those conditions. One primary limitation on the growth of oysters and salt marsh plants is the availability of fresh water. These species require a suitable blend of fresh and marine water. Oysters cannot reproduce if waters are too fresh.

If waters are too salty, predatory snails will arrive and eat them. Oysters also feed on planktonic species that need the right blend of fresh and marine waters. Marsh plants also require fresh water for the nutrients and sediments it provides. Without new sediments, a marsh will drown as its soils settle and compact. Without nutrients, it will starve.

Thus, if the Southeast wants to restore coastal ecosystems to prepare for sea level rise, then it must also restore rivers. Salt marshes, oyster reefs, and mangroves depend on the supply of fresh water, sediment, and nutrients delivered to the coast by rivers. Furthermore, if not enough river water is delivered to the coast by those managing rivers, our existing salt marshes, oyster reefs, and barrier islands may simply drown beneath the rapidly rising waters.[8] This is one of the most profoundly important reasons why we need to finish the river revival and restore the ecology of southeastern rivers. And as we restore rivers, we must also restore coastal marshes.

The United States has lost over 50 percent of its marshes since the dawn of European colonization.[9] We've converted them to agriculture, filled them in to create new land, or dredged them for navigation. We drained marshes to reduce mosquito populations. Our roads cut others off from the tides. Marshes died back due to rising salinities when large dams began withholding fresh water from the coast. More recently we've been losing marshes to an overabundance of nutrients.[10] Excess nutrients make marsh grasses tastier and easier to eat by marsh snails. These herbivores are more abundant now than ever due to our overharvest of the terrapins, crabs, and fishes that once kept their populations in check.[11]

Fortunately, there's hope for marshes. Now that we realize the value of the ecosystem services that they offer us, salt marsh restoration in the United States is on the rise.[12] In some places the marshes are restored to promote habitat conservation. More commonly, coastal engineers are building marshes to protect adjacent urbanized areas. Marsh restoration is in an early phase of development. There's still plenty of trial and error, missteps, and surprises. But ecologists and engineers are developing better approaches. Efforts include planting salt marsh grasses, plugging drainage ditches, and restoring tidal access to marshes trapped behind roads. In some areas, we are creating new marshes by pumping sediment into shallow waters and then hand-planting plugs of marsh grass.

Salt marsh restoration is lagging in the Southeast despite the region's vulnerability to coastal flooding, but efforts are underway to catch up. State governments and conservation organizations are partnering to test different restoration methods on both the Gulf and Atlantic Coasts. One of the largest projects thus far is at Lightning Point in Bayou La Batre, a small town at the center of Alabama's commercial seafood industry. The Nature Conservancy coordinated the restoration of salt marsh and shoreline habitats to help protect the town from flooding and restore wildlife habitat. Project partners built one and a half miles of breakwaters to suppress wave energy, then used over 365,452 cubic yards of dredge material to create forty acres of salt marsh. The project was completed in 2020, and biologists expect that the marsh will need five years to mature. The Lightning Point project should inspire other large-scale marsh restoration projects along the southeastern coastline.[13]

Like salt marshes, oyster reefs are also in bad shape. Oyster reefs were abundant in nearly every estuary of both southeastern coasts before the arrival of the Europeans. Due to a lack of early survey data, biologists don't know how much was lost in the Southeast, but they estimate that oyster reefs are down to between 10 and 15 percent of their former extent globally. For example, on the James River oyster reefs once lined the banks for a dozen miles inland, but now are nearly absent.[14] Given the size and geographic positions of Native American shell middens along both coasts of the Southeast, archaeologists know that oysters were far more abundant in pre-Columbian times than they are today.

Three problems caused the decline of oyster reefs. The largest problem was overharvesting. Oysters became a popular food for European and American colonists. Fisheries from New England to the western Gulf of Mexico supplied jobs and income for many generations. But most fisheries eventually collapsed due to overharvest. Reefs along the Gulf Coast fared better than those on the Atlantic Coast, due to protections on the extensive fisheries in Louisiana and Texas.

Another threat to oysters was water pollution. As filter feeders, oysters are sensitive to water quality declines. Eutrophication from sewage discharge and fertilizer runoff caused algal blooms and periods of low dissolved oxygen that suffocated young oysters. Some blooms produced toxins that killed or sickened oysters. And a third factor contributing to the loss of oyster reefs was sediment pollution. Before farmers practiced soil conservation, excessive sediment delivered by rivers to the coast smothered many reefs with mud.[15]

Because of oysters' economic value, for many decades oyster fishers have tried to restore or sustain reefs by depositing hard materials (such as oyster shell or rock) on the estuary floor so that young oysters have a place to settle. Sometimes they disperse hatchery-raised oyster larvae (spat) over reefs where recruitment has been insufficient. These days conservationists are restoring reefs for a different reason. Reefs in the shallow, intertidal zone are effective breakwaters that protect salt marshes and estuary beaches. They also offer habitat for wildlife. To build these reefs, restorationists pile discarded shells from seafood processors in domes just offshore or place them in mesh bags that are bound together and anchored in the shallows. Restorationists construct more elaborate reefs of concrete structures or long walls of oyster shells held together with wire mesh. We are still on the learning curve for creating oyster reefs. A recent review calculated that only 75 to 90 percent of created reefs survive in the Southeast.[16] Clearly, we need more innovation and research.

Economic studies have bolstered arguments for oyster reef restoration. Biologists estimate that the ecosystem services provided by oyster reefs range from $2,200 to $39,600 per acre per year. These services include providing habitat for commercially valuable species such as Blue Crab and Red Snapper and cleaning the water of sediment and nutrients through filter feeding. And coastal communities recoup the costs of reef restoration where new reefs help sustain the harvest of wild oysters.[17]

Our craving for oysters may help them recover. The mollusks have become trendy as upscale oyster bars have opened in cities across the United States. What is new about our old obsession is that farmers are raising many of these oysters. Like those who tour wineries, oyster connoisseurs delight in sampling varieties with different tastes and provenance. Farmers grow them in the shallows of estuaries. Each estuary produces a unique blend of texture and flavor ratios of salt, sweet, and umami. This variation is a product of the estuary-specific blend of fresh and marine waters, the mixture of plankton on which the oysters feed, and the chemistry of the water supplied by the estuary's streams. If the popularity of oyster consumption expands, entrepreneurs throughout the Southeast could farm oysters and market varieties unique to their community. Other benefits would include reduced coastal erosion, improved water quality, and more habitat for wildlife, including other commercially valuable species.

The future of the southeastern coast could be one of working *with* nature, not against it. We can restore barrier islands, dunes, beaches, salt marshes, and oyster reefs to prepare for sea level rise. These ecosystems are beautiful and were ubiquitous before urbanization. They supported wildlife in quantities and varieties that would seem unbelievable by today's standards. Historical accounts describe estuaries and offshore waters teeming with shellfish and finfish including snapper, grouper, mullet, drum, sturgeon, flounder, shrimp, and crabs. Widespread restoration of these ecosystems would sustain thriving tourism, seafood, and sport fishing industries, even as sea level rises and the climate changes.

A vision of the future like this is an antidote to fears that we'll finally push nature fully aside in a scramble to adjust to climate change. Marshes, reefs, and mangroves and their denizens will be our allies with whom we cooperate for mutual benefit. We'll value ecosystems for the services they provide, and we'll value their species as our neighbors that get the job done. But to make this future happen, the Southeast will need to manage its rivers differently. We will need to send enough fresh water, nutrients, and sediment—not too much, not too little—to help restore coastal wetlands and oyster reefs and protect the coast from flooding. To do this, we must rethink how we live with and manage our rivers and our coasts. Much depends on whether the region can control its growing thirst for fresh water during a time when it is becoming scarcer.

31
Drought

THE WARNING SIGNS BEGAN IN January 2016. The preceding November and December had been exceptionally warm and wet for the Southeast, with rainfall totals nearly double the historical average. Creeks and rivers were swollen. The Mobile River Delta became a thirteen-mile-wide river, a degree of flooding not seen in over a quarter century. But in January the rain abruptly stopped across the region. Not just the unusually heavy rains, but all rain. By the end of the month, rainfall levels were far below average.[1] Still, it was no cause for alarm. Reservoirs were full, and soils were amply saturated for spring planting.

By March it was clear that a problem was brewing. Farmers watched as young corn in the fields of North Georgia and Alabama struggled. Ranchers saw their pasture and rangeland grasses dying. Dam managers nervously eyed declining reservoir levels at a time when they are normally peaking for the year. The anomalous conditions were also being tracked by the US Drought Monitor, a weekly report on drought conditions across the United States compiled by federal agencies and the University of Nebraska–Lincoln. The monitor assigns categories ranging from "abnormally dry" to levels of increasingly intense drought coded as "moderate," "severe," "extreme," and "exceptional." In March, the monitor showed that abnormally dry weather—the precursor to drought—had spread across the interiors of Georgia and the Carolinas.[2] Most residents of the region thought the weather was idyllic. Skies were clear, temperatures were slightly above average, and people crowded beaches and parks. But those whose livelihoods directly depended on rainfall and the water supply anxiously awaited the long, sweltering summer ahead.

By mid-July 2016, over two dozen counties in northern Georgia and Alabama, and southern Tennessee and South Carolina, were in extreme drought, and much of the area in surrounding states was in moderate or severe drought.[3] The signs of drought were obvious across the region. The sun was punishing, and temperatures rose far above average. Clouds formed briefly, but then disintegrated. The dry air sponged the last remaining moisture from soils. Crop failures were widespread on farms lacking irrigation. Ranchers imported hay from outside the region. Farm ponds used for watering crops and livestock evaporated. In the forests, the flush of new growth spurred by the winter's abundant rain was drying out. Grasses crunched underfoot. Tree roots snapped as soils dried, hardened, and cracked. Insect populations crashed, sending ripples of hunger through food webs.[4]

This was the fourth drought for the Southeast since 2000. The cause was the Bermuda High, which had overreached its typical position and was hovering over the Southeast. We learned earlier that the Bermuda High is a broad area of elevated atmospheric pressure in the Atlantic caused by dry air falling to Earth's surface from high in the atmosphere. During a typical summer, the Bermuda High moves from a position closer to Africa to a position off the southeastern US coast. While there, its clockwise-rotating winds sweep moisture from the Gulf and Caribbean into the Southeast. But due to global warming, during summer the Bermuda High is now more likely to extend over the Southeast, causing dry conditions by interrupting the flow of moisture into the region and preventing cloud and storm formation. This is what triggered the drought in 2016.[5]

Because droughts are a normal climate phenomenon in the Southeast, most southeastern states have a plan to deal with it. Good drought plans delineate short-term measures to manage water supplies during periods of water scarcity. Such plans identify essential and nonessential uses of water and develop response measures calibrated to drought levels. None of the southeastern states have great plans, but North Carolina and Georgia have the two best plans within the region.[6] North Carolina doesn't specify mandatory actions for counties, cities, or public utilities during a drought, but it does require each of them to have and follow their own drought plan. Each week, the state government issues a summary of drought conditions based on data from the US Drought Monitor, then local governments and utilities enact their plans. This strategy is good because it values local knowledge and adaptability. For example, during an extreme drought a county lacking large reserves of water would need to begin rationing water sooner than a neighboring county that has sizable water reserves. The downside of North Carolina's plan is that many local governments lack the expertise and resources to develop and implement robust drought plans.

In contrast to North Carolina's emphasis on local response, Georgia has a top-down approach to drought planning.[7] Its plan mandates actions for various levels of water scarcity, ranging from public education to limits or bans on forms of outdoor irrigation such as lawn watering. The director of the state's Environmental Protection Division must decide to activate the drought plan. Before they can officially declare a drought, the director must gather information and opinions from constituents around the state who would be affected by the activation of the drought plan. This might include, for example, farmers, city managers, and manufacturers. While Georgia's top-down approach to managing water supplies during a drought allows for a rapid and comprehensive response, the involvement of constituents who will suffer economic hardship with a drought declaration can undermine effective decision-making.[8] Criticisms aside, both North Carolina and Georgia have developed drought plans that are more advanced than those of other states in the region. Alabama and South Carolina, for example, leave the planning—which is optional—to local governments. The drought plans for these two states are primarily guidelines for local governments to consider if they try to develop a drought response.[9]

September 2016 was brutally hot and bone dry. Moderate and severe drought had spread, and several counties in northern Georgia and adjacent Tennessee were in an exceptional drought—the highest drought classification possible. Western North Carolina was abnormally dry or in moderate drought, and water conservation measures were in effect. Georgia declared a level 1 drought in September, but this only required that officials in affected areas inform the public they were in a drought. States with weak drought plans were wholly unprepared and did little to nothing.[10]

DROUGHT 255

While state and local governments responded—or didn't—conditions in streams across the region deteriorated. By early autumn, most smaller creeks were dry. The flow in larger creeks and small rivers was a hot trickle or came to a halt. The most desperate streams were those in agricultural areas where farmers extracted groundwater to irrigate crops. Each day the pumps would turn on, and each day streamflow would drop a notch further.

The dying streams across the region elicited heartbreak from anyone who visited to bear witness. Playful creeks were now tracks of mud and silt-covered rock. Pools of water left isolated by dropping water levels were death traps, where temperatures rose and dissolved oxygen levels plummeted. Fish spent their last days gulping air at the surface before going belly-up. The sun broiled mussels to their death in the hot mud. The smell of a dying ecosystem hung in the stifling air.

Even larger streams suffered in the fall of 2016. Groundwater feeding into them tapered off, currents slowed, and waters became overheated in the sun. In some streams, permitted wastewater discharged from industry and sewage treatment systems was a rising fraction of the river's volume. This could be a mixed blessing. On the one hand, this discharge helped some streams keep flowing. On the other hand, this exposed wildlife to abnormally high concentrations of contaminants. This pollution became even more concentrated downstream as water evaporated from the rivers. Nutrient concentrations rose high enough in some rivers to trigger algal blooms, which accelerated water quality decline.

The reduced flow from the region's rivers altered coastal ecosystems. The rivers feeding Mobile Bay, for example, were so weak that Gulf water —clear and salty—pushed higher than usual into the estuary. Anglers in Mobile Bay caught fishes normally found far offshore. Fishers saw crabs, shrimp, and saltwater fishes over a dozen miles up the Mobile River Delta. Nearshore waters of the open Gulf, normally tainted with sediments and plankton blooms, were unusually clear. Though anglers and swimmers enjoyed these conditions, the delta's vegetation got a taste of the high salinity that sea level rise promises to bring. Consequently, marsh and swamp vegetation died back in some areas.[11]

The reduced flow in rivers also affected water supplies for inland communities across the region. Water levels in reservoirs dropped steadily as surface waters rapidly evaporated and inputs from creeks and rivers declined. Public water supply systems struggled to meet demand. Lake Purdy, Birmingham's drinking water impoundment, dried out completely. The lake, normally a serene, blue expanse nestled between forest-green hills, became a pit of orange, polygon-fissured mud. Birmingham's water utility bought water from a neighboring city for twice what it normally spent to produce its own water.[12] The water bills spiked, and the utility imposed extreme surcharges for above-average water use. Comparable stories played out across the region.[13]

At the region's big reservoirs, dam operators followed their water control manuals, the guidelines for how much water to withhold or discharge depending on season and impoundment levels. Dam owners must write the manuals with input from stakeholders, including state and federal agencies obliged to ensure enough water flows past the dam to sustain wildlife and the water supplies of communities downstream. Due to the drought, the manuals allowed many dams to hold back more water than normal to ensure there was enough for local public water supplies. Consequently, power production declined at hydropower dams. Buford Dam, which creates Lake Lanier, Atlanta's primary source of water, cut power production by 65 percent. The Tennessee Valley Authority (TVA) cut production by 50 percent across its network of dams.[14]

The dwindling water supplies of autumn 2016 highlighted unresolved legal battles over water rights in the region. Among these was the dispute over the waters of the Apalachicola-Chattahoochee-Flint River Basin, which extends over western Georgia, eastern Alabama, and the Florida Panhandle. The competing demands on these waters by the states have entangled them in one of the longest-running water conflicts in the eastern United States. Much of the demand on the basin's water is from the Atlanta metropolitan area, which sprawls across the basin's headwaters. Atlanta has been growing rapidly, and the metro area is now home to over five million people. Because the underlying bedrock is impermeable to water infiltration, Atlanta meets its water needs by drawing from local rivers, especially the Chattahoochee. In addition, there are twelve dams on the Chattahoochee and sixteen power plants. The latter constitute the greatest use of the basin's water.[15] There's also heavy demand on the Flint River from thousands of Georgia farmers. They draw irrigation water from the river or nearby shallow wells hydrologically connected to the river. The Chattahoochee and Flint Rivers merge to form the Apalachicola River near the Florida state line. Florida gets the dregs of the basin. During droughts, especially, water quantity is low and water quality is poor.

The legal battles between the three states, known as the Tri-State Water Wars, have raged since 1990. Much of the dispute initially focused on Lake Lanier, the large reservoir on the Chattahoochee near Atlanta.[16] The US Army Corps of Engineers built and manages the reservoir. To satisfy metropolitan Atlanta's growing demand for water, in 1989 the corps announced it would begin allowing Atlanta communities to purchase water stored in Lake Lanier. Both Alabama and Florida sued, claiming that the corps had no right to allocate Lanier's water for this purpose. After a lengthy battle the federal courts decided that the corps could withhold and sell water from Lake Lanier to Atlanta.

Meanwhile, Florida's oyster fishery in Apalachicola Bay collapsed. It had been one of the most productive oyster fisheries in the United States. In 2013, Florida sued Georgia again, claiming that its overuse of water had contributed to the fishery's demise.[17] Georgia argued that overharvesting and climate change were to blame. The states fought their case all the way to the US Supreme Court. In spring 2021, the high court decided in favor of Georgia, stating that Florida had failed to provide enough convincing evidence that Georgia was responsible for the fishery's collapse, and that Florida's own mismanagement of the fishery could not be ruled out as a cause.[18]

As for Alabama, it has been unable to muster a successful legal challenge to Georgia's use of the Chattahoochee and other watersheds the two states share. Because Alabama lacks a statewide water policy plan (it's the only southeastern state without one), it has been unable to build a strong legal case for why it needs access to these waters. Though stakeholders and state agencies began working on a plan several years ago, the state government in Montgomery has failed to prioritize its completion. A detailed drought plan would be part of a good statewide water plan, and the absence of both hindered Alabama's response to the 2016 drought.

Some parts of the Southeast got relief from the drought by late summer 2016. The northern Gulf Coast began getting rain in August, and the southeastern Atlantic Coast enjoyed rainfall from Hurricane Hermine and Tropical Storm Julia in September. Hurricane Matthew in October delivered intense amounts of rain and flooding into the Coastal Plain of Florida, Alabama, and Georgia.[19] Though this helped with the drought, Matthew caused catastrophic levels of destruction in some areas. And, the Bermuda High prevented these three tropical storms from reaching deeper into the region, where people and the rivers needed rainfall the most.

As the days grew shorter in fall, some hoped that a shift to cooler temperatures would force the drought to end. No such luck. Temperatures remained 6–8°F above normal, and a La Niña weather pattern established. This was the climate counterpart to the El Niño that had caused above-average rainfall the prior winter. Because La Niñas bring dry winters to the Southeast, hope faded that the drought would soon end.[20]

The late fall forests were eerily calm. There was no chatter of insects. Birds were absent from the forest canopy. Instead of the progression of color from greens to yellows, golds, and reds, leaves turned brown, died, and hung from the branches. Trees on shallow soils died. Lush young growth in the understory spurred by the heavy rains of the previous winter also died.

The unusually dry, warm, and sunny conditions of late autumn 2016 encouraged outdoor recreation. Despite burn bans issued by most states, visitors to forests lit barbeques, campfires, and fireworks. Predictably, hundreds of wildfires broke out, including dozens of major blazes. Fires were the worst in the mountains, where steep terrains prevented crews from containing the fire. Smoke columns loomed over burning ridgelines. An acrid, smoky haze settled over portions of the region, causing air quality to decline and threatening public health. An especially tragic fire in Gatlinburg, Tennessee, killed fourteen people and destroyed over 2,460 structures. Altogether, it was the worst fire season the region had endured in over a century.[21]

By the time the drought peaked, in late November 2016, 92 percent

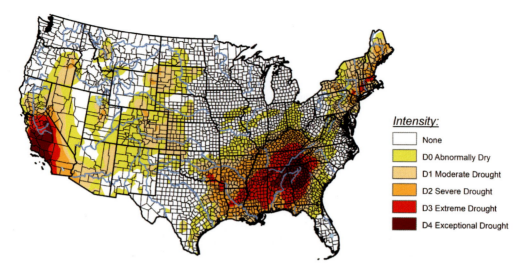

One of the Southeast's most damaging droughts peaked in November 2016. The rapid onset of intense droughts is becoming the new normal and threatens the region's biodiversity and water supplies. Map from the US Drought Monitor, an online tool jointly produced by the National Drought Mitigation Center at the University of Nebraska-Lincoln, the US Department of Agriculture, and the National Oceanic and Atmospheric Administration. Map courtesy of the National Drought Mitigation Center.

of Alabama, Georgia, Mississippi, and Tennessee was in a condition of severe drought or worse. Relief finally came with the arrival of a storm system in late November that brought widespread rain to the region, but the rest of the winter was abnormally dry, and the drought lingered until March 2017, ending a year after it began.[22]

Climate scientists say that the 2016–17 drought offered a preview of the region's future droughts because it hit fast and hard. The drought began quickly, just weeks after a period of record-breaking rainfall. And it was severe: soils were drier than at any time since 1895, when record keeping began. Climatologists expect that future droughts will begin quickly and be much more intense than in the past because the warming atmosphere forces much faster evapotranspiration rates.

The 2016–17 drought also illustrated that the Southeast is not ready for a future of declining water supply punctuated with powerful droughts. During the 2016–17 drought, the region suffered agricultural losses in the hundreds of millions of dollars and water supplies reached critically low levels in many communities. If southeastern states were prepared with better water plans, they would have avoided some of the disruption and had a faster recovery. The region was lucky because the 2016–17 drought only lasted a year. Most southeastern droughts last for several years.[23] Had the drought continued through the summer and fall of 2017, many more communities would have experienced water supply emergencies. They would have had to ration water and buy water from outside their area.

In this new era of population growth and rapid climate change, southeastern states need to step up preparations for declines in water availability and periods of scarcity. As they do, it is inevitable that most communities will seek out new sources of water. Unfortunately, the first possibility many will consider is building dams to create more reservoirs. Yet a new wave of dam building would be devastating to southeastern rivers. The region's remaining aquatic biodiversity would suffer, and the coasts would not receive the river water needed to mitigate sea level rise. Is there any reason to hope that the region can have enough water while avoiding new dams and reservoirs?

32

Reservoir Reservations

MORE DAMS. THIS IS THE solution many southeastern communities will consider first when they awaken to the region's water supply crisis. We've already built nearly a thousand dams in the Southeast for public water supply over the past two centuries, and dozens more have been proposed in recent years.[1] So, dams might seem to be a logical solution to the water scarcity challenge that the region will face this century. We can use them to capture water when it's abundant, then use it as needed. Problem solved, right?

Wrong. As we have seen already, dams create more problems than they solve. Dams broke the migrations of sturgeon and other anadromous fishes. They caused the extinction of mussels, snails, and fishes, and fragmented the populations of remaining river species. Dam reservoirs drowned sacred Native American sites beneath cold water and mud, and they displaced rural Americans from family farms and sleepy riverside towns. Reservoirs drowned river wonders such as the Muscle Shoals on the Tennessee River and the Devil's Staircase on the Coosa River. And dams warped the natural flow regime of rivers, thus diminishing the ecosystem services that free-flowing rivers provide to us. The emerging wisdom is that dams are an imperfect nineteenth-century solution to a twenty-first-century problem. Dams should be the last choice, never the first.[2]

Another compelling reason we should avoid new water storage dams is their expense. The cost of building a dam varies with size and circumstance, but it ranges from tens to hundreds of millions of dollars. There is land acquisition, design, permitting, construction, conservation mitigation, and new water treatment plants. Sometimes there is costly litigation over land acquisition and environmental damage. Budget overruns are the norm, not the exception. These problems have bedeviled recently built dams, throughout the world and in the

Southeast. Developers estimated that Duck River Dam in Alabama would cost $42 million, but its final cost was $68 million. They predicted that the cost of Hickory Log Creek Dam in Georgia would be $19 million, but its costs quintupled to $100 million. Most egregious was Hard Labor Creek Dam in Georgia. Developers originally estimated that the dam would cost $41 million, but the final bill was for $350 million, or 854 percent more than they projected.[3]

Where does this money come from? Usually it is in the form of municipal bonds approved and issued by local governments. The burden of repaying bondholders, plus paying the interest on the bond and the cost of bond insurance, falls on the local government issuing the bond. Governments collect this money by raising the price of water for ratepayers and by collecting taxes. Because local residents pay for them, dams are only affordable by communities with enough water customers and taxpayers. Small communities planning a dam rely on estimates of future population growth to justify issuing municipal bonds. But local population growth estimates are often wrong. Economic downturns suppress immigration, other communities become more desirable to potential new residents, and large employers in the area can downsize, close, or move. Thus, justifying a costly dam project on a gamble of future population growth is risky, and has led communities into significant debt that residents will be paying off for decades to come.[4]

Hydrologists have shown that reservoirs are crude tools for addressing water insecurity because reservoirs enhance evaporation. While river water stays cool due to groundwater inputs and mixing, impoundments expose their waters to the sun and mix poorly. As surface waters warm, evaporation rates increase. In the Southeast, for every acre of reservoir surface, one million gallons of water is lost to evaporation per year. That's the equivalent of a cube of water fifty-one feet wide being lost to evaporation per acre per year. Evaporation rates are greatest in warm weather—and especially during droughts—when our demand for water peaks. During a hot summer day, 200 million gallons can evaporate from the thirty-eight thousand acres of Lake Lanier. That's a lot of water, considering that the Atlanta metropolitan area currently uses about 560 million gallons a day.[5] The problem of evaporative loss from reservoirs will worsen as the atmosphere warms.

Another problem with reservoirs is that they don't just capture water, they capture mud. Very few dams allow sediments to flow downstream. Instead, silt accumulates in reservoirs at the expense of water storage capacity. Because of the siltation problem, engineers design dams and reservoirs to be much larger than they would otherwise need to be. With periods of heavy rainfall and flooding on the rise in the Southeast due to climate change, the US Army Corps of Engineers predicts that impoundment siltation rates will worsen this century.

Meanwhile, the sediments trapped in reservoirs are desperately needed downstream to support river ecosystems and build land at the coast, especially in river deltas and estuaries. Coastal geologists have raised the alarm that deltas around the world may disappear this century due to sea level rise, sediment starvation, and land subsidence. Over five hundred million people live in the world's deltaic cities, which include Bangkok, Cairo, Tokyo, and New Orleans. In the Southeast, deltaic cities include Savannah, Norfolk, Jacksonville, and Mobile. Delta marshes and swamps protecting these cities from flooding need new river sediment to build elevation as sea level rises. Otherwise, these coastlines will disappear much faster than other coastal areas.[6]

Reservoirs also contribute to climate change. Dam construction requires the use of substantial amounts of fossil fuels, from creating concrete and steel, to powering the machines needed in construction. Moreover, reservoirs pump greenhouse gasses into the atmosphere long after completion. Microbial decomposition of dead plant material in the low-oxygen conditions of the reservoir produces methane, a greenhouse gas that traps heat in our atmosphere over twenty-five times more efficiently than does carbon dioxide. Methane from dam reservoirs already contributes 1.3 percent of the world's greenhouse gas emissions.[7] This is one of the strongest reasons why we should avoid building more dams.

Dams and reservoirs should be a solution of last resort to address water insecurity. They are expensive and inefficient, threaten biodiversity, hinder coastal land-building, and contribute to climate change. But if dams are not a reasonable solution, where can the Southeast turn for more water?

33

Water from the Rock and Sea

"Water, water everywhere, but not a drop to drink." This common misquotation from Samuel Taylor Coleridge's *The Rime of the Ancient Mariner* aptly describes the Southeast's predicament. Oceans border the region on two sides, yet the Southeast is facing a century of declining water supply. Could salty marine waters provide the Southeast with the water it needs?

The extraction of fresh water from saline water, or desalination, is already sustaining communities in arid regions of the world, such as the Middle East, where freshwater supplies are almost nonexistent.[1] In the United States, desalination for municipal and industrial uses accounts for only 2 percent of freshwater production, but the practice is spreading, especially in California, Texas, and Florida, where populations are large and freshwater resources are limited. Within the Southeast, Florida leads the pack, with 167 desalination plants. North Carolina has eighteen, South Carolina and Virginia have six each, and the remaining southeastern states have just one plant.[2] More cities would use desalination to provide drinking water, but the technology is prohibitively expensive in all but a few circumstances.

The problem is reverse osmosis, the process of pressurizing source water and forcing water molecules through a membrane that prevents the passage of salt and other contaminants. The energy required for this process inflates the cost of producing drinking water by a factor of three. Some cities reduce costs by processing brackish water, which has less salt. It is also possible to build desalination plants alongside fossil fuel or nuclear power plants and use their waste heat to improve desalination efficiencies.[3]

But, even with these cost-saving approaches, desalination plants are consistently affordable only where there are no other sources of water to easily exploit. Not even multiyear droughts ensure these facilities are economically sustainable. Multiple plants built in Australia and California during recent long droughts were forced to close when rainfall resumed.[4] For similar reasons, producing fresh water at the coast and pumping it to inland communities will not be economically reasonable, especially when factoring in the cost of building and maintaining pipelines.

Desalination plants also create environmental challenges. Communities must carefully dispose of the brine produced because high concentrations of salt are toxic to most aquatic life. Engineers must design offshore pipelines that minimize injury and death of wildlife through impingement on intake screens. Desalination facilities using electricity produced from fossil fuel combustion contribute significantly more to global warming than do typical water treatment plants. A few desalination plants in Australia use solar energy, but this practice is still rare in the United States.[5] But setting environmental concerns aside, the economics simply do not favor desalination as a widespread solution for the Southeast. Until the technology improves, desalination is only a reasonable solution for large coastal cities that have exploited all other options.

If not the ocean, another place to look for new water is underground. The geology beneath the Southeast supplies groundwater that already helps the region meet its water needs. Each day, hundreds of millions of gallons are pumped from wells to be used for public water supplies and irrigation. For example, aquifers provide 40 percent of the public water supply in Alabama, 25 percent in Georgia, and 52 percent in North Carolina.[6] Groundwater is available year-round and is cleaner than surface water. Could groundwater supply what the region needs during the coming century of declining surface water supply? Maybe. The answer depends on local geology and the way we manage aquifers.

Rainwater becomes groundwater by collecting in low places on Earth's surface, such as streams, ponds, and wetlands, then seeping belowground. The topmost layer of groundwater is the water table, a zone that varies in depth depending on rainfall inputs. Below the water table are aquifers, layers of sediment or rock with internal spaces that are consistently saturated. Shallow aquifers can fill rapidly, on the scale of weeks to years, but water must seep downward for thousands of years to infiltrate aquifers hundreds to thousands of feet belowground.

The availability of aquifers as sources of drinking water varies widely across the Southeast due to the region's complex geology. In the hills and mountains of the southeastern uplands, good groundwater is hard to find. Rock layers of many varieties make up Earth's crust in this zone. Some are porous and readily store water, while others are impervious. Some leach minerals into the water that can be toxic to us. When drillers tap into clean aquifer water, wells typically yield only enough water to support a few homes or a small community. Groundwater is especially scarce across the Piedmont, a large swath of the southeastern uplands between the mountains and the Coastal Plain. Granite, a dense volcanic rock that cannot store water except in faults that have split the rock, composes the crust beneath the Piedmont. With rare exception, Piedmont cities have no groundwater supply and rely on surface water from rivers. These cities are among the Southeast's largest, including Charlotte, Raleigh, and Atlanta. Taking all this into consideration, we can see that groundwater will not make up for water shortages in the southeastern uplands.[7]

Groundwater availability is quite different on the Coastal Plain. The ground beneath the region consists of porous layers of sand and soft limestone that readily store groundwater. The water fills the interstitial spaces between sand grains or the holes that riddle limestone layers. Most cities and many farms on the Coastal Plain rely on these aquifers.[8] Does this mean that the Coastal Plain will avoid future water shortages? The answer depends on how many straws cities and farms punch into these aquifers, and how much water they extract.

Communities across the United States are withdrawing groundwater at an unsustainable rate. Regions like the California Central Valley and the Midwest are infamous for rapidly depleting their aquifers for agricultural irrigation. Many farmers in both regions have abandoned lands because their wells ran dry. But many communities and farms in the southeastern Coastal Plain also extract groundwater unsustainably. They've extracted so much that NASA satellites detected a decline in the gravitational pull of Earth's crust in the region during the first decade of this century.[9] Groundwater availability in these regions and across the United States will worsen because aquifer recharge rates will decline as atmospheric temperatures and evapotranspiration rates rise.

Groundwater extraction is especially high along the southeastern coast, where populations are concentrated and fresh water at the surface is scarce. Near Jamestown, groundwater in wells is dropping seven feet per decade despite the region's high rainfall. In south-central Georgia, well levels have declined ten feet per decade since the early 1980s. In parts of the Florida aquifer (the aquifer beneath all of Florida and parts of southern Alabama, Georgia, and South Carolina), well water levels are falling thirty-five feet per decade.[10] If these aquifers were being used sustainably, well water levels wouldn't drop at all.

These unsustainable rates of withdrawal have already come at a great cost. Sand aquifers have permanently lost water storage capacity as sand layers collapse when water is extracted.[11] In coastal areas, voids in shallow aquifers created by extraction are filling with salt water seeping in from the ocean. This problem, known as saltwater intrusion, raises the cost of water treatment and forces wells to close. Some communities affected this way include Fernandina Beach, Florida; Brunswick, Georgia; and Hilton Head, South Carolina. And, as we learned earlier, groundwater depletion along the southeastern Atlantic Coast has doubled the rate of land subsidence and accelerated relative sea level rise by one inch per decade.[12]

In the Florida aquifer, where porous limestone is at or near the surface, groundwater overextraction has caused underground caverns to collapse and holes, known as sinkholes, to open at the surface.[13] Unsustainable use of groundwater has caused wetlands to dry out and shallow lakes to drop in volume or disappear entirely. Elsewhere, springs flow with less volume, and a few stop flowing during the dry season. Overextraction can also cause streamflow to decline in streams fed by groundwater. This is a nasty problem for Georgia's Flint River and its tributaries, many of which still retain diverse mussel communities with endemic and rare species. New irrigation systems tapping into the shallow aquifer in the Flint watershed have caused streamflow to decline by 30 to 40 percent. During droughts, some creeks dry up completely. Consequently, mussel populations, including those of

Two of 110 sinkholes forming in January 2010 in Central Florida due to overuse of groundwater by local farmers. Photo by Ann Tihansky, US Geological Survey.

endangered species, have plunged.[14] These problems will spread and intensify as human populations rise and irrigation increases.

Some might argue that the unsustainable use of groundwater is worth the benefit of enjoying a temporary abundance of water for homes, businesses, and farms. Sinkholes and the losses of streamflow, wetlands, and springs are just unfortunate collateral damage. But such rationalization is blind to the future value of groundwater reserves. Future generations will need that water, too. And the Southeast will need full aquifers when the next multidecadal drought strikes. Southeastern groundwater reserves are also important to the food security of the country. If agricultural productivity declines in California or the Midwest due to climate change or groundwater depletion, the nation may need the Southeast to increase its agricultural output to make up the difference. If that day comes, having enough groundwater for irrigation will be critically important.

Given the value of groundwater to the region's future, some southeastern states have taken steps to protect their groundwater resources. South Carolina limits withdrawal in areas with less aquifer capacity. Florida's regulations protect minimum streamflows, lakes, and wetlands. Virginia's permitting process protects other beneficial uses of water and fish and wildlife habitat.[15] In contrast, Tennessee and

Georgia give wide latitude to agricultural withdrawals, and neither Alabama nor North Carolina limits the timing or volume of extraction. Those states with weak or no regulations should develop groundwater extraction regulations soon in this age of climate change and population growth.

Fortunately, several southeastern Atlantic Coast communities are taking a proactive approach to managing their groundwater. Some cities began reducing their reliance on groundwater in the 2000s, and their aquifers began refilling. Other cities are experimenting with techniques to recharge aquifers with stormwater or treated wastewater. These include pumping it into groundwater injection wells or spreading this water across infiltration fields and allowing it to passively seep belowground. Some cities do this to counteract saltwater intrusion and subsidence, while others do it to store water for later use. Because these are novel water management strategies, the associated regulations, best practices, and technologies are still evolving.[16]

The proactive initiatives above are good steps, but the Southeast needs much more groundwater conservation and innovation to avoid the water shortages that the combination of climate change and population growth will bring. This is especially true in the rocky uplands, where exploitable aquifers are absent, and on the Coastal Plain, where cities and farms are extracting groundwater at unsustainable rates.

These proactive groundwater conservation strategies are encouraging. But even if we scale up such efforts, much of the Southeast will still face water insecurity as the climate warms and the population grows. Most communities in the rocky uplands of the region lack exploitable aquifers. Small coastal communities will be unable to build desalination plants unless there is a technological breakthrough lowering the production cost of fresh water. And reservoirs, as we saw in the previous chapter, are a poor choice because they are expensive, wasteful, and ecologically destructive. But although it may seem we have run out of options, there is one last source of new fresh water in the Southeast, one that's been with us all along.

34

The Source Within

FACED WITH DECLINING WATER SUPPLIES due to climate change and a growing population, the Southeast must find new sources of water this century. We've considered new dams and water reservoirs, but given their expense, inefficiency, and ecological destructiveness, they should be a solution of last resort. We've looked to the ocean, but desalination is far too expensive for most coastal cities. There's water belowground, and it's enough to help many Coastal Plain communities and farms if they use it sustainably, but upland communities of the Piedmont and the montane regions don't have access to big aquifers. Though it may seem we have run out of options, there's one last source of water we haven't yet considered, and we can learn about it from Atlanta.

Challenged with supporting a large and growing population, and sitting atop impermeable Piedmont granite, the Atlanta metropolitan area has struggled to find enough water. Historically, the city has relied on reservoirs built by damming local creeks and rivers, some of which it shares with Alabama and Florida. At the turn of this century, with the Tri-State Water Wars still unresolved and having already impounded most of the rivers and creeks in the region that could be easily dammed, Georgia initiated a new approach to managing the Atlanta water problem. In 2001 the state created the Metropolitan North Georgia Water Planning District, with the goal of securing the water supply for the metro area.[1] Since that time over one million people have moved to the region. But instead of the water supply declining over that period, the district has seen the availability of water increase. Where did Atlanta find this water?

The source of water that Atlanta found, and the water the Southeast needs, is the water flowing through our cities and our bodies. It's the water we've had all along. Water policy experts calculate that conservation and efficient use of existing water sources can supply the water the region needs, even as populations grow and water availability declines. Water conservation is about using less water by changing what we do. For example, one can replace a thirsty lawn with gardens of drought-tolerant plants that can endure the hot southern summer with minimal watering. Water efficiency, on the other hand, is about doing what we are already doing, but doing it with less water. This could be installing a low-volume flush toilet instead of keeping the water-guzzling thrones on which most of us sit.

Water efficiency is the easier strategy of the two. It's hard to convince people to practice conservation because it requires them to modify goals and form new habits. Not only have we become conditioned to an abundance of cheap, clean, and convenient water, but in today's busy world few people have time to rethink how they live with water. Instead, water efficiency policies can facilitate change at a systemic level without the public having to change their behavior. When scaled up to the entire community, water efficiency policies reduce water demand and leave more water in our reservoirs, aquifers, and rivers. While communities put efficiency policies into play, education and leadership can guide the public to being more receptive to water conservation.

For many, the best part about water efficiency policies will be their affordability. While new dams and desalination plants inflate the price of water for consumers, water efficiency can reduce production expenses, and utilities can pass the savings on to ratepayers. For example, a 2008 report from water policy experts at American Rivers compared the cost of water efficiency policies with that of new dam construction for cities in the Southeast. They estimated that Columbia, South Carolina, could meet its water needs and save $45–100 million by implementing water efficiency policies instead of constructing new dams. Charlotte, North Carolina, could save $75–160 million. Metropolitan Atlanta could save $300–700 million and save enough water to fill another Lake Lanier. In sum, new dams and reservoirs can cost up to 8,500 times more than water efficiency policies per gallon of water made available.[2]

Water efficiency policies should first address fixing leaks in water supply systems. The US water distribution system leaks over six billion gallons of drinking water each day, an equivalent of 14 to 18 percent of our total daily water use.[3] In addition, governments can fund rebate programs to encourage the installation of water-efficient appliances. Building codes can require new homes and businesses to have efficient fixtures and appliances and require owners to upgrade existing buildings before their sale.

Water conservation policies can also play a key role, and some of the greatest strides can happen outside the home. No less than 30 percent of household water use in the Southeast is for outdoor watering.[4] Policies can encourage homeowners and businesses to plant lawn grasses, wildflowers, shrubs, and trees that require far less water than what is typically planted in the region. Many are native species used by wildlife such as butterflies and birds. Additionally, lawn sprinkler systems can have moisture sensors that prevent use when unnecessary. If needed, cities can promote conservation by requiring businesses with substantial landscaping to pay a higher rate for irrigation water.

In the Atlanta metropolitan area, the water efficiency and conservation policies of the Metropolitan North Georgia Water Planning District led to a 10 percent decline in total water use and a 30 percent decline in per person water use in the years following its establishment. This is even more impressive given that the metro area's population grew by 25 percent during this period. The district achieved this by offering a toilet rebate program, doubling down on finding and repairing leaks, and requiring drive-through car washes to recycle their water. The district also implemented conservation pricing—where water users pay more as consumption increases—to incentivize conservation. Residents were encouraged to minimize outdoor water use. Those with dry lawns were given yard signs to show concerned neighbors that they were taking part in a sanctioned water conservation program. The district is rolling out new initiatives in the next few years and expects continued declines in water use.[5]

Fixing leaks, encouraging the use of water-efficient appliances and plumbing, and incentivizing water conservation will be enough for many southeastern cities needing to increase their water availability or reduce the expense of supplying drinking water this century. However, some communities may need to adopt another water efficiency approach, known as water recycling or water reclamation. This is the cleaning and reuse of our wastewater. Many people instinctively recoil in response to the idea. We have strong cultural taboos against reusing water. They arose from hard lessons learned from disease outbreaks over past millennia. Reinforced by the rise of modern science and germ theory, we treat wastewater like it is a source of plague. But modern technology has changed this equation. We can now transform used water, even nasty sewer water, into clean water suitable for all needs, including drinking.

Whether in our home communities or when traveling, most of us routinely consume water that's recently passed through a wastewater treatment plant. Communities whose water source is surface water are often downstream from another community discharging their treated wastewater into the watershed. This reuse of wastewater discharged into the environment is called de facto water reuse. As a society, we are okay with this, perhaps because we assume that nature cleanses the water. While there is some truth to that notion, it is also true that surface water is often laden with contaminants, even if it is not downstream from a wastewater treatment plant. Two federal laws shield us from these contaminants. The Safe Drinking Water Act (1974) sets lofty standards for cleaning the water our utilities provide, and the Clean Water Act (1972) requires that wastewater discharged from treatment plants is clean.[6]

Some US communities facing frequent water shortages are improving water efficiencies by recycling water without sending it into the environment first. One approach is recycling graywater—all the wastewater from a household except what we flush down the toilet. The average US resident creates twenty-seven gallons of graywater each day. With filtering and minor chemical treatment, homeowners can use graywater for toilet flushing and yard irrigation, the two largest household uses of water. If residential use of graywater was widespread, a city such as Birmingham, Alabama, could reduce its production of potable water by 25 percent. Graywater use can extend the longevity of a community's water supply and reduce wear and tear on wastewater treatment plants. And unlike surface water, graywater is available all year long, even during droughts.[7]

Graywater use is still uncommon in the United States. Until recent decades, water has been abundant enough to preclude a strong incentive for using graywater. Graywater use also requires the installation of new plumbing systems and their periodic maintenance, though households can eventually recoup these costs with lower water bills. Local governments and utilities can promote graywater by funding installation rebate programs. Such costs are trivial compared to the price tag on a new dam, well, or desalination plant.

The greatest obstacle to using graywater is the scarcity of research on implementation at the community scale. As a result, local governments lack scientifically validated management practices to adopt. This lack of research contributes to why many states treat graywater as a health hazard and prohibit its use. A few states allow graywater use, but laws vary widely, tend to restrict usage, and are rarely updated by policymakers. Fortunately, loyal activists in some communities are working diligently to mainstream graywater usage. They have replumbed their homes, pioneered laundry-to-landscape irrigation systems, and pressured local governments to update ordinances. Given their persistence and the rising frequency of water shortages in the United States, we may see a much-needed graywater revolution in the coming years.[8]

Graywater reuse is a decentralized water efficiency approach because it happens in the household or neighborhood. There are much greater water efficiency gains available through the recycling of wastewater sent to the treatment plant. For years, a few communities have used treated wastewater that is nonpotable (unsuitable for drinking) for industrial purposes, crop irrigation, or artificial lakes. But this is a clunky antidote to water scarcity because separate distribution systems are needed, and demand is low and often seasonal.[9]

Increasingly, US communities are adopting or considering potable reuse, where water managers pipe treated wastewater back into the water supply and put it through advanced treatment processes to meet or exceed standards set by the Safe Drinking Water Act. Redundant systems are employed to purge the water of contaminants, including reverse osmosis, nanofiltration, advanced oxidation, biologically active filtration, electrodialysis, and electrodialysis reversal. And there is testing —lots of testing—to ensure that the water meets safety standards. The water produced is cleaner than any surface water.[10]

The most common procedure with potable reuse water is to pump it (after cleaning) into the reservoir or aquifer used for a community's drinking water supply. After this water has mingled with the receiving waters, water managers extract the mixture, process it at the water treatment plant, and pipe it to customers. Alternatively, some coastal communities use this water to replenish extracted groundwater to prevent saltwater intrusion and subsidence. Because these uses require time in the environment, such systems are known as indirect potable reuse. The process supplies water during droughts, conserves the local water supply, reduces discharge of treated wastewater into rivers, and recaptures dollars invested into standard wastewater treatment.

Although indirect potable reuse is used around the world, in the United States it is only common in California, where populations are large and water is frequently scarce. In the Southeast there are two plants each in Virginia and Georgia, but in 2017 about half a dozen southeastern communities were considering or building indirect potable reuse facilities.[11]

The other way to recycle treated wastewater is direct potable reuse, where water managers skip the step of laundering it in the environment and pipe the wastewater directly to the water treatment plant. The wastewater is then refined into fully potable water and sold to customers.[12] Public perception has been the greatest barrier to its adoption. Promoters must conduct a considerable amount of educational outreach to overcome the public's gag reflex at a system that has borne the unfortunate moniker of toilet-to-tap. But with rising populations, the spread of water scarcity, and improved technology, direct potable reuse may soon receive long-overdue attention.

The only two direct potable reuse plants in the United States are in Texas. The cities of Big Spring and Wichita Falls installed their plants during a severe drought in 2013. Water supplies were low, and there were no other tenable options. An energetic public education campaign and sheer desperation won the public over. In Wichita Falls, the former included T-shirts touting with Texan pride "Wee Recycle" and "We Put the No. 2 in H2O." After the drought, Wichita Falls switched to an indirect potable reuse system, but Big Spring kept direct potable reuse.[13]

One challenge to potable reuse is that the purification and testing technologies are expensive and drive up the price of water. Water conservation and efficiency policies are far more affordable, and when dry spells or droughts end, surface and groundwater become cheaper alternatives. Thus, potable reuse in the Southeast will remain uncommon in the immediate future. But this will change if the costs of production decline or water shortages reach a sustained threshold of severity, a likely scenario for large cities of the immediate coast and the Piedmont.[14]

As water shortages increase in the Southeast, state water plans will influence whether and how local governments pursue water efficiency and conservation programs, drill for groundwater, or build dams, desalination plants, or potable reuse facilities. When done well, these plans and their supporting policies are based on a model of a state's water budget. Just as good personal financial planning requires that one develops a budget, good statewide water planning begins with a water budget. Water budgets track how much water enters a state (precipitation and river inflow), how much water people need, and how much water is unavailable for direct use (e.g., lost through evapotranspiration and river outflow).

When these inputs, uses, and outputs are known, states can develop water plans to balance the needs of stakeholders and prepare for crises such as droughts. Plans can coordinate activities among state agencies, ensure the sustainability of water withdrawals and pollution discharges, provide adequate water monitoring, resolve disputes among stakeholders, and facilitate cooperation between the state, local governments, and other stakeholders. Governments should regularly update water plans to adjust for changes in water inputs, population size, water usage patterns, and available technologies. States with strong water plans know what they are doing, and why they are doing it. They enjoy economic and environmental security because they are prepared for threats and opportunities. Unfortunately, most southeastern states have woefully underdeveloped water plans.[15] Only Georgia, North Carolina, and South Carolina have mapped out their water budgets. Alabama has no water plan; it is the only state in the Southeast without one.

Though state and local governments have much work ahead to prepare for managing their water resources in the era of climate change, the good news is that with water conservation and efficiency, the Southeast can have all the water it needs this century, and possibly beyond. As states develop or upgrade their water policies, they can find inspiration in communities across the United States that have already implemented successful programs. And, by pursuing water conservation and efficiency programs, communities can avoid investing in expensive alternatives, some of which—especially new dams and reservoirs—cause significant environmental harm. Furthermore, communities that minimize their need to extract more water from the environment will leave more water in rivers for other vital uses, such as irrigation and the sustaining of riverine and estuarine ecosystems.

We finally have an answer to one of the big questions about the Southeast's future: whether the region will have enough water this century and beyond. It will, provided that states and local governments make smart choices. Now it's time to address the other big question about the region's future: can southeastern river biodiversity survive the future we are creating?

Some are cynical about biodiversity's future. They point out that we've already lost dozens of aquatic species to extinction, and hundreds of others are endangered. Environmental problems from last century burden us and our rivers, and now climate change is causing new challenges.

Nevertheless, there are reasons to be hopeful about the future of biodiversity. More people than ever before want to save other species because it's the right thing to do. But for those who need more practical motivations, the science of ecology has revealed that we are inextricably dependent on biodiversity and natural ecosystems for our survival and well-being. Other species provide us with food and keep pests and diseases in check. Coastal ecosystems protect us from storm surges and sustain fisheries. Plants supply the oxygen we breathe, control erosion, and draw carbon dioxide from the atmosphere. Biodiversity provides us with these and many other ecosystem services that our best technologies cannot replicate affordably or at meaningful scales. People around the world are increasingly aware of this, either due to education or due to firsthand experience as the ecosystems around them unravel. As more people—especially leaders and policymakers—come to understand the importance of nature to their economy and well-being, protecting biodiversity will become a greater priority.

Meanwhile, we are now in an era when humanity is managing the planetary systems that sustain life on Earth. We have reshaped the lands, the oceans, and the sky, and repurposed the planet's biogeochemical cycles. It wasn't our intention to become planetary managers. But due to our large population, patterns of natural resource use, and technologies, our harmful impact on the environment is now measurable at the planetary scale. Importantly, because we are not managing these Earth systems wisely, the future of biodiversity depends on what we choose to do next. If we continue living on the planet as we do now, the rate of other species going extinct will climb sharply this century.

So, our future depends on biodiversity, and now biodiversity's future depends on us. If we are to thrive (or perhaps merely survive) in the Anthropocene, then we must live in a way that also allows biodiversity to thrive. This is true for our oceans, continents, islands, and rivers. Here in the Southeast, we need the unseen thousands of river species that provide us with clean drinking water. We need forests to supply groundwater to sustain river flows between rains. We need predators to keep invasive aquatic species in check. We need salt marshes, oyster reefs, and mangrove swamps to sustain coastal fisheries and absorb storm surges. And rivers and other ecosystems of the region need us to manage them wisely. This confluence of fates—a mutual interdependency of us and nature for survival—gives me hope that humanity will make the right choices to save biodiversity, and itself. If we are to ensure that southeastern aquatic biodiversity flourishes this century, then we must restore biodiversity to our rivers where it has been lost.

35

Mussel Power

IN 1972 THE US CONGRESS passed the Clean Water Act. Its goal was to restore water quality in all US waters to the point they are safe for swimming and fishing. Fifty years later we still have a long way to go. Though regulation of point source pollution late in the twentieth century led to modest improvements in water quality, most rivers in the Southeast and across the United States still carry harmful levels of nutrients, sediment, bacteria, and/or toxins. For this reason, some biologists and conservationists suggest that we rebuild mussel populations to help us restore and maintain water quality in rivers. In other words, we need a mussel revival.

We know that mussels, as filter feeders, can clear river water of contaminants. But how realistic is the idea that mussels could help us keep our rivers clean? A first challenge to this idea is that many mussel species are endangered, and even widespread species have declined in some rivers and disappeared from others. Given this, have enough species survived for us to repopulate rivers with mussels?

The answer is yes. For every species lost to extinction in the Southeast, eleven others have survived. That leaves us with many species to work with. Of the region's 270 mussel species, just over a hundred are doing well in our rivers, and a few are even thriving in reservoirs.[1] Furthermore, we can help endangered mussel species rebuild their populations in the wild through captive breeding and release programs. Saving these endangered species is not just the right thing to do; the US Endangered Species Act (1973) requires it. So, when we help populations of endangered species grow, they can help us keep our rivers clean.

Another question is whether mussels can do enough to improve river water quality in a consequential way. Compared to the size of a river, mussels are small creatures. Are they up to the task? We can begin to answer this by considering how much water a mussel can filter.

Water filtration rates vary by species and size, the quality and quantity of food in the water, temperature, current speed, and season.[2] For these reasons, filtration rate estimates among mussel species typically range from five to fifteen gallons of water per day. Though this seems like a lot of water for a small creature to filter, a large creek will, even during a low-flow period, discharge millions of gallons per day. Thus, the power of mussels to measurably improve water quality will depend on their collective weight, or biomass, in the river.

The potential impact of mussel biomass on water quality is made clear by way of an ecological catastrophe in the Great Lakes. Remember the Zebra Mussel, that small invasive species from Europe brought to the Great Lakes in ballast water carried by freight ships? Both the Zebra Mussel and the Quagga Mussel (a similar species arriving in the Great Lakes about the same time) are small and fast growing, and encrust any solid surface. In the absence of predators, competitors, diseases, and parasites, the two species took over the Great Lake ecosystems. They carpet the lake bottom and any firm surface, growing to densities exceeding 100,000 per square yard.[3] Though each mussel can only filter a quarter gallon of water a day, their collective impact has nearly eliminated plankton in the lakes. Unfortunately, plankton are the basis for the food web in the Great Lakes, and plankton-eating minnows and their predators have all but disappeared in the heavily infested regions.[4] And, the inordinate number of mussels and increased water clarity have caused other ecological problems, such as the proliferation of nuisance aquatic weeds growing on the lake bottom. The situation is a nasty mess, with no hope currently of any reversal. But it illustrates how a large biomass of small filter feeders can clean huge volumes of water.

In contrast to the Great Lakes, where plankton historically supported the food chain and mussels were never plentiful, in southeastern rivers in the pre-industrial era mussels were often most of the biomass. Based on the few rivers where mussels are still abundant, we can surmise that small to midsize rivers sustained ten to twenty mussels per square yard, while larger rivers supported densities up to several hundred per square yard.[5] Could densities like these be enough to improve water quality in rivers today?

Several research teams have tackled this question by studying rivers with large mussel populations. They have discovered that mussels can filter a substantial fraction of a river's volume and improve water quality. Caryn C. Vaughn and colleagues found that in an Oklahoma river during the summer, mussels filter nearly 100 percent of river water per day.[6] Similarly, native mussels in the Hudson River, New York, filter the equivalent of the river's discharge each day. German scientists found that mussels filter 15 to 82 percent of river water per day over the course of a summer and dramatically reduce plankton levels. And in the Upper Mississippi River (much larger than the Oklahoma or New York rivers just mentioned), mussels filtered 12 percent of water during normal flows.[7] In all these studies, biologists determined that during winter, the season when river volume is highest, the portion of a river's discharge filtered by mussels became an insignificant fraction. But even here there is good news. Summer is when river water quality is the worst because flows are low, contaminants are more concentrated, and plankton levels spike.

So, if we help mussel populations rebuild to levels like they were in pre-industrial rivers, we should see significant improvements in water quality. Where mussel populations are strong, we'll spend less to convert river water into drinking water. Rivers will be safer for fishing and recreation. And the work done by mussels will help restore other components of the river ecosystem. More translucent waters will allow aquatic wildlife to find food, mates, or spawning sites more easily. Clearer rivers stay cooler and, thus, hold more dissolved oxygen than waters clouded with mud and algae. And the light reaching the stream bottom will fuel the growth of beneficial algae that sustain snails and other stream creatures. Will the impact of mussels be enough to attain the Clean Water Act's goals of having swimmable and fishable waters in our rivers? That's hard to know at this point: the answer will vary among rivers based on the degree of pollution and the amount of mussel biomass. But, there should be no doubt that mussels can be a powerful component of the green infrastructure we need to help us through the coming years.

The mussel revival—the era of restoring mussel species to the point where they are thriving again in southeastern rivers—won't happen overnight. A daunting challenge is that 65 percent of southeastern mussels are endangered or are showing vulnerability to endangerment. We'll need many years to rebuild the populations of all these species. For some, we will need to begin by restoring conditions in their rivers to levels where mussels and their host fishes can flourish again. Many species will need captive breeding and release programs to resuscitate their populations. This even includes some of the more common species, like the Elephantear, which have populations trapped behind dams. Saving these declining populations and endangered species not only prevents biodiversity loss, but builds resiliency in the mussel community. The more species that are prospering in a river, the more quickly a river ecosystem can recover from a catastrophe such as a strong flood, drought, or chemical spill.

The potential benefit of a mussel revival is a sign of the mutual interdependency that we now have with biodiversity. Many mussel species need quick and deliberate action from us to avoid extinction. Meanwhile we are struggling to improve water quality in our rivers, and we could use some mussel power to restore river ecosystems. We have a lot of work ahead of us to achieve this, but, fortunately, the mussel revival has already begun.

36

Back from the Brink

THE ALABAMA LAMPMUSSEL WAS ONCE one of the most endangered species in North America. The diminutive, golden mollusk is endemic to the Tennessee River system. Its original distribution spanned over 250 miles as the crow flies, from the mountainous valleys of eastern Tennessee to rocky creeks near the Mississippi-Alabama state line. Across this range, watersheds formerly home to the mussel have endured the usual sequence of deforestation, siltation, and minor channelization. But the most disastrous change for the mussel was the damming of the Tennessee River and almost every one of its major tributaries by the Tennessee Valley Authority (TVA). This led to the extinction of several species, and left dozens of others like the Alabama Lampmussel scattered in small, vulnerable populations.[1]

The bleak outlook for the Alabama Lampmussel began to change in spring 2010 as a small team of biologists searched for it and other rare mussels in the Paint Rock River, a tributary of the Tennessee River. The biologists were from the Alabama Aquatic Biodiversity Center, a facility that opened in 2005 and is supervised by Dr. Paul Johnson. The center's mission is to conserve and restore rare freshwater species in Alabama's rivers and creeks. The biologists in the Paint Rock were on a mission to inventory the mussel community of the Paint Rock, but a specific goal was to find one or more gravid Alabama Lampmussels and return them to the center so the young could be raised in the safety of the laboratory. The center would then release young mussels back into the Paint Rock and other suitable streams in the species' former range to rebuild populations.

The Paint Rock River was the only place biologists had seen the Alabama Lampmussel in a half century. Other surveyors had found a handful of them in the Paint Rock in the 1980s, and a biologist captured a live female in 2004. During the first three years after the

Alabama Aquatic Biodiversity Center opened, teams from the center had made many trips to the Paint Rock in hopes of finding an Alabama Lampmussel. Though each time they returned empty handed, in April 2010 the team was optimistic. They had recently found shells of deceased lampmussels in the river's shallows. The biologists knew the mussels were hiding out somewhere near where they were searching, and felt it was only a matter of time before they found them.[2]

The prospect of finding an Alabama Lampmussel wasn't the only reason why Alabama Aquatic Biodiversity Center teams were surveying the Paint Rock River. Nor were they there because the river winds through an exquisite labyrinth of densely forested mountain slopes and narrow fertile valleys. The biologists were on the Paint Rock because it was never dammed. And, as one of the few free-flowing rivers of the Southeast near the epicenter of the region's aquatic biodiversity, the Paint Rock shelters an extraordinary number of species, including 100 fishes and 45 mussels, many of which are imperiled.[3] Part of the survey team's mission was to document this biodiversity.

The Paint Rock is a refugium river, a stream where many river species survived while river industrialization wiped out the stream fauna of other rivers in the area. Other examples in the Southeast include the Cahaba, Upper Clinch, Duck, Etowah, and Conasauga.[4] Most refugium rivers don't include the entire river—and none are pristine—but they have dodged enough industrialization to be a haven for species that have had bad luck elsewhere. Paradoxically, some small refugia are immediately below dams, where fast tailwaters reduce siltation and dissolved oxygen levels are high. Wilson Dam's tailwaters on the Tennessee River, for example, are critical habitat for ten imperiled mussels.[5]

Although the Paint Rock River has been a refugium ever since the damming of the Tennessee River system, its watershed has seen better days. While forested mountains border the river and its tributaries, settlers have farmed and ranched the valleys for over two centuries.[6] Long ago, some farmers redirected sections of the river into canals to free more land for agriculture. Stormwater floods laden with sediment and nutrients hit the river after major rains. Riverbanks are eroding,

The Paint Rock is a refugium river, a stream where river species survived while river industrialization devastated the stream fauna of other rivers in the region. Photo by R. Scot Duncan.

and silt smothers some stream habitats. Left unchecked, these degradations could cause the rare species of the Paint Rock to vanish.

Fortunately, the Alabama Chapter of the Nature Conservancy has been helping the Paint Rock. The Nature Conservancy is a nongovernmental organization (NGO) whose mission is "to conserve the lands and waters on which all life depends." The Nature Conservancy was founded in the mid-twentieth century by scientists who foresaw the importance of protecting ecologically unique and sensitive areas. For its first decades, its focus was land conservation in the United States and abroad. But as the century waned, it became clear that land preservation was insufficient to ensure that ecosystems, even protected ones, would survive. For conservation to succeed, it must be practiced beyond the preserve's borders.[7] Today, the conservancy continues to protect what it calls "the last great places" through land preservation, but it also works in communities around the world to help them improve how they live with nature.

This dual strategy is evident in the Paint Rock watershed. The Nature Conservancy has been instrumental in protecting over forty thousand acres of intact, forested headwaters and floodplain surrounding the river. It has secured funding to pay for more than forty stream restoration projects on the properties of landowners.[8] To address sediment pollution and improve water quality and habitat, the conservancy has stabilized crumbling riverbanks to stop erosion, capture new sediment, and help the growth of riparian vegetation. It has removed low-water stream crossings that were barriers to the migration of fishes. The Nature Conservancy has worked with ranchers to fence cattle out of the river and install water tanks. And on former pasture purchased by the conservancy, a tributary to the Paint Rock that had been straightened and channelized was returned to a more natural course and riparian trees were planted in the adjacent floodplain.[9] Boots-on-the-ground stream restoration work like this is essential to save aquatic biodiversity, restore ecosystem services, and improve the resiliency of watersheds.

Like other environmental NGOs, the Nature Conservancy helps society achieve worthwhile goals unmet by laws and the free market. There are many other NGOs working to protect southeastern rivers. Some are land trusts, like the conservancy, that secure land for watershed protection. This is an important service because there is very little federal- and state-protected land in the region compared to the rest of the nation.[10] Some NGOs, like the Waterkeeper Alliance and the Southern Environmental Law Center, are watchdog groups ensuring that environmental laws are enforced. Many river NGOs try to influence policies and practices at the watershed level, while others, like American Rivers and International Rivers, have a much broader focus. Some work with businesses to encourage environmentally friendly practices, or work with lawmakers to shape regulations. Many NGOs educate the public about the value of healthy rivers, and larger NGOs have staff scientists who do conservation research.

None of this is easy work. While NGOs tend to be far more flexible and adaptable than government, they have no direct authority over environmental laws or their enforcement. They often advocate for change in situations where powerful stakeholders are empowered and enriched by the status quo. NGO budgets are small and reliant on memberships, donations, and grants. Securing these funds consumes considerable time.

State and federal agencies tasked with river protection and wildlife conservation have similar financial constraints as NGOs. Lean budgets hamper law enforcement, research, restoration efforts, and recruitment and retention of talent. The financial challenges are especially bleak in the Southeast. Despite the region's vast number of freshwater species on the endangered species list, there is scant federal and state funding allocated for their conservation. Governmental spending on endangered species outside the region is two to three times higher per species of mussel, four to thirteen times higher per species of crayfish, and thirty-five to fifty-two times higher per species of freshwater fish. These disparities may grow because the US Fish and Wildlife Service is currently reviewing dozens of proposals to protect new southeastern species under the Endangered Species Act.[11] Many, if not all, these species need protection, but unless the service's southeastern region gets more funding, the expenditure per endangered species will shrink even further.

One remedy for the funding gap has been for state and federal agencies to partner with NGOs by sharing expertise, equipment, and funding. Partnerships are now the norm for stream restoration because these conservation projects usually cost more than what any single entity can afford. The partnership approach is also effective because it encourages buy-in from a much wider swath of society.

Ideally, there would be enough funding available to rebuild the populations of all endangered species and to restore damaged river ecosystems. Unfortunately, resource scarcity has led many NGOs and governmental agencies to concentrate their efforts on protecting and restoring selected rivers or portions of rivers. The focus is on defending biodiverse refugia such as the Paint Rock River, and their endemic and imperiled species. While this means that conservationists neglect some endangered species and rivers, this is a necessary strategy given the scarcity of resources available for conservation in the region.[12]

Efforts by the Nature Conservancy and its partners to protect the Paint Rock River paid off on the morning of April 22, 2010—Earth Day, as it happened—as the team from the Alabama Aquatic Biodiversity Center surveyed the river near the town of Hollytree. Finding mussels can be demanding work. It requires long hours of snorkeling in shallow water while wearing thick wetsuits or drysuits to stay warm. Visibility is low, and one must concentrate to spot mussels hidden on the bottom. But on that day persistence paid off with the discovery of seven adult Alabama Lampmussels. Most importantly, two of them were gravid females and were taken back to the center.[13] The mussels' new home at the center was a large room with a drain-lined concrete floor and an elaborate network of pipes conveying water, pressurized air, and electricity to row on row of tanks. As unnatural as this might seem, the facility had everything needed to create a nurturing environment for the new guests.

The practice of propagating freshwater mussels and other southeastern aquatic species is relatively new, and the challenges can be daunting. Many rare aquatic species need the water chemistry, temperature, oxygen level, and food availability to be exactly right. And as with the Purple Bankclimber, a significant obstacle to propagating mussels is discovering which fish species is a suitable host for its larvae. Fortunately, the center biologists had a head start. Recall that lone female Alabama Lampmussel captured on the Paint Rock in 2004? She was gravid, and that allowed biologists to identify two bass species as suitable fish hosts for her species.[14]

There are over a dozen facilities in the United States that propagate rare mussels for release into the wild. The US Fish and Wildlife Service runs several, but state governments operate most. In the Southeast, there's one each in Alabama, Tennessee, North Carolina, and Kentucky, and two in Virginia. Some of these facilities also raise endangered snails and fishes. Though these facilities operate on shoestring budgets, their work is vital to preventing extinctions and restoring river ecosystems.

The efforts of the Alabama Aquatic Biodiversity Center for the Alabama Lampmussel were a remarkable success. The first batch of young lampmussels raised from the two females collected in 2010 yielded 10,769 juveniles. The young mussels lived in captivity until they were

Young Alabama Lampmussels raised in captivity and readied for release into the wild by the Alabama Aquatic Biodiversity Center. Programs such as these are slowing the loss of southeastern aquatic species to extinction. Courtesy of the Alabama Aquatic Biodiversity Center.

large enough for survival trials in the Paint Rock and other streams within the species' historical range. If mussels survive and grow during such trials, then the biologists place more into the stream. Follow-up monitoring reveals how well the new population fares.

As of 2019, the Alabama Aquatic Biodiversity Center was propagating fourteen other mussel species and four snail species for similar rescue efforts. For the Alabama Lampmussel, the center has released 26,894 mussels across six different creeks or rivers, including the Paint Rock, Elk, and Sequatchie. The released mussels are growing well, and some are reproducing.[15]

The work by the Alabama Aquatic Biodiversity Center and others to rescue threatened and endangered mollusks from extinction is a reason to be hopeful. But while these species get a short-term boost in their chances of survival, these restored populations are still isolated and vulnerable to many hazards. Their ultimate survival depends on whether we can restore enough ecological integrity to river ecosystems so populations can grow and spread. If we want to restore mussels to improve water quality and rehabilitate degraded river ecosystems, then we've got a lot to learn and a lot of work to do. Fortunately, what may be the world's largest river restoration experiment is happening on the East Coast, and it is teaching us a lot about restoring watersheds.

37

Restoration Blueprint

THERE'S NOTHING QUITE LIKE CHESAPEAKE Bay. This, the nation's largest estuary, is 200 miles in length and is almost three times the size of Rhode Island. The bay is the lower part of the Susquehanna River Valley and was drowned by the Atlantic after the peak of the most recent ice age. The bay's watershed is twice the size of South Carolina and spans six states. Over 150 major streams feed the bay, including such legendary rivers as the James, Potomac, Patuxent, and Rappahannock. Portions of the watershed are heavily urbanized. Most of the seventeen million residents of the watershed reside in heavily urbanized areas that include the metropolitan areas of Norfolk, Richmond, Baltimore, and Washington, DC. Despite the urban influence, agriculture covers over 30 percent of the watershed.[1]

Chesapeake Bay is also noteworthy because it is the site of one of the largest, most ambitious watershed restoration projects in the world. Due to the estuary's size, population density, and intensive agriculture, Chesapeake Bay has suffered epic levels of sediment and nutrient pollution. State efforts to clean up the pollution during the first decade after the passage of the Clean Water Act in 1972 were woefully inadequate. The bay became muddier and more eutrophic (polluted with nutrients) by the year. Seagrass meadows, oyster reefs, and other estuarine ecosystems vanished, and dead zones devoid of oxygen and life tormented portions of the bay. In 1983 the US Environmental Protection Agency (EPA) began working with states in the watershed to clean up the bay by reducing river pollution across the watershed. But attempts to tame pollution over the next twenty-five years were unsuccessful. In 2009, the Chesapeake Bay Foundation and other environmental groups sued the EPA for not enforcing the terms of the agreement the agency and states had committed to and signed. The next year, the plaintiffs and the EPA settled the lawsuit. The new legally binding agreement, known as the Chesapeake Bay Clean Water Blueprint, set new pollution goals for sediment, phosphorus, and nitrogen. The six states and the District of Columbia must each do their part to reach these goals by 2025. Though most of the Chesapeake Bay watershed lies just beyond the Southeast, the outcomes of this massive effort will influence stream restoration in the United States for decades to come.[2]

Chesapeake Bay may be the most intensely studied body of water on the planet. Though efforts did not limit pollution during the first decades, scientists identified the source of the pollution. Most was nonpoint source pollution coming from the eighty-three thousand farms in the watershed. Farms contributed 42 percent of the nitrogen, 55 percent of the phosphorus, and 60 percent of the sediment pollution in the bay.[3] Subsequent efforts to control this pollution included a combination of mandatory and voluntary programs. Some farmers used conservation tillage, cover crops, forest buffers, and livestock fencing to protect streams and reduce erosion. They also adopted better techniques for spreading fertilizer and manure to reduce nutrient pollution and minimize their own expenditures on these strategies.

Scientists found that much of the remaining pollution in Chesapeake Bay came from urban areas. A substantial part of this pollution was discharge from wastewater treatment plants. As a form of point source pollution, this was a relatively easy issue to fix. States upgraded their plants with modern technology and infrastructure, and their contribution to nutrient pollution declined. Combined with agricultural reforms and policies such as lawn fertilizer bans, this led to an 11 percent reduction in nitrogen, a 21 percent reduction in phosphorus, and 10 percent reduction in sediment in Chesapeake Bay from 2009 to 2017.[4] This is good progress, but it is not enough to meet the 2025 goals of the blueprint agreement.

Watershed scientists also identified another major form of urban pollution affecting the bay: stormwater runoff.[5] Urban stormwater—the rainwater that flows across a city's landscape without sinking into the soil—is one of the most difficult forms of water pollution to control. Its most damaging feature isn't the garbage, nutrients, bacteria, or toxins that it gathers from the urban setting; it's the volume and speed of the water itself. Cities create so much stormwater that receiving streams cannot handle the volume or velocity of stormwater they receive. So much water surges through these streams that it changes their shape through a process known as stream evolution.

Stream evolution can be a natural process occurring whenever there is a change in a watershed's hydrologic cycle, the pattern of water flowing through a watershed. Hydrologic changes occur naturally as a region's geology or climate shifts and alters levels of precipitation, erosion, groundwater availability, or surface flow. In response, streams change in depth, width, sinuosity, and slope until they reach a new equilibrium. Because natural hydrologic changes are usually slow, stream evolution toward a new equilibrium occurs over centuries and millennia. This is ample time for stream biodiversity to adjust to the changes.

In contrast, human modifications to a watershed take immediate effect and often cause rapid stream evolution that few aquatic species can tolerate. Urban landscapes covered with hard, impervious surfaces such as roofs, streets, and parking lots cause drastic hydrologic changes. The volume of stormwater created overwhelms an urban stream's capacity to assimilate it. Stormwater floods scour away mussel beds and wildlife habitats. Streams cut deeply into the landscape, and banks collapse. Winding streams straighten as high-velocity floods carve a more efficient path downstream. This process continues for decades, until the river carves a deeper and wider channel that can handle the higher volumes of water and sediment it now carries. Without assistance, wildlife may need decades to recolonize the stream, if it even can.

Unchecked urbanization during the latter half of the twentieth century initiated rapid stream evolution in many Chesapeake Bay tributaries and streams across the nation. Erosion from urban streams prompted by each rainstorm sent fine particles—silts and clays—through creeks and rivers into the bay, where they raised water temperatures, blocked light for seagrasses, silted over oyster reefs, and reduced visibility for wildlife. Because these particles also carry nutrients—including nitrogen and phosphorus—sediment pollution leads to eutrophication. Ordinarily, river sediments in low doses are a healthy contribution to estuarine ecosystems. They bring nutrients to seagrass meadows and marshes, and help the latter keep up with subsidence and sea level rise. But the amount of sediment pollution from urban and agricultural stormwater has been detrimental to the entire estuary. The problem is especially bad in Maryland, the state bordering the top half of the bay. Much of the state is a hilly region of loose sand and clay that easily erode from urban streams.[6]

Civil engineers, urban planners, and stream restoration scientists in the Chesapeake Bay watershed and across the United States are currently trying to curb the gush of water and sediments from urban areas, but they are battling a stormwater problem that has vexed humanity for millennia. Urban dwellers typically reduce stormwater flooding in cities by lining streets with ditches and pipes that collect stormwater from roofs and pavement and divert it into streams or nearby water bodies. They also straighten urban streams, sometimes paving them, to speed the flow of water out of the city. While all this can reduce urban flooding, urban creeks and rivers are usually the most degraded streams in a watershed. In the Southeast, for example, nearly every urban creek or river is in the process of stream evolution as it adjusts to urban stormwater. Where cities haven't paved or widened streams, trees fall into the creek as banks collapse and the channel widens. Street litter swept

Across the Southeast we have transformed creeks, such as this one in Raleigh, North Carolina, into stormwater ditches. This destroys stream ecosystems and damages the rivers that receive the stormwater. Courtesy of Alan Cressler.

up by flash floods hangs from trees and marks floods that reached high above the stream's surface at normal flow. Swift currents have blasted away the natural streambed and replaced it with fragments of brick and concrete, discarded tires, and other urban artifacts.

Urban stormwater isn't just a problem for wildlife. Many cities now produce so much stormwater that it routinely overwhelms stormwater systems and floods low-lying neighborhoods. Property is destroyed

Urban streams in the Southeast are hammered by stormwater floods, trash, and toxins, and all of it winds up in the rivers and coasts downstream. Courtesy of Nelson Brooke, Black Warrior Riverkeeper.

The stormwater floods now battering urban creeks can reach epic levels. Note the wooden beam above Birmingham-Southern College graduate Allison Justus Brown. The beam was deposited in the tree during a winter flood in Turkey Creek Nature Preserve, Pinson, Alabama. Photo by R. Scot Duncan.

and damaged. Transportation is disrupted. People are caught unaware. Because of urban stormwater, a freshwater flood now occurs every two to three days in the United States. Each year these floods cause dozens of injuries and deaths and over $8 billion in damages.[7]

Urban flooding will worsen this century. Demographers predict that the US urban footprint—the total extent of the urban landscape—will more than double between 2000 and 2050. Meanwhile, climate change has already caused a rise in the frequency and intensity of rainfall events, and climatologists say this trend will continue. This spells danger for communities with aging stormwater systems, which are ill equipped to handle both the expanding urban footprint and storm intensification. Making matters worse, issues of social inequity are entangled with urban flooding. In many US cities, white policymakers used zoning laws and home loan policies such as redlining to steer minorities and the poor into less desirable neighborhoods. Many of these were on low-lying floodplains, next to creeks and rivers that now flood regularly.[8]

While it's been obvious for many decades to scientists and city managers that we should do something about urban stormwater, neither the Clean Water Act of 1972 nor its later amendments authorized the EPA to regulate the amount of water discharged from urban or agricultural areas. However, the 1987 amendments did empower the EPA to regulate contaminants in stormwater.[9] This created an opening to address the stormwater problem. Hydrologists considering the problem knew that aquatic ecosystems trap sediment and absorb nutrients and toxins. Thus, their thinking went, if we could restore streams blighted by urban stormwater so they behaved like natural ecosystems again, then the streams themselves could do some of the heavy lifting needed to decontaminate stormwater. Unfortunately, no widely accepted theoretical frameworks or techniques for restoring streams existed at the time. Into this void stepped David Rosgen.

In the early 1980s, after working as a hydrologist for the US Forest Service for two decades, David Rosgen quit his job and began work as a researcher and consultant in the private sector. Seeing the need for a standardized approach to stream restoration, he developed a simple system for diagnosing the problems of degraded stream channels and designing solutions to repair them. His approach, now known as natural channel design, entails considerable reengineering of degraded stream sections to restore normal stream function.

Rosgen's prescription for restoring a degraded stream starts with classifying the stream into one of several categories based on its position within the stream valley. Then, practitioners must identify a healthy stream of the same category to use as a reference. Features such as slope, width, depth, curvature, streambed sediment type, and channel dimensions are measured in the reference stream and used to design the new structure of the degraded channel. Restoration often involves reshaping the stream channel with heavy equipment to have a structure like the reference stream. Rocks placed in the stream divert the current away from eroding banks, which themselves are armored with logs and tree root wads. The result is a stream section with a more natural ecology than the traditional stream stabilization approach of lining a creekbank with concrete.[10]

Rosgen quickly built a reputation as a leader in stream restoration. He offered classes to teach his methods to professionals from around the United States tasked with taming unstable streams. He translates the science of fluvial geomorphology (the study of a stream's form and function in the landscape) into simple concepts his students can apply even if they have little or no academic training in the field. He has trained well over fourteen thousand students, who use his method across the United States. Many work for federal agencies, or state and local governments. Rosgen's method found success because it filled a vacant niche in the young field of stream restoration at a time when demand was rising for a more natural approach to stream management. Today, the United States spends over $1 billion a year on stream restoration, with most of that spent hiring engineers using Rosgen's method.[11]

But this isn't a clear-cut success story. Many fluvial geomorphologists who specialize on river restoration are critical of natural channel design. Among them are some of the world's most prominent river scientists. They claim that natural channel design has a high failure rate because of its simplistic approach and use by practitioners with insufficient training. In 2014 Margaret A. Palmer and colleagues conducted a scientific review of 644 projects and found that over half of the natural channel design project managers surveyed reported no improvements in restored stream stability, and very low improvements in water quality or biodiversity. Critics also argue that natural channel design projects interrupt a stream's evolution toward a new equilibrium that matches the hydrology of the new landscape, and the projects do nothing to address the upstream watershed conditions

The United States spends over $1 billion a year on stream restoration projects. Courtesy of Alan Cressler.

ultimately responsible for stream degradation. Because funding is rarely allocated for monitoring, bad restoration techniques are perpetuated and stream restoration science doesn't advance as rapidly as it should. Critics also complain about the adoption of natural channel design by state and federal agencies, and the requirement that practitioners receiving government grants for restoration projects have completed Rosgen training. This, they claim, stifles scientific innovation and means that hundreds of millions of taxpayer dollars are spent on a technique that has yet to be validated scientifically.[12]

These and other criticisms ignited a fiery debate in the stream restoration community that has raged for a quarter century. Points and counterpoints have been exchanged in the literature and at scientific conferences. Rosgen and many of his followers shrug off most of the criticism. They point out that streams need restoration now and cannot wait for academia to produce enough science to fashion a different approach.

Despite the controversy, local and state governments have completed hundreds of stream restoration projects in the Chesapeake Bay watershed over the past decade, and more are on tap. With the blueprint's 2025 deadline looming, states are throwing money at stream restoration, hoping it will reduce urban stormwater pollution. The stakes are high because the EPA could penalize states not meeting pollution reduction goals. They might, for example, lose opportunities to obtain new permits for wastewater treatment plants or new housing developments in the watershed. So, in addition to the agricultural reforms and wastewater treatment plant upgrades, bay states have planned seven hundred miles of stream restoration projects in the watershed.[13]

Are these stream restoration projects working? Most of the projects are new, so it's not yet clear. It takes time for the fresh ecosystems to adjust. Sediments must settle, plants must grow, and wildlife must return. The answer is also unclear because there's little monitoring of individual projects. At this point, whether the projects are working depends on whom you ask. Scientists tracking a small subset of the projects report some successes in reducing nutrient and sediment pollution, but improvements are modest. Advocates of stream restoration projects defend them as works in progress—each one is a unique experiment. Others say these projects are cosmetic and do nothing to solve the real problem, which is the creation of too much urban stormwater.[14]

Physical geographer Rebecca Lave has systematically evaluated the claims made by the opposing sides of the Rosgen wars, and interviewed David Rosgen and his followers and critics. She's concluded that the truth on many features of the debate is somewhere in the muddy middle. Does Rosgen's approach ignore upstream conditions? No, but practitioners often do ignore these in their projects. Is there a high project failure rate? Hard to say. Restorationists rarely monitor projects, and publications on the Rosgen method are scarce. Furthermore, failed projects may be attributable to incorrect application of natural channel design methods, not the method itself. Lave found that some critics have unfairly focused on a few extreme failures that are unrepresentative. Lave does agree with Rosgen's critics who say that natural channel design emphasizes stream stability instead of allowing streams to adjust to their new hydrology, and that federal and state policies curtail scientific advancement.[15]

Lave has offered suggestions for how the stream restoration field can move beyond the rift. She points out that the sustained barrage of scientific criticisms has neither slowed nor reversed the use of natural channel design. She suggests that critics should accept that Rosgen's method is here to stay for a while, and that both sides should abandon their harbored misconceptions and cooperate on more research and innovation. Finally, Lave suggests that Rosgen should continually update his recommended approaches as the science of restoration advances.[16]

Novel approaches to stream restoration are emerging. Monitoring of completed projects has increased, and some states require it. While natural channel design emphasizes stream form, stream restoration scientists are studying the restoration of ecosystem services and river functions such as energy flow, nutrient cycling, and connectivity with the landscape. Another trend is to focus restoration efforts on the ultimate causes of stream degradation instead of the symptoms. Proponents of this latter strategy argue that when restorationists address the root causes of stream degradation, then stream ecosystems will do much of the restoration on their own. In urban watersheds, this would require reducing stormwater runoff.[17] While it could seem fanciful to imagine transforming our concrete jungles into a landscape that can help restore streams, there's a scientific field dedicated to this goal.

Low-impact development (LID) is a set of design practices to mitigate urban flooding and protect streams. The goal is to keep rainwater as close as possible to where it falls by capturing stormwater and allowing it to evaporate or seep into the ground. Some LID strategies detain stormwater long enough for pollutants to be filtered out and reduce flash flooding in urban creeks.[18] All these approaches help local governments follow pollution regulations.

Because LID strategies bring a bit of nature back into the city, they are another form of green infrastructure. Trees planted along urban streets absorb rain and release water through photosynthesis. So-called green roofs intercept rainfall and use it to sustain rooftop gardens of grasses and wildflowers. Retention ponds detain stormwater to later evaporate or recharge groundwater supplies. Small depressions known as bioswales capture stormwater runoff for use by plants or ground infiltration. Streets, driveways, and parking lots paved with porous materials allow stormwater to recharge the groundwater. Downspouts on homes and businesses route rainwater into gardens or rain barrels and cisterns for later use. Cities are creating public spaces that function as parks in fair weather and floodplains and stormwater retention zones during heavy rains. Because many LID techniques rely on the planting of trees and gardens, these approaches improve the aesthetics of neighborhoods otherwise dominated by pavement.[19] Finally, most LID strategies are relatively cheap and easy to install. In contrast, the conventional solution for addressing urban stormwater problems is to replace old and inadequate stormwater pipes, which are usually beneath busy city streets.

LID helps solve another problem climate change causes in cities: the urban heat island effect. The materials used to build our cities—especially concrete and asphalt—absorb sunlight, then radiate heat into the urban setting long after the sun has gone down. During the summer, southeastern cities are often more than ten degrees hotter than adjacent rural or suburban areas. It's a problem that will intensify as global warming continues. This extra heat increases the threat of heat-related health problems, which can be life threatening. Warming urban temperatures also increase the spread of mosquito-borne diseases, intensify air pollution, and drive up the use of energy to keep buildings cool. Six of every ten southeastern cities have already seen a worsening in one of the four measures of heat waves (timing, frequency, intensity, and duration), and Birmingham and Raleigh are two of only five US cities exceeding the national average for all four measures.[20]

Urban heating also harms streams. When rain falls on hot surfaces in the city, stormwater absorbs this heat and carries it to streams. Subsequent water temperature spikes cause physiological stress and death of aquatic animals. Warmer waters can also trigger harmful algal blooms in rivers and reservoirs. A large urban area can shed enough water during a storm to affect river temperature for miles downstream.

LID can mitigate all these problems significantly. Plants, bioswales, and retention ponds do not absorb as much heat as do buildings and pavement. Trees planted along streets reflect sunlight back into the atmosphere and keep roads and sidewalks cooler. And the evaporation of water and the pumping of it into the air by plants during photosynthesis pull heat off the landscape and disperse it into the atmosphere. Thus, we can think of plants as nature's air-conditioning units.

Despite these benefits, very few cities use LID solutions. None are part of standard building or landscaping practices in the Southeast. Progress is stymied by societal norms, lack of awareness and training of professionals, misconceptions about the effectiveness and cost of LID, and outdated building codes. But these are social challenges, not engineering or economic barriers. And as such, communities can quickly overcome them and get to work mitigating flooding, stormwater pollution, stream degradation, and urban heating. Given that these problems are worsening, that day cannot come soon enough.

The Chesapeake Bay Clean Water Blueprint is an enormous socioecological restoration experiment nearly four decades old. It's an innovative approach to guiding cooperation among federal, state, and local governments to restore and manage a shared resource: the nation's largest estuary. The landscapes and stakeholders involved are diverse and include fishers in Maryland, coal miners in Pennsylvania, farmers in New York, and legislators in Washington. States are trying novel approaches to reducing agricultural pollution. They are testing new ways of rebuilding stream sections, including regenerative stormwater conveyances and stream-wetland complexes. Some of these restoration attempts will fail, others will succeed—we will learn from all of them. The lessons learned here will influence ecological restoration across America for years to come.

Progress has been mixed. The midway point to the 2025 blueprint deadline to reach pollution reduction targets was 2017. The EPA's midpoint review reported that bay waters were cleaner than they had been in recent decades. Sediment and phosphorus mitigation had achieved over 60 percent of the reduction targets agreed to in 2010, and the EPA credited this for healthier streams and lakes and an expansion of seagrass meadows in the bay. Nitrogen reductions, however, were at only 30 percent of the targeted goals. While nitrogen pollution from agricultural areas had declined and been nearly eliminated from wastewater, nitrogen from urban stormwater had increased since 2010.

States will have to work hard to reach the blueprint's goals by 2025. They've already done the easy fixes, such as addressing point source pollution at wastewater treatment plants. It's the nonpoint sources of pollution that are most problematic. Urban stormwater continues to defy taming, and little time remains to adopt widespread LID practices across the vast metro areas of the watershed. Agriculture is still the biggest source of pollution. States must do a lot to either encourage or require farmers to adopt soil and nutrient conservation practices. Many farmers have embraced these practices, but there has been stiff resistance from others. In 2011, the Pennsylvania Farm Bureau, the American Farm Bureau Federation, and other national agricultural industry groups joined with the National Association of Home Builders to file a lawsuit against the EPA, challenging the blueprint. Though a federal court dismissed the suit, this resistance is indicative of the attitudes many farmers, developers, and builders have toward conservation reforms.

Even if the EPA, the six bay states, and the District of Columbia do not meet pollution targets, these efforts have already improved conditions in the bay and its tributaries. Chances are that if the 2025 targets are unmet, the EPA and the regional governments will develop a new blueprint to guide watershed restoration the rest of the way. After forty years of restoration attempts, it's clear that saving Chesapeake Bay is a marathon, not a sprint.

The scientific and political lessons we are learning in the Chesapeake watershed are important to the Southeast because land use change in the region—the conversion of forests and prairies to farms and cities—has tremendously disfigured the hydrologic cycle of the southeastern stream. The widespread adoption of new wastewater technologies, urban stormwater mitigation strategies, and agricultural conservation reforms would do much to bring about a river revival in the Southeast. Water would be cleaner, stream habitats would be rebuilt, and ecological recovery would be measurable far downstream.

But as beneficial as these changes would be for both people and wildlife, they would still fall short of fully restoring ecological function in southeastern streams. Most importantly, none of these improvements would change the fact that damming has distorted the flow and connectivity of the region's rivers and large creeks. Unless we do something to address this, dozens of aquatic wildlife species in the Southeast will remain endangered, and many will go extinct this century. This is why scientists and river advocates have been studying the problem of what to do about dams.

38

Let 'Em Flow

The Elk River is a midsize stream flowing through southern Tennessee and northern Alabama. It's a tributary of the Tennessee River, joining it just upstream from Muscle Shoals. The Elk is popular with paddlers and fishers and offers refuge for eight endangered species. Most of these rarities are mussels, including the Cracking Pearlymussel, Shiny Pigtoe, and Cumberland Monkeyface. Also among the eight is the tiny Boulder Darter. It is one of the rarest fishes in North America, and the Elk River is its last secure stronghold.

For many decades the outlook for these and other aquatic wildlife in the river was bleak. The problem was the Tennessee Valley Authority (TVA) and the unnatural flows it created in the Elk River through its operation of Tims Ford Dam. Today, however, most Elk River populations of these rare species are stable, and some have grown. Their prognosis improved with the emergence of a new conservation strategy that addressed how dam operators release water from reservoirs.

The TVA built Tims Ford Dam in the late 1960s to create a recreational reservoir. Secondary purposes included storing floodwater and generating power. It was the power production that caused the most trouble for the river. Starting in 1970, when the TVA completed dam construction, for half the year the authority didn't release enough water from the dam to keep the river flowing. During these times the riverbed was dry for up to ten miles downstream, and a decline in water quality was measurable for another fifty miles.

The problem was Tims Ford Dam's oversize turbine. With a 45-megawatt capacity, the turbine was too large for a river this small.[1] To understand why this caused problems for aquatic wildlife, it helps to know a bit about the design of hydropower dams. The power to spin a turbine comes from the release of water at the bottom of the forebay, the water immediately upstream from the dam. This water is under high pressure because it supports the weight of water stacked above it. The deeper the reservoir, the greater the pressure, and the more power available to spin a turbine. Tims Ford Dam is tall (175 feet) and has a deep forebay (125–40 feet). This combination supplied plenty of force to spin the dam's turbine during optimal conditions. However, this design created three problems for the aquatic wildlife downstream from the dam. First, because the released water was from the bottom of the reservoir, oxygen levels in the tailwaters were near zero, and this caused aquatic life to suffocate. Second, the water released from the bottom of the reservoir was extremely cold. During winters, temperatures averaged 41°F. In summers, released waters were 61°F, about 20 degrees below what they would be normally and far too cold to support most aquatic species native to the river.[2]

The third and most severe problem created by the big turbine was that the TVA couldn't run it often enough to keep the river flowing. Because the turbine was so large, running it continuously would quickly drain the water from the reservoir. This would bring an end to power generation and curtail recreation in the reservoir until it refilled. Thus, during the summer and early fall, the season when rainfall was low and evaporation was high, the TVA limited turbine operation to an hour in the morning and an hour in the afternoon. When the turbine was off, the dam released only a trickle of water from a small side sluice, sending water forty miles downstream to the town of Fayetteville for its water supply.[3] But twice a day—Monday through Friday—the turbine spun to life, a frigid flash flood roared down the river channel, and the river flowed for an hour before running dry again. On weekends the TVA didn't run the turbine at all; this was to allow the reservoir to refill and to avoid endangering canoeists dozens of miles downstream. This is how it was for 183 days per year. This operational plan for the dam worked well for the TVA, reservoir boaters, river canoeists, and the folks of Fayetteville, but it was devastating for the river's ecosystem.

By the late 1970s, the TVA openly acknowledged that many of its dams were causing severe water quality problems like those on the Elk River, affecting 340 miles of river within its purview. For the next several decades the TVA took steps to improve conditions by modifying dam operations. Eventually, its efforts helped repair the river ecosystems it had damaged. However, the TVA's initial motivations to improve water quality on the Elk River had little to do with saving endangered species or protecting native biodiversity. Instead, the TVA wanted to establish populations of non-native trout.

Trout are cold-water fish and can endure the chilly waters discharged by a dam. In the 1970s, many state wildlife agencies in the Southeast were raising Rainbow and Brown Trout in hatcheries and releasing them below dams to create sport fishing opportunities for anglers. After the TVA built Tims Ford Dam, Tennessee's wildlife agency released lots of trout (seventy thousand a year) into the Elk River. The fish survived well during the winter and spring, but those not taken home by anglers died during the summer because the TVA released so little water through the dam, and the water released was nearly devoid of oxygen.[4]

In 1984 the TVA installed aerators in the turbine shaft at Tims Ford Dam to improve oxygen levels, but the trout still died over the summer. Then, in 1987, the TVA installed more aeration devices and a small turbine to run whenever the big turbine was not in operation. The modifications created a continuous stream below the dam during the summer, but only a few Brown Trout survived to the next winter and the fishery still depended on the release of hatchery fish.

It was also in 1987 that ichthyologists proposed the Boulder Darter for federal protection. The US Fish and Wildlife Service listed it as an endangered species the next year. Dams had caused a lot of hardship for the darter. The species' original range was a hundred-mile section (as the crow flies) of the Tennessee River and its tributaries across northern Alabama. Within this expanse the fish inhabited rock outcrops in strong currents, where it sheltered in crevices that offered protection from predators and the current. Most of these outcrops now rest in the mud at the bottom of reservoirs behind Wilson and Wheeler Dams on the Tennessee River. Though much of the Elk River escaped the influence of these reservoirs, as of 1987 the fish was only known to inhabit nine locations, all of which were in the Elk River. These remnant populations were suffering from poor water quality in the river due to the TVA's management of Tims Ford Dam. Cold, low-oxygen water released from the dam probably caused the loss of the two populations closest to the dam within a year of the species' listing.[5]

During the fifteen years following its receiving federal protection, biologists took steps to rebuild the darter's population. They learned more about the darter's habitat requirements and reproductive needs. Surveyors found several new populations on the Elk, and a tiny population was discovered near the mouth of Shoal Creek, another Tennessee River tributary. They propagated the darter in captivity and released the fish to augment exiting populations and to start new ones in patches of suitable habitat within the Elk River and Shoal Creek.

While conservationists struggled to save the Boulder Darter, and the TVA sought to establish trout populations in the Elk River, river scientists elsewhere in the world were developing an entirely new approach to restoring ecosystems in streams that were being degraded by unnatural river flows caused by dams.[6] These new strategies would eventually help the darter and dozens of mussel species on the Elk River.

In the 1990s, scientists began studying how to manipulate the timing, quantity, and quality of reservoir water released from large dams to restore river ecosystems and biodiversity. Such restorative releases are known as environmental flows, instream flows, or ecological flows. The overarching strategy of environmental flows is to mimic what a river would do naturally while still fulfilling the dam's primary purposes.

The amount of water naturally flowing through an undammed river varies across the seasons. In the Southeast, a river's typical baseflow (the amount of water it carries between periods of rainfall) is low in the summer and fall and high during the winter and spring. High-rainfall events and droughts periodically cause floods or unusually low baseflows, respectively. River scientists refer to a river's pattern of flows—including seasonal variation and periodic irregularities—as its natural flow regime.[7]

Dams impose artificial flow regimes on rivers. For example, managers of flood-control dams in the Southeast will withhold spring floodwater in the reservoir and release it in late summer. This imposes consistency on an ecosystem that is naturally variable. In extreme cases of disruption, as happened at Tims Ford Dam during the summer, artificial flow regimes bear little resemblance to a river's natural flow regime. In general, artificial flow regimes disrupt the life cycles of river plants and animals, whose behavior and reproduction evolved to respond to natural variation in water depth, temperature, chemical composition, clarity, and current speed. Artificial flow regimes also distort physical processes that sustain river habitats. Floods, for example, build sandbars, distribute nutrient-rich sediment onto floodplains, clean out gravel bars of fine sediments, and carry sediments to the coast. Meanwhile, low flows allow plants to colonize riverbanks and islands, and their gentle currents help populations of some wildlife species grow.

Dam operators can restore some of a river's natural flow regime by releasing environmental flows from the dam. River scientists help develop the guidelines, or prescriptions, for environmental flows for managers to use throughout the year. A prescription might call for releasing more water in winter and less during summer. And it might prescribe the release of surges of water to simulate flood events. A well-developed and well-implemented environmental flow prescription can preserve river biodiversity and restore some of the ecosystem services that free-flowing rivers provide.

Environmental flow prescriptions are not yet in widespread use by river managers in the Southeast. However, there are many success stories of their use from around the world that illustrate the benefits they could bring to the region. For example, farmers in many countries depend on seasonal floods to raise the water table, irrigate crops, and deliver fresh silt and nutrients to agricultural floodplains. Fishers downstream from a dam need flooded forests to supply nursery habitat for young fishes. To meet these and other needs, hydrologists model the river's natural flow regime and design environmental flow prescriptions. Such efforts to restore and maintain these and other ecosystem services have in recent years improved the ecological health of major rivers in South Africa, Australia, and parts of Europe and sub-Saharan Africa.[8]

A barrier to the adoption of environmental flows is that developing a new flow prescription for a river is not easy. Changing the timing and release of water from a dam can help river ecosystems and stakeholders, but if not developed or implemented properly, the prescription can cause ecological or social harm. In addition, implementing new flows can help some stakeholders but hurt others. At Tims Ford Dam, for instance, releasing more water during the summer to maintain a flowing river and sustain trout populations would reduce water levels for boaters in the reservoir. Given these challenges, managers have been understandably cautious about testing new management approaches such as releasing massive amounts of water to simulate floods.

Another impediment to the use of environmental flows is that there are few opportunities to introduce changes into a flow prescription for hydropower dams owned by public utilities. Such opportunities arise only when these utilities apply to the Federal Energy Regulatory Commission (FERC) for dam relicensing. The relicensing process allows new science and public input to help shape flow prescriptions. However, because relicensing for a dam occurs only every thirty to fifty years, this process might be a once-in-a-lifetime opportunity for stakeholders to have a say in how dam operations affect the river or reservoir. The long intervals between relicensing is also a problem because the climate is changing rapidly, as is our scientific understanding of how climate change is affecting rivers.[9]

Sometimes a dam's design imposes limits on how much good environmental flow prescriptions can do. At Tims Ford Dam, for example, the oversize turbine led to problems of balancing the competing needs of keeping water in the reservoir and sending water downstream for trout and the people of Fayetteville. Some dams lack the design to release warm surface water from reservoirs or lack the reservoir capacity to mimic a large flood. For these and other reasons, most environmental flow prescriptions in the United States have simply been to ensure that dams release a minimum baseflow for supporting fish and wildlife populations or for sufficiently diluting pollution legally discharged into rivers.

Though most environmental flow prescriptions in the United States were written for dams in the Pacific Northwest and the Northeast, one of the most influential environmental flow restoration projects in US history was for the Savannah River in the early 2000s. The Nature Conservancy and the US Army Corps of Engineers partnered to lead the way in developing an environmental flow prescription for the river. Up to that point in time, there was no widely accepted method for developing environmental flow prescriptions. Early attempts in the previous decade did not always use the best science available or involve all the stakeholders who would be affected by changes in river management. So, the conservancy and the corps designed a new process of developing an environmental flow prescription and applied it to the corps' management of its three dams on the Savannah River. The process brought together river science experts, state and federal agencies, nongovernmental organizations, community leaders, and others. The outcome was a science-based environmental flow prescription that was responsive to environmental needs and the needs of stakeholders. Notably, the new prescription was not a static plan. Instead, as aspects of the prescription were implemented, scientists (including my brother, Will Duncan) monitored and evaluated the impacts, then used the findings to revise the prescription to better achieve its goals. The collaboration, known as the Savannah Process, was so successful that restorationists have used it across the United States and around the world.[10]

A logistical challenge to the greater use of environmental flows is that there are many dams needing these prescriptions, yet so few scientists who can produce the research needed to draft them. In the United States an average of thirty dams will be up for relicensing each year during the 2020s. Each should have its own environmental flow prescription, but scientists typically need years to gather the necessary data. To overcome these challenges, some of the world's leading river hydrologists developed a method in the mid-2000s for developing flow prescriptions for entire regions. The approach, known as the ecological limits of hydrologic alteration, or ELOHA, is a compromise between the need for intensive science and the need for timely action to improve river management. ELOHA practitioners use available data to model at the regional scale the natural flow regime for distinct types of river segments (e.g., headwaters, river deltas). They then apply these models to rivers when an environmental flow prescription is needed, such as during FERC relicensing. As they develop the prescription, the scientists factor in the river's degree of hydrologic alteration, plus the needs of stakeholders and the ecosystem services on which stakeholders depend. If the plan is adopted, the effects of the new environmental flow prescription are evaluated and river managers modify the prescription as needed.[11] River scientists have used ELOHA in several US states, including Maine, Massachusetts, and Connecticut. In Virginia scientists developed an environmental flow prescription for the Potomac River, a tributary to Chesapeake Bay. Approaches such as this and the Savannah Process will play an increasingly significant role in restoration of our rivers.

Back on the Elk River, the environmental flows implemented by the TVA in 1987 at Tims Ford Dam did not help the trout populations. Nor did the flow improvements help the Boulder Darter or the dozens of mussel species struggling in the lower section of the river. In 1993 the TVA added more aerators to oxygenate the tailwaters and scheduled the release of even more water to help the trout survive the summer. Again, the attempts failed to help the trout. By the early 2000s pressure was mounting on the TVA to mitigate the damage its dams had been causing to endangered aquatic species throughout its sphere of operation, including on the Elk River.[12] Then, in 2004, a flood destroyed the small turbine that had been installed in Tims Ford Dam to provide a continuous flow of water during the warm months. This was a setback for the TVA, but the event sparked a turnaround for the ecological health of the Elk River. Concerned about its liability for the endangered species downstream from its dams, the TVA reached out to the US Fish and Wildlife Service for advice on managing its troublesome dams.

In 2008, under the guidance of the US Fish and Wildlife Service, the TVA implemented a robust environmental flow prescription for Tims Ford Dam that included releasing more water from the sluice, running the large turbine more frequently, adding new aeration devices, and releasing warm surface water through the spillway. Finally, after two decades of missteps, the water released from Tims Ford Dam improved in quality. Water temperatures, oxygen levels, water volumes, and current speeds rose enough to allow Elk River biodiversity to begin a recovery. Today, mussel populations below Fayetteville are booming. The river supports a new population of endangered

Alabama Lampmussels that biologists from the Alabama Aquatic Biodiversity Center released in 2008. Even trout populations are faring better during the summer, though hatcheries still release some to maintain the fishery. In addition, the TVA is using environmental flow prescriptions at other dams, including Wilson Dam, and these efforts are sustaining dozens of imperiled aquatic species.[13]

As for the Boulder Darter, the water quality improvements and other recovery strategies—including the release of over 2,200 captive bred darters into the Elk River and Shoal Creek—have helped the little fish. Though biologists struggle to get an accurate population count in the swift, deep, and muddy waters of the river, the consensus is that the fish is more secure today than at any time since Tims Ford Dam was completed over fifty years ago.[14]

The story of the Elk River illustrates how far river ecosystem management has progressed since the 1970s and the beginnings of the American river revival. During the golden age of dam building, we only considered species conservation for those rivers sustaining economically important fisheries. These days, when water managers make decisions about dam operations, supporting biodiversity with environmental flows is an important consideration.

Will environmental flow prescriptions be enough to prevent widespread extinctions caused by the damming of southeastern rivers? It might for some species, but there is only so much that environmental flows and other restoration efforts can do for a dammed river ecosystem. Dams and reservoirs have fragmented the populations of formerly widespread species like the Boulder Darter and Alabama Lampmussel. Even if these population fragments and their rivers are well managed, these populations are still small and isolated. This leaves them vulnerable to catastrophic events, which can wipe them out. And while the odds of a population-killing catastrophe may be rare in any given year, these remnant populations must survive alongside humanity indefinitely if they are to avoid extinction. Furthermore, as long as dams remain on the river, they will block important ecological processes. Not even a perfect flow prescription will allow migrating river fishes like shad and sturgeon, nor the mussel larvae they carry, to swim past a dam.

39
Struggling Sturgeon

IT'S LATE WINTER, AND MANY sturgeon are on edge. For months they have patrolled the waters of the southeastern Gulf and Atlantic Coasts, vacuuming up invertebrates from their mud burrows. There are the rare hazards of boat collisions and sharks, but it's an otherwise easy life, and the high-fat, high-protein diet pays off. Whereas most in the population have converted this food energy into fat reserves and growth, some sturgeon have been producing copious amounts of egg and sperm. They are the ones that are restless. It's their year to spawn, and their bodies are swollen with sex cells. Now, as they days have lengthened and the waters have begun to warm, their natal river is calling them home and they are anxious to heed the call.

When the time is right, the migrants begin swimming upriver—sometimes alone, sometimes in small groups. If a male can find a migrating female, he will escort her up the river in hopes he will be the one to mate with her. Other males swim ahead, then wait for females just below spawning areas. Spawning takes place at night, often with multiple males jostling for position next to the female. By the time spawning is over, both sexes are exhausted. Adult sturgeon do not feed in rivers, swimming upstream for weeks is taxing, and mating is a rough-and-tumble affair. Spawning is so grueling that males and females need two to three years to recover before they can spawn again. During their recuperation, others will make the journey upriver to spawn.[1]

The sturgeon are doing what they have done for millions of years: feed, avoid dangers, grow, migrate, spawn, rest, then repeat. Evolution has honed these instincts to maximize one outcome: an individual's reproductive success. And although a sturgeon population is an aggregation of individuals, the net outcome of their efforts is a population struggling to rebuild itself. To be clear, there's no coordinated effort.

Sturgeon have no idea that they are endangered, no memory of overfishing, and no awareness that dams block them from reaching most of their former spawning habitat. But through the collective effort of individuals following their instincts, each sturgeon population has the capacity to rebuild itself. The question is, are we doing enough to give them that chance?

Our efforts to revive the Gulf and Atlantic Sturgeon subspecies were nonexistent until the 1970s. The situation for the Gulf Sturgeon improved first. Between 1972 and 1990, Gulf Coast states one by one banned all forms of sturgeon fishing. Then, in 1991, the US Fish and Wildlife Service listed the Gulf Sturgeon as threatened. This brought the fish federal protection and required the service to restore its populations. However, it wasn't possible to design population restoration initiatives because scientists knew so little about the fish. Thus, the first conservation priority became studying the Gulf Sturgeon's basic biology.[2] Biologists learned how to net sturgeon as fishers had done decades before. This allowed them to estimate population sizes for each river in the species' historical range. They measured, weighed, and tagged each captured fish to estimate growth and survival rates and population sizes. They drew blood for genetic analysis, clipped fins to estimate age, and identified favorite foods and preferred spawning habitats. They adopted innovative technologies as they became available. Most recently, they began using sonar-equipped boats to see fuzzy images of sturgeon swimming through murky river and estuarine waters.[3]

The surge of science gradually paid off. Netting surveys showed which habitats were most important for the sturgeon at different life stages and needed special protection. They also revealed which populations were growing, remaining stable, or declining. Genetics research confirmed what had been previously suspected, that each river's population was genetically distinct. The discovery of unique genetic markers for each population empowered biologists to identify the natal river for any sturgeon captured within the species' range. This allowed them to map the range of each population.

Such discoveries now inform regulations and policies that help protect the Gulf Sturgeon. For example, the endangered listing requires that businesses and state and local governments consult the US Fish and Wildlife Service when planning the use of federal funds, permits, or licensing for projects that may affect the sturgeon or its habitats. If a state is proposing to replace or build a bridge across a river used by sturgeon, say, then construction might be allowed only during times when sturgeon are not using the river, and only if the bridge design does not impede sturgeon migration.

Biologists have made other discoveries suggesting that the Gulf Sturgeon has more capability to rebound than previously known. For example, genetics researchers found that a very few sturgeon in some river populations had immigrated from a different natal river. This could be important because most sturgeon populations are small and, therefore, have low genetic diversity. A population's genetic diversity is like a toolbox: the more tools (genes) it has, the more capacity the population has to cope with environmental change. Thus, if these rare immigrants successfully breed in their adopted populations, they can enrich the gene pool.

Another discovery is that while most Gulf Sturgeon spawn in the spring, several populations have an autumn spawning run. Biologists suspect that more sturgeon breed in the fall than in the past due to intense fishing pressure during the heyday of the sturgeon fishery. Because nearly all commercial fishing occurred in spring, the few sturgeon with genes driving them to spawn in the fall evaded capture, successfully reproduced, and contributed a disproportionately higher number of offspring to the next generation than those genetically wired to breed in the spring.[4] Over time, this would result in an increase in fall spawning. A population with autumn spawning should grow faster and be more resilient to catastrophes such as unusually strong spring floods that would otherwise wipe out a year's worth of offspring.

A third way Gulf Sturgeon have more capacity to rebuild their populations than previously appreciated involves a shift toward earlier reproduction. Researchers comparing modern survey data to historical landing data noticed that biologists were not catching very large sturgeon like those caught during the fishery's early years. And yet, after decades of protection, the oldest fish in any of the river populations have had time to grow to these large sizes. Biologists think we "fished down" the population, a phenomenon seen in fisheries around the world. It happens when commercial fishers target the largest fish in the population because they are more valuable than smaller fish. This removes those genes in the population that caused some individuals to grow faster and larger than others. Sturgeon genetically wired for reproduction at a younger age and smaller size were more likely to slip past the nets and reproduce than were sturgeon with genes for delayed reproduction and larger growth. The result is that modern sturgeon populations have a smaller average size and reproduce at a younger age than the original population.[5] This is another change that can help sturgeon populations rebuild quickly.

So, what will it take to help sturgeon populations rebuild to the point where they number in the thousands and their future is secure? Sturgeon scientists are trying to answer that question as they update the recovery plan for the Gulf Sturgeon and write a recovery plan for the Atlantic Sturgeon. Based on the science thus far, the most important thing sturgeon need right now is access to enough good-quality habitat for them to rebuild their populations. Furthermore, river industrialization—principally damming, industrial pollution, and freshwater extraction—seems to be the chief impediment to supplying the habitat the sturgeon need to recover.

The impact of river industrialization becomes clear when we compare the rivers in which the Gulf Sturgeon has disappeared to those in which it has survived, and we examine population trends among the latter. By studying historical data, biologists have discovered that the Gulf Sturgeon once had populations from the Suwannee River in northern Florida to the Rio Grande River in South Texas. Across this range at least eleven distinct river populations exist or once existed. Of these, one is gone, and three are disappearing. The Rio Grande population was overfished and never recovered, and the presence of dams and low water levels due to extraction for irrigation means that the river could not support a sturgeon population today (eight other Texas rivers seem suitable for supporting a sturgeon population, but there are no historical records of sturgeon from these rivers). Three rivers—the Tombigbee, Alabama, and Ochlockonee—once had large populations, but dams prevent the surviving sturgeon from reaching spawning habitats. Without intervention, these three populations are doomed.

This leaves the Gulf Sturgeon with just seven spawning populations. Three of these, the Pearl, Pascagoula, and Escambia river populations, are struggling to survive. These populations are at less than 2 percent of their prefishery size due to periodic industrial chemical spills and catastrophic floods caused by hurricanes (large floods cause dissolved oxygen levels to drop, and this kills sturgeon and other fishes).[6]

Of the remaining four populations, the Apalachicola River population has only five hundred to a thousand sturgeon, and dams block

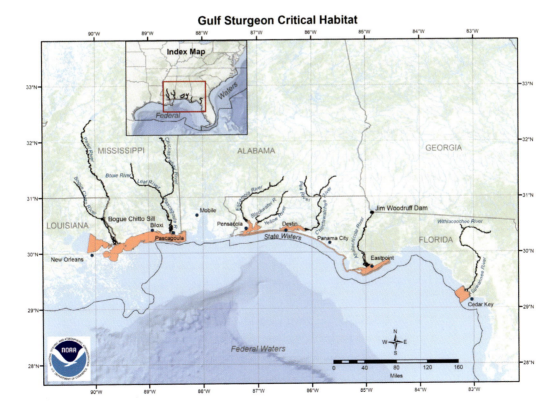

Critical habitat designated for the Gulf Sturgeon under the US Endangered Species Act. Only seven spawning populations remain. Map by the National Oceanic and Atmospheric Administration (NOAA), with assistance from Esri, Garmin, the General Bathymetric Chart of the Oceans (GEBCO), the NOAA National Centers for Environmental Information (NCEI), and other contributors.

The Suwannee River supports the healthiest Gulf Sturgeon spawning population because the river has never been dammed. CC BY 3.0 Mike Tilley/Flickr.

the fish from reaching most of the population's former spawning habitat. Modeling suggests that the population will need nearly a century to make a full recovery, given its slow rate of growth.[7] The remaining three river populations that still successfully spawn—the Suwannee, Yellow/Blackwater, and Choctawhatchee—are doing substantially better than the others. The chief reason is that all three rivers are undammed or minimally dammed. Taken altogether, the statuses of the surviving, declining, and former populations of the Gulf Sturgeon illustrate how much river damming and industrial pollution are preventing the species from recovering.

Blocking Gulf Sturgeon from reaching their spawning habitats is just one way dams are preventing them from rebuilding their populations. Hatchling sturgeon cannot tolerate the high salinities of undiluted marine water and must remain in the river for the first nine months of their life. Because dams prevent access to upstream habitat, dams shorten the length of river available to hatchling sturgeon.[8]

Dams also work together with water extraction to reduce the amount of habitat available to Gulf Sturgeon at the coast. Because sturgeon between one and six years of age cannot tolerate high salinities, they inhabit estuaries near the mouth of their natal river, where marine waters are diluted by river water. However, because of river water extraction and reservoir dams, less water flows through most southeastern rivers than in the past. This allows salty marine water to creep higher into estuaries and constrict the foraging habitat available to young sturgeon. The Suwannee River, for example, flows at a mere 40 percent of its former capacity due to water withdrawal. Sturgeon biologists have calculated that this flow reduction will limit the sturgeon population to less than half its original size.[9] The same is undoubtedly true for other southeastern rivers harboring Gulf Sturgeon.

Climate change is also hastening the loss of estuarine habitats for Gulf Sturgeon. Due to greater evapotranspiration rates in a warmer climate, southeastern rivers will carry less water to the coast in the coming decades and beyond. And as if that were not bad enough, rising sea level and salty tides are taking away brackish water habitats in estuaries. Environmental flows from dams can help in some rivers, but dams will always withhold substantial amounts of fresh water and block access to critically important habitats needed by hatchlings and spawning adults.

The outlook for the Atlantic Sturgeon is also grim. After much dispute and delay, the US Fish and Wildlife Service placed the Atlantic Sturgeon on the endangered species list in 2012, two decades after listing the Gulf Sturgeon. Whereas the Gulf Sturgeon was listed as threatened, the service classified the Atlantic Sturgeon as fully endangered. Experts believe the overall population is at an all-time low.[10]

Due to the delay in listing, research on the Atlantic Sturgeon is two decades behind that for the Gulf Sturgeon. Many ecological research findings about the Gulf Sturgeon likely apply to its East Coast counterpart, but this needs confirmation. Other important conservation data specific to Atlantic Sturgeon conservation remains completely unknown, even basic information, such as which populations have survived.[11]

For example, until recently biologists thought that the James River population (the one that saved the Jamestown colony from starvation in 1609) had been lost to overfishing. But sturgeon sightings by the public in 2007 sparked hope that some had survived. When biologists surveyed the river, they discovered a sturgeon population numbering over three hundred adults.[12] While this is far short of the original population—estimated to be at twenty thousand adults—it's enough fish that the population could make a comeback if we provide the resources the population needs to grow.

But despite the dogged persistence of sturgeon populations like the one in the James River, the long-term outlooks for the Gulf and Atlantic Sturgeon remain bleak. Industrial pollution, reservoir dams, and freshwater extraction are causing the loss of estuarine habitats needed by young sturgeon. The inescapable conclusion is that the future of both sturgeon subspecies depends on how the Southeast decides to manage dams and its use of river water.

40
A Shad Story

THE SOUTHEAST HAS STRUGGLED WITH the problem of managing dams and fish migrations for several centuries. It all began with three species of East Coast fishes in the genus *Alosa*, the Blueback Herring, Alewife, and American Shad. Each spring, massive schools of these fishes migrated from the ocean to spawn in the headwaters of rivers from Florida to Canada. These alosines became an important source of protein and revenue for many Euro-American residents soon after colonization. However, the spawning runs were destined to falter in the era of river industrialization, and now these species are at all-time population lows. Efforts are underway to save these species. One of the foremost strategies that biologists favor for restoring them—and populations of migratory river fish around the world—was first used on East Coast rivers several hundred years ago, to settle disputes between fishers and milldam owners.

The alosine spawning runs up East Coast rivers were colossal. Fish filled the river channel, and their collective struggle against the current generated a wake that sloshed against the shoreline. Where rock outcrops, islands, or logjams created a bottleneck, fishes streamed through narrow passages as a solid, wriggling mass. Like with sturgeon, the spring spawning runs of these species supplied nutrition to Native Americans during the lean days when winter food stores were waning. Native Americans taught Euro-American settlers how to catch the fish, and the annual migrations soon became part of the seasonal rhythm of early colonial life.

By far, the most coveted of the three alosines was the American Shad. Shad are larger than the other alosines, and a young adult provides a pair of good-sized fillets. They are also tasty. Their scientific name, *Alosa sapidissima*, roughly translates as "savory herring."[1] American Shad migrated much farther inland than did river herring

(Blueback Herring and Alewife feed and migrate together and are so similar in appearance that biologists usually refer to them as river herring) and became an especially important food source for many inland farming families. Shad reached 450 miles up the Yadkin and Great Pee Dee Rivers, 380 miles into the Savannah River watershed, and over 200 miles on the Saint Johns, Altamaha, Edisto, Santee, Neuse, and James Rivers.[2] Those who fished for "poor man's salmon" were mostly white landowners who strove for self-sufficiency on small farms. Most lived in the Piedmont, where land was rocky and steep, but much cheaper than the coveted soils of the Coastal Plain. Like two-legged shad, farmers would migrate to the rivers each spring: camping, socializing, and scooping fish from the river with dip nets. The atmosphere was often festive and occasionally rowdy. Afterward, they loaded their wagons and returned home with crates of pickled fish and the occasional hangover.[3]

It wasn't long before industrialization began affecting the shad. A few entrepreneurs in the eighteenth and nineteenth centuries saw the market potential in the predictable, voluminous migrations of a tasty fish. They hired laborers to build weirs of stone or brush to corral fish for easy harvest and process the thousands of fish caught. They packed fillets in salt to remove the moisture, then stuffed them in barrels with vinegar. Pickled shad was then sold in local markets. Shad were so important to colonial America and the early US economy that author John McPhee dubbed it the Founding Fish.[4]

These commercial fishers were not the only ones viewing southeastern rivers with an entrepreneurial eye. Other industrialists began

Millions of American Shad once migrated up most Atlantic Coast rivers of the Southeast each spring to spawn. Overfishing and river dams caused the fishery to crash, but conservationists are endeavoring to restore their populations. Courtesy of Fritz Rohde, National Oceanic and Atmospheric Administration.

harnessing rivers to grind grain, saw lumber, and gin cotton. They built dams to raise the river level so that falling water would spin a waterwheel, whose shaft supplied rotational energy for running the mill's machinery. But milldams and fish migrations were incompatible uses of the river, and this put poor fisher-farmers at odds with wealthy industrialists.[5]

The ensuing disputes were more than a struggle over resource use; the controversies embodied the clash of two conflicting economic visions for America at the time. The fisher-farmers valued self-sufficiency and had little use for industrialization. They spent the little they earned from selling surplus grain on necessities they couldn't make on the farm. The miller's vision, in contrast, was one of connecting rural agricultural productivity to the broader market economy. Millers catered to the wealthier farmers, those who owned more land or more productive land, and who managed their farms and plantations as businesses. Most fisher-farmers couldn't even afford the miller's services and instead ground their grain with handmills at home. The dams and gristmills had no clear benefit to them; these structures merely took away a valued food resource and social event.[6]

Heated controversies over milldams and fishing rights arose in communities across the Atlantic seaboard in the eighteenth and nineteenth centuries. Dam sabotage and mob violence erupted in a few instances, but most fisher-farmers peacefully banded together and submitted petitions against milldams to state legislatures, arguing that the dams were taking away their rightful access to fish migrations. Many also voiced the concern that dams blocked river navigation, which was still crucial in this era before railroads. Legislators, typically from the wealthier classes, might have dismissed these petitions if it were not for the fact that they needed the votes of the pro-fishing poor to keep their elected posts.[7]

As a compromise, many state legislatures in the nineteenth century mandated new approaches for managing dams. They required dam owners to facilitate boat passage with locks, a technology that was already in wide use. They also mandated that dam owners accommodate migrating fishes. People had been tinkering with building fishways (structural features designed to help fish past a dam or other barrier) in New England and Europe as early as the seventeenth century. But two centuries later, fishway technology was still primitive, primarily consisting of simple ramps or sluices in the dam, through which water flowed and fishes could theoretically swim. The efficacy of these fishways was hotly debated. Dam operators and their supporters argued, without evidence, that fish migrations continued as they always had. Fishers with empty nets argued otherwise. Eventually, even the dam owners agreed that fishways at their dams were not working.[8]

The milldams were just one of several reasons the shad stopped migrating up the rivers. Shad prefer to spawn in clear waters over gravel and rock.[9] But many East Coast rivers became choked with mud and sand as settlers converted more forest to agriculture, and this drove shad away. Even more dire for the migrating shad was the rise of large commercial fisheries lower in the watershed on the Coastal Plain. Previously, small-scale commercial fishers high in the watershed sold their pickled shad locally. But in the nineteenth century more ambitious fishers began harvesting enormous amounts of shad. These fishers captured shad en masse using teams of laborers, who strung large nets across the river and captured entire schools of migrating shad at once. As one net was pulled ashore for clearing, another was already in place to catch the next school. The fishers sold some of the shad to plantation owners, who fed pickled fish to their enslaved workers, but they also shipped shad to major East Coast cities, Great Britain, and Europe.

Some state legislatures required that fishers allow a fraction of the fish to continue upstream, but fishers simply moved to the estuaries, intercepting shad before they entered the river. By the mid-nineteenth century, one could count shad nets by the thousands in the shallow waters of Atlantic Coast estuaries, some stretching for over a mile in length.[10]

The estuarine shad fishery survived through the first decades of the twentieth century, but eventually river populations couldn't produce enough shad to sustain commercial harvest. Like sturgeon, most Atlantic Shad (and river herring) spawn in their natal rivers, so several years of intensive fishing could wipe out a river's fishery. Some East Coast river populations survived, but at much smaller sizes. Small schools of shad trickling out of these rivers and into the ocean found

African American workers haul a spawning school of American Shad to the shore in April 1915. Commercial fishing operations such as this one decimated shad populations along the Atlantic Coast. George Grantham Bain Collection, Library of Congress.

one another and joined with schools of river herring, just as their ancestors had done for thousands of years prior. Together, the species migrated seasonally along the coast. During summer they ventured northward along the Atlantic Coast to the Gulf of Maine and beyond to take advantage of the abundant food in these cooler waters. When waters began to chill in autumn, the schools would migrate to waters farther south, much closer to the rivers into which they would migrate to spawn each spring. Those from the southern portion of their range might travel over 1,200 miles in a year.[11]

After the estuarine fisheries for shad and river herring collapsed, these large offshore schools attracted the attention of commercial fishers. In the early twentieth century, total landings in the United States were around 33,000 US tons of shad. But as nets, boats, and fishing fleets grew larger, the migrating schools of shad and river herring shrank. By the end of the twentieth century, offshore shad landings had dropped 98 percent.[12]

It was not until recently that fishery regulators began attempting to preserve the American Shad fishery. In the late twentieth century, some Atlantic Coast states, including the Carolinas and Virginia, began augmenting their shad populations through hatchery programs. The number of these programs peaked around two thousand, although many have since closed due to a combination of budget cuts, limited success in rebuilding populations, and the difficulty of capturing enough wild shad to harvest egg and sperm.[13] Oceanic shad fishing was banned in 2005, but fifteen years later there's no indication that shad are recovering as a result. It doesn't help that shad are still caught as bycatch in offshore fishing operations targeting other species, or that many states allow recreational or commercial harvest in rivers.[14] In some rivers, predatory fishes introduced for sport fishing are believed to consume many juvenile shad as they migrate to the ocean. Today seventy rivers host spawning populations, but this is little more than half the number of rivers that historically supported shad.

As for the river herring, when shad populations declined, commercial fishers began intensively harvesting Alewife and Blueback Herring. Populations of both species are now at all-time lows. In 2019, the National Marine Fisheries Service determined that despite the trends, protecting either species under the Endangered Species Act was unwarranted. There's good and bad news here. On the one hand, neither species is at risk of extinction in the foreseeable future. On the other hand, neither species will get the attention it needs to fully recover.[15] It's conservation purgatory.

Though many populations of shad and river herring have died out because of damming and overfishing, the idea of helping alosines and other migratory fishes bypass dams survived. Fish biologists resurrected the idea in the 1930s, during the first decade of the golden age of dam building. At that time, their motivation was to save the lucrative salmon spawning runs on West Coast rivers. The technologies developed and lessons learned for managing salmon were later exported back to the East Coast, in a desperate attempt to save American Shad, river herring, and other commercially valuable migratory fishes on East Coast rivers.

41
Safe, Timely, and Effective

TO UNDERSTAND THE HOPES AND challenges for fish passage at today's southeastern dams, we must begin by venturing to the Columbia River in the Pacific Northwest, where modern fish passage practices in the United States began, during the first years of the golden age of dam building. The Columbia has few rivals in terms of extent. Its basin is the size of Texas and spans portions of British Columbia and seven US states. The river is 1,200 miles long, cuts through four mountain ranges, and spills more water into the Pacific Ocean than any other river in the western hemisphere. Given its magnitude and power, the Columbia attracted considerable attention during the golden age of dam building. By the end of the twentieth century, there were over 150 hydroelectric dams in the basin, including eighteen on the Columbia and its main tributary, the Snake River.

In the 1930s, as the federal government and regional utilities planned and built the Columbia River's first dams, there was considerable resistance from a coalition of powerful stakeholders advocating to keep the Columbia flowing freely. Elsewhere in the United States during the Great Depression, residents welcomed large dam projects for the job creation, flood control, and inexpensive electricity they would provide. But many wanted to keep the Columbia River flowing freely because it produced extraordinary numbers of salmon.[1]

Each year in the Pacific Northwest multiple species of salmon make exhausting treks upriver to spawn in the final days of their life. Salmon are large fish (even the smallest species can grow to thirty inches in length), and a single school can include thousands. Many Native American tribes held lands along the Columbia River and its tributaries, and they relied on salmon runs for sustenance and the continuation of their cultural practices. The basin also supported profitable salmon canning operations, employing thousands. Fishing tourism

A fish ladder at Bonneville Dam on the Columbia River, at the border of Washington and Oregon. The Bonneville ladders were the first ever deployed at dams constructed by the US Army Corps of Engineers. CC BY 2.0 US Army Corps of Engineers.

was also a growing industry. Altogether, salmon played a leading role in the geographic identity of both Native Americans and the region's newcomers, and salmon generated $10 million annually in the 1930s ($115 million in today's dollars). Proposals in the early twentieth century for dams on the Columbia threatened to end all that, and Native Americans, canners, and conservationists formed coalitions to advocate for salmon protection.[2] It was a clash of economic and cultural values, much like the conflict between shad fishers and the first dam builders on the East Coast.

Because of salmon's value, federal and state legislators had by the 1930s already protected the migrations with laws regulating dam construction and operation. However, engineers had not yet developed effective methods for helping adult Pacific salmon navigate past dams on their upstream spawning runs, or methods for helping young salmon past dams on their journey to the ocean. This had salmon advocates particularly worried about construction of Bonneville Dam by

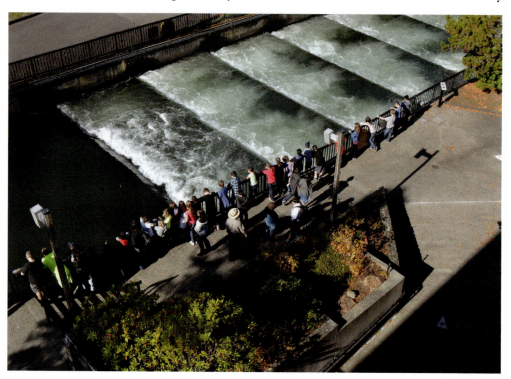

the US Army Corps of Engineers on the Columbia River. At just 146 river miles from the coast, the dam would become the gateway to most of the watershed, and it therefore posed a great threat to the future of the Columbia's salmon.

Understanding the importance of salmon to the region, and being legally required to maintain the salmon runs, the corps trialed several fishways at Bonneville at its completion in 1937. Engineers equipped the dam with three fish ladders, concrete channels through which water flows from upstream to downstream. The channels have a gentle slope, so fish can swim against the current created by the descending water. Bonneville's fish ladders were pool-and-weir systems, where partial walls (weirs) in the channel slowed the current and created pockets where fish could rest between bouts of swimming upstream.[3]

The corps also built four fish locks at Bonneville to move fishes upstream. These worked like boat locks to move fishes from below the dam to the reservoir above the dam. The locks were tall chambers inside which a current and splashing water attracted fishes. When enough fishes gathered, a technician closed the chamber's entrance and flooded the chamber. Fishes followed the rising surface of the water (nudged upward with a basket) and then were released into the reservoir behind the dam.[4]

As an added precaution, the corps built a fish hatchery at Bonneville. The federal government had used fish hatcheries since the 1870s to augment populations of overfished species and establish new populations of valuable species wherever possible. At the Bonneville hatchery, the corps captured salmon during spawning runs, harvested their egg and sperm, and raised the young in captivity before releasing them below the dam at an age when they were likely to survive.

The ultimate success of salmon conservation at Bonneville and elsewhere on the Columbia River was mixed. Though thousands of salmon used the fishways and the hatchery released tens of thousands of young fish, these strategies did not sustain the scale of the salmon migrations from before the corps built the dam. Additionally, within two decades of Bonneville's completion, federal agencies and utility companies built another forty dams on the Columbia and its tributaries, each further hindering the salmon migrations.

Although the Columbia's salmon migrations today are a fraction of what they once were, they would have disappeared entirely if not for the efforts of dam engineers and fisheries biologists. And, the salmon's use of the fishways at Bonneville inspired more fish passage innovation. Most of this work was to support salmon fisheries on the West Coast, but fishery managers in the Northeast imported some fishway technology to help the struggling Atlantic Salmon fishery. Since fishways on the continent had first appeared on the East Coast, in hopes of sustaining American Shad spawning runs, this return of fish passage practices to the Northeast completed a round-trip journey.

Many innovations that scientists tried after Bonneville involved modifying fish ladders. Fishway engineers made some wider to allow more fish to migrate through at once. Other ladders had weirs with notches at the top for surface-swimming species and openings at the base for bottom swimmers. Some ladders had weirs with vertical slots, through which fish could swim at any depth, while others had weirs staggered left and right to create a serpentine pathway within the channel.[5] Fishway engineers discovered that they could easily deploy narrow metal chutes at small dams on headwater creeks to pass some species. Sometimes, they simply had to cut out a section in the top of a small dam to allow migrating fish to pass upstream.

At tall dams, where fish ladders were impractical because of how long they would need to be, engineers designed lifts to move fish up and over the dam. Also known as fish elevators, lifts were more practical than fish locks because they did not require a vertical shaft filled with water to move fish upward. At some large dams, technicians captured migrating fish from the tailwaters and released them into the upstream reservoir. Dam operators also used existing boat locks to pass migrating fish upstream or downstream, a technique known as conservation locking. Some of these strategies worked in reverse to help fishes migrate to the ocean, but engineers sometimes built separate fishways for safe downstream passage.[6]

But despite these innovations and the legal protections, many fish passage strategies and technologies deployed in the twentieth century proved to be unable to sustain migrating fish populations. As scientists studied the problem, it became clear that dams were inherently unsafe environments for fishes. For downstream migrants, including young headed to the ocean, whirling turbine blades at hydropower dams mortally wounded them. Those dodging the blades endured extreme and rapid changes in water pressure and temperature that injured or killed them. At some dams, fishes tumbled over spillways and collided with rocks in shallow plunge pools at the bottom. Sucking currents at water intakes impinged other migrants on screens. Still other fishes entered water diversion passages leading to dead-end channels, from which there was no escape.[7]

Sometimes fishways were the threat. Poorly designed fish ladders produced currents that were too swift or confusing for fishes. Fishes emerged upstream and were too exhausted and bewildered to evade awaiting predators, which could include large fishes, seabirds, or even bears. Fishways could also delay migrations. Migratory fishes are on tight schedules because their movements and behaviors must coincide with seasonal changes, and delays cost fishes valuable time and energy. Delays at dams can occur when fishes cannot readily find the entrance to a fishway or when they become confused or fatigued in fish ladders.

By late in the twentieth century, it was clear that fishways were only occasionally successful, and stakeholders were frustrated. Fishers, scientists, and environmentalists pointed to decades of steeply declining migratory fish populations. Legal requirements to accommodate migrating fishes angered dam owners because the fishways they installed often didn't work. Furthermore, most fishways—including locks, lifts, and ladders—were expensive to build, modify, or replace. And there were accountability issues because funding for most fish passage projects came from taxpayers and utility ratepayers and shareholders.

These concerns led Congress to address fish passage in the National Energy Policy Act of 1992, signed into law by President George H. W. Bush. Based on the failings of existing fish passage strategies and technologies, a provision of the act mandated that fish passage must be safe, timely, and effective. Safe: fishes must be able to migrate upstream or downstream at dams without stress, injury, or death. Timely: fish passage at dams must proceed without significant delay. Effective: the outcome of fish passage must be the successful movement of targeted fishes past dams.[8]

The 1992 act was a shot in the arm for fish passage science. During the heyday of large dam construction, earlier in the twentieth century, the exorbitant cost of building fishways had prohibited true experimentation, which ideally would involve repeated cycles of hypothesis testing, field trials with replication, and data analysis. Instead, engineers designed fishways based on past success, failures, and educated guesswork. After the passage of the 1992 act, federal agencies supplied more funding for research on how to improve fish passage. Project managers recruited scientists from more disciplines—including physiology, behaviorism, and economics—to examine the fish passage problem in new ways.[9] Laboratories were built to study how fast and for how long different fish species and fish of different ages could swim in the range of currents encountered in a fish ladder. Engineers built computer models to match the swimming abilities of fishes to the hydraulics of potential fish ladders. Fish biologists experimented with finding the best attraction flows and splashing sounds to lure different target species into fishways. Other biologists experimented with lights, sounds, and screens to herd fishes into fishways and away from danger. And as they deployed new fishways, biologists marked fishes to track their success at migrating past dams. Economists compared the costs and benefits of different strategies over the short and long term. The practice of fish passage had finally become a fully fledged science.[10]

Despite how important fisheries for sturgeon, shad, and river herring once were in the Southeast, there has never been a concerted, region-wide effort to revive populations of these and other commercially valuable species with fish passage on the region's large dams. Very few dams in the region have fishways, and most that do have poor track records. A principal reason for this is that engineers mostly adopted the designs of successful fishways from other regions rather than developing new designs to accommodate the behaviors and swimming abilities of southeastern fishes.

A prime example of this is the Striped Bass. This is a large, anadromous, predatory fish that like sturgeon and shad was an important staple in the diets of Native Americans and colonists. "Striper"

populations ranged from Canada down to the Saint Johns River in Florida, and along the Gulf Coast from the Suwannee River to Lake Pontchartrain. Commercial harvesting began in the nineteenth century and accelerated during much of the twentieth century. Eventually, overharvest caused populations to crash, and fishery managers imposed a fishing moratorium in the early 1980s.

Striped Bass are more ecologically flexible than sturgeon and shad, and this has helped the species recover since the fishing ban. For example, while some striper populations spawn only in their natal rivers, others that spawn across many rivers helped restore overfished populations. Such versatility helped Striped Bass populations partially rebuild, and they now support a popular sport fishery and a limited commercial fishery.[11]

Because Striped Bass migrate upriver to spawn, dams are a major obstacle to them making a full recovery. But fisheries biologists know very little about stripers' use of fishways. It's unknown, for example, what type of fish ladder would work best for them. Stripers avoid

A school of river herring migrating upstream to their spawning grounds. Throngs of river herring and American Shad once churned the creeks of the Southeast's Atlantic Coast each spring. Photo by Ryan Hagerty, US Fish and Wildlife Service.

typical fish ladder designs imported from the West Coast, which generally force them to move in single file, because they migrate in schools. The fish use lifts on some dams in the Northeast, but there are so few fish lifts on southeastern dams that scientists don't know whether lifts would help the region's populations.[12] Fisheries scientists must learn more about migratory Striped Bass before fishways at dams can help the species recover.

As for the American Shad, many biologists believe that fish passage at dams could help them recover by improving access to spawning habitats. However, shad are notoriously picky about what fishways they'll use (river herring are much less finnicky). Because shad migrate in schools, they avoid chutes and fish ladders that are steep, long, and narrow. Shad easily fatigue and get confused on fish ladder turns, where currents swirl in unexpected ways. Short, straight chutes work fine, but only for small schools on small dams. Shad use fish lifts in the Northeast, but as is true for Striped Bass, it's unknown whether lifts would work well for southeastern river populations.[13]

The fish passage prospects for Gulf and Atlantic Sturgeon are even more troubling. Conventional fishways do not accommodate sturgeon, which are huge, schooling bottom-dwellers. No dams in the eastern United States have routinely passed either subspecies. From what is known about White Sturgeon use of fish ladders in the Pacific Northwest, outfitting southeastern dams for sturgeon would require construction of oversize fish ladders with no sharp turns and enough depth for several sturgeon to pass at once. A more vexing challenge would be designing a fishway to entice sturgeon away from the streambed, where the river is deep and swift, to a shallower and gentler fishway. Alternatively, fish lifts might work on the region's tall dams. Thousands of White Sturgeon once used lifts on Bonneville Dam. But though lifts are less expensive to build than fish ladders, they are still quite expensive, and it's unknown whether southeastern sturgeon would use them.[14]

―⁕―

With migratory fish populations crashing along the East and Gulf Coasts and conventional fishways coming up short, fish passage scientists are developing alternative fishways. The corps is testing one new approach on the Cape Fear River, the largest river contained entirely within North Carolina.[15] In the early twentieth century, the corps built three lock-and-dam structures on the river for commercial navigation. However, these dams plus overfishing caused a steep decline in migratory fish populations. Sturgeon lost over 60 percent of their freshwater habitat, and shad populations dropped to 13 percent of historical estimates. The corps began conservation locking in the 1960s, but this was only marginally successful. While dam removal would be best for restoring migratory fish populations, the reservoirs on the Cape Fear provide water for cities and industries and navigation for recreational boaters.[16]

In 2000, the corps decided to test a rock-ramp at the lowermost lock-and-dam structure on the Cape Fear River. A rock-ramp is an elongated rocky slope spanning all or part of a river and leading fishes up and over the crest of a dam. In theory, rock-ramps allow the passage of many fish species while maintaining an upstream reservoir. Boulders on the ramp dissipate the falling water's energy and offer a variety of channel types for fishes of assorted sizes, strengths, and schooling behaviors. Because rock-ramps resemble whitewater rapids of natural streams, they are classified as a form of nature-like fishway. Fishery managers in Europe have used rock-ramps for decades, and several have been built in the Midwest, but they are untested for North American fishes migrating between rivers and the coast.[17]

The corps began construction on the rock-ramp in 2009, when funding became available. Three years later, the Cape Fear became the first Atlantic Coast river to have a rock-ramp fishway.[18] It's still unclear whether the rock-ramp will work. The corps' goal is that at least 80 percent of the population for each migratory fish species will ascend the fishway. To test the rock-ramp's effectiveness, biologists tagged Striped Bass and American Shad with sonic transmitters and released them below the rock-ramp from 2013 to 2015. Unfortunately, only 19–25 percent of bass and 53–65 percent of shad ascended the rock-ramp.[19] These results fell far short of the goal, and passage for both species was less than or equal to the efficiency of conservation locking at the dam. This was a major disappointment, but there's still reason to hope for a better outcome. Engineers can move boulders within a rock-ramp to create new channels and flow patterns that might improve fish passage. This adjustability is one of the big advantages of rock-ramps over conventional fishways like ladders and lifts. The corps is planning to adjust the Cape Fear rock-ramp early this decade. If it can get the rock-ramp to work, the Southeast will have a way to maintain some of its essential reservoirs while providing effective fish passage.

The corps wants to replace another southeastern dam with a rock-ramp but has faced fierce local opposition. The project is related to the corps' ongoing work to expand the harbor in Savannah, Georgia. Because the expansion is damaging estuarine habitat for the endangered Atlantic and Shortnose Sturgeon, federal law requires the corps to complete a mitigation project to compensate for the harm. The project selected was to replace the dilapidated New Savannah Bluff Lock and Dam on the Savannah River with a new structure allowing sturgeon

This innovative rock-ramp on the Cape Fear River replaced a dam that blocked the passage of migrating Striped Bass, sturgeon, herring, and shad. Nature-like fishways such as this over or around dams may help us revive fish migrations. Courtesy of Alan Cradick.

to access historical spawning habitats. After reviewing several options, the corps proposed replacing the dam with a rock-ramp. However, this would cause the upstream reservoir to drop by six feet. With current technology, a loss of reservoir capacity is unavoidable when replacing tall dams with rock-ramps. If the Corps designed a taller rock-ramp to maintain the current reservoir level, the fishway would either be too steep or too long for sturgeon and other fishes to navigate.[20]

The proposed loss of reservoir height had residents and governments in the area concerned because the reservoir has created valuable waterfront property in Augusta, Georgia, and its sister city across the river, North Augusta, South Carolina. In winter 2019 the corps simulated the drawdown by releasing water from the dam for a week. Residents watched as the reservoir dropped by six feet and left waterfront docks stranded over mud. Within weeks both the State of South Carolina and the City of Augusta sued the corps in federal court to stop the rock-ramp project. A focal point of the suit was the interpretation of language in federal legislation from 2016 requiring the corps to maintain the reservoir at levels supporting navigation, water supply, and recreational activities. The plaintiffs argued that the language means that there can be no lowering of the reservoir height. The corps interpreted the language more broadly to mean that the reservoir must provide enough depth to support those uses, though not necessarily as they are supported now. In November 2020, a federal judge ruled in favor of the plaintiffs, thus barring the corps from proceeding with its plan for the rock-ramp. However, the corps filed with the Fourth Circuit Court of Appeals to overturn the judge's ruling. In April 2023 the appeals court sided with the corps, but stipulated that the corps' plan for the rock-ramp must ensure that reservoir water levels are high enough for both water supply and recreation in Augusta.[21] This was a win for the corps, but it now faces a daunting challenge. To meet the mandates of the new ruling, it must replace the lock and dam with a fish passage structure of a design or size that has never been attempted. What happens next is unclear. The only thing known for sure is that if the corps does nothing, a flood will eventually wipe out what's left of the dam, and the reservoir will dewater on its own.

∽

There has been an entirely different outcome in a small rural town in Maine, where scientists are testing an alternative fishway with the support of residents. The project is in downtown Howland, at the confluence of the Piscataquis and Penobscot Rivers. A hydroelectric dam spans the mouth of the Piscataquis and since 1916 has blocked endangered Atlantic Salmon and other migratory fishes from reaching over a thousand miles of spawning habitat in headwater streams. In the early 2000s, the utility that owned the dam began cooperating with environmentalists, the Penobscot Indian Nation, and wildlife agencies to find a way to restore fish migrations. The utility offered to give up its ownership of the dam, and project planners considered removing the structure. But Howland residents wanted to maintain the reservoir for fishing and other forms of recreation.[22]

Fishway engineers proposed building a bypass channel next to the dam for fish to use. Bypasses are constructed of rock and other sediments to resemble a natural stream and are another form of nature-like fishway. Like with rock-ramps, engineers strategically place rocks of varied sizes in the channel to create currents and resting pools suitable for fishes of different sizes and swimming strengths.

The bypass concept isn't new, but fishway designers have avoided them for several reasons. Bypasses require substantial space in the adjacent landscape to offer a gentle slope. If a bypass is too short and steep, the current will be too strong, cause erosion, and deter fish from swimming upstream. Another drawback is that bypasses usually don't offer enough signal to attract migrating fishes. The strong current and loud noise of the tailwaters below a dam's spillway can easily obscure the outflow from a bypass.[23] Finally, as with rock-ramps, bypasses can necessitate a drop in reservoir height to shorten the length of the bypass.

Fortunately, enough land was available at Howland to build a bypass large enough to surmount most of the above challenges. Completed in 2016, the bypass is over a thousand feet in length and a hundred feet wide. Thus, the channel is stable enough to avoid erosion and carries enough flow to attract migratory fishes. Furthermore, designers

anticipate that the bypass can accommodate strong swimmers like salmon and weaker swimmers like shad. The bypass did require the reservoir to drop by several feet, but the residents agreed to the change and funding was made available to lengthen public boat ramps at the reservoir.

Within days of the bypass's opening, the first salmon were passing upstream into the Piscataquis River. Despite early signs of success, the bypass is still an unproven technology on the East Coast and biologists are studying its effectiveness. As part of the agreement among stakeholders, if the bypass is not providing enough passage for salmon and other migratory fishes after fifteen years, then the dam will be removed. But if the bypass works, scientists will have successfully tested a fishway design that accommodates a much wider range of migratory fish species than a conventional fishway.[24]

We won't know until after more experimentation whether rock-ramps and bypasses can restore fish migrations in the Southeast and elsewhere. Hopefully scientists can design them to offer safe, timely, and effective passage because nature-like fishways could potentially accommodate dozens of species that once freely migrated though the

This bypass on the Piscataquis River in Howland, Maine, allows Atlantic Salmon to migrate past a dam that blocked spawning runs for a century. CC BY 2.0 US Fish and Wildlife Service.

region's rivers. Among them were basses, bowfins, buffalos, catfishes, chubs, crappie, darters, drums, flounders, gars, hogchokers, lampreys, logperches, menhaden, mooneyes, mullets, needlefishes, paddlefishes, redhorses, shiners, suckers, sunfishes, topminnows, and quillbacks. However, a few of the region's migratory species have unusual anatomy, physiology, or behavior, and they may need custom-made fishways. Fish passage scientists are developing this for one such species, and their efforts are beginning to pay off. The species they are saving is the strangest and most talented North American fish, the American Eel.

42

American Eel, Superhero

"In a world where dams and overfishing threaten the survival of migratory fish species, one fish uses special powers to save its kind: American Eel . . . fish superhero." If any southeastern fish deserves a movie trailer introduction like this, it is the American Eel. The eel has the most mind-blowing migration of all fishes on the continent. Yet due to dams and the collapse of eel fisheries around the world, the species has never been more imperiled than it is now. Fortunately, the eel has superpowers—a dozen in all—that help it survive. And at large dams, where eels face dangers they can't overcome, fish passage scientists are finding ways to help.

The eel is a remarkably long fish, with a tubelike body trimmed with a continuous fin. Its slender design gives it an uncanny ability to squeeze into narrow spaces beyond the reach of other predatory fishes, allowing it to capture or avoid becoming prey.[1] Tight space maneuvering: superpower one.

Eels are catadromous, meaning they spend their lives as adults in fresh water, then migrate to salt water to spawn. We already know that Gulf and Atlantic Sturgeon and shad are unusual for living in marine environments and breeding in fresh water. But in the fish universe, it is exceptionally rare to do the reverse. Unfortunately, catadromy makes the eel more vulnerable than other migratory fishes because its young must swim upstream against the current, while young anadromous fish can drift with the current. Nevertheless, catadromy is a special ability, and is superpower two.

American eels range from southern Greenland to northern South America and inhabit a greater range of marine and freshwater environments than any other fish on the planet.[2] They were often the most common fish in rivers throughout their range. Habitat adaptability: superpower three.

American Eels make a grand migration from their spawning grounds to the streams where they grow to adulthood. Those from the periphery of their range travel nearly ten thousand miles during their lifetime. Epic migration ability: superpower four.

Eel experts believe that all eels spawn in the Sargasso Sea, a gyre of warm water spanning over one thousand miles in diameter off the southeastern Atlantic Coast. To reach land and rivers, the tiny eel larvae do a lot of passive traveling on currents. When the current is favorable, they drift; when it is not, they swim. They don't seem to aim for a specific site on the coast, they just get what they get. But as weak swimmers, they are vulnerable to predators patrolling oceanic waters. Since there's no place for them to hide in the open ocean, their ribbon-like bodies are transparent to help them evade detection. They look like a piece of transparent tape with a dot for an eye and a tiny mouth. Power of invisibility: superpower five.

As growing eels migrate to their freshwater destinations, they must transition from living in the ocean's pelagic zone, to estuaries, to big rivers, and finally to the small streams and ponds where they mature into full adults. To adjust to these different environments, eels transform the shape, structure, and color of their bodies: shape-shifting is superpower six.

Approaching the coast, larval eels morph into a shape more like the adults. At two to three inches long and still transparent, they are known as glass eels.[3] Glass eels arrive in coastal waters at about a year old and begin venturing up estuaries and into rivers. As glass eels travel inland, they shape-shift again. This time they produce skin pigments, ranging from olive to gray, for better camouflage as their physiology changes and transparency is no longer possible. These eels, known as elvers, eventually begin the upriver migration, though a few remain behind and complete their maturation in estuaries. Some elvers may find places to settle lower in the watershed, while others push higher, sometimes hundreds of miles inland. Eventually, elvers morph into the yellow eel phase, where they have a longer body form and a yellowish hue. Yellow eels continue to migrate upstream and eventually find places to settle, usually where nooks and crannies are abundant for hiding in the day and hunting at night.

The yellow eel phase is the longest of the fish's life cycle; it can be as short as two years or as long as forty.[4] Such a life span is far longer than that of most fishes and longer than the average human for most of our history. Unusual longevity: superpower seven.

When the time for the spawning migration is nigh, yellow eels shape-shift a final time, into the silver eel phase. This segment of their life involves the most dramatic physical changes and demands the adoption of new powers. By this point females are up to five feet in length, and males about three feet.[5] Both sexes develop a bronze-black back and silver belly for better camouflage in the ocean. Their sex organs mature and begin manufacturing huge quantities of sperm or egg. They cease to feed, absorb their own gut for its nutritional value, and rely only on stored fat for fuel. Their eyes double in size and develop an enhanced sensitivity to the blue wavelengths of light that penetrate ocean waters. They also refurbish their swim bladder so they can remain buoyant during the migration. Fasting ability and enhanced vision: superpowers eight and nine.

Like some superheroes, the American Eel has a mysterious backstory. After it morphs into the silver eel phase, the details of what happens next are one of ichthyology's greatest puzzles: the adult eels simply disappear. Eel experts suspect they complete a reverse migration to the Sargasso Sea, traveling a different route because they can't ride the currents like they did as larvae. It's only rare captures in a research or fishing trawl that hint of a migration toward the Sargasso Sea—the rest is deduction and guesswork. No one has ever seen them spawn there, or anywhere for that matter.[6] The migration must take weeks to months, depending on where they start, but again, scientists simply don't know. Because they presumably all spawn in the same place, an eel from a steamy tropical creek in Venezuela might mate with one that lived in a glacier-fed stream in Greenland. Superior navigation skills allowing eels from faraway streams to reach the same destination: superpower ten.

Ichthyologists believe that eels die after spawning because no one has ever found an adult migrating back upriver, and without a gut they wouldn't survive long. So, like most species of salmon, eels die for their young.

All superhero stories have villains, and one of the eel's two archnemeses wields knives with superior precision: it's the sushi chef. Asian cuisine has featured eels for centuries, especially in Japan, where eel, or *unagi*, is a staple. Commercial exports of yellow and silver eels from the United States to Asia were high through the late 1980s but declined substantially thereafter due to overfishing. Even as sushi became popular in the United States, commercial harvest remained at low levels due to population declines and caps on harvesting to protect the stock. Today, the American Eel population is at a historically low level, and the National Marine Fisheries Service considers the fishery depleted. The US Fish and Wildlife Service evaluated whether the American Eel was endangered in 2007 and 2015, and both times concluded that while the population has declined drastically, it is stable and does not warrant listing. Still, the eel remains a species of concern due to its reduced population and commercial importance.[7]

Despite these concerns, exports of glass eel from the United States have skyrocketed in recent years to support eel aquaculture in Japan. Eel farming is a thriving industry in Japan, but scientists haven't discovered how to entice freshwater eels to reproduce in captivity. As a work-around, the Japanese import live glass eels and raise them until they reach harvestable size. The interest in US glass eels arose due to bans on the harvest of Japanese and European eels because these species are now endangered.[8]

Most glass eels in the United States are harvested in Maine during the upriver migration, which coincides with the time of year when Japanese aquaculturalists need eel stock. Fishers in Maine can get well over $2,000 per pound of glass eels, and in 2018, a pound of glass eels could fetch $2,800. A small glass eel fishery operates on the Cooper River in South Carolina, but that migration occurs during a time of year when demand is low and fishers make far less money than those in Maine. Just as happened with freshwater mussels, exorbitant prices like this have led to poaching and illegal marketing. A recent bust by the US Fish and Wildlife Service resulted in the conviction of nineteen smugglers responsible for over $7 million in eel trafficking.[9]

The rising demand for American Eel, at a time when fishery stocks are at an all-time low, has prompted Maine fishery managers to restrict the harvest of glass eels. Understandably, this has upset glass eel fishers, who argue that overfishing isn't the problem. Instead, they blame the numerous dams in the region, which block upstream eel migration to habitats where they can mature. Although fishers rarely accept blame for declining fish populations, they have a good point. Dams are the other archnemesis of the American Eel. Migration past dams is a challenge for juvenile eels, and typical fishways do little to help them. Elvers and yellow eels rarely enter fish lifts and locks. Glass eels and elvers cannot swim against the fast currents in chutes, and their success in ascending fish ladders is unknown but thought to be poor.[10]

Fortunately, eels have two more superpowers, and each helps them with upstream dam passage. First, they can breathe through their skin. This allows eels to travel short distances across land or minor stream barriers. Second, although they have slippery skin and weak pectoral fins, young eels can climb walls. Biologists have found glass eels and elvers trying to climb over dams by wedging their way up crevices and rough surfaces and squeezing through small openings. All they need is a little moisture and a rough surface or structure against which they can brace their bodies. For this reason, a subset of creeks and rivers have supported small eel populations despite the ubiquity of dams. Amphibious skin and superior climbing abilities: that rounds out a dozen superpowers.

Even the most able of superheroes occasionally needs a little help, and biologists have developed upstream fishways at larger dams for eels at various stages of maturity. These so-called eelways involve a ramp that looks something like a pegboard, which eels climb up by weaving their way through the pegs. A thin sheet of flowing water on the ramp keeps the eels moist. At dams blocking smaller eels, closely spaced tufts of plastic bristles replace the pegs.

Though eelways can provide complete passage over small dams, young eels can fatigue with excessive climbing at large dams. Thus, eelways at large dams lead to fish lifts or traps that dam operators check regularly and empty above the dam. One technical challenge for biologists is placing ramps where eels can find them. Some eelways are highly successful and pass tens of thousands of eels each season, while other eelways are rarely used.[11] Despite the mixed results, eelway

Young American Eels leave the ocean and swim up rivers to find habitats in which they can mature. Fisheries biologists have designed ramps that help them climb over dams and other barriers to migration. Photo by Alexander Haro, US Geological Survey.

technology is a major breakthrough in the conservation of the species.

Unfortunately, helping large silver eels get past dams when headed downriver on their spawning migration is proving to be more difficult for biologists. A lot is at stake for these fish. Unlike for shad or sturgeon, the spawning run is their one-and-only chance to reproduce after decades of survival and growth. Dam spillways that release water to keep the river flowing are dangerous because of rough landings, surging water, and sudden changes in water temperature and pressure. Even more dangerous is the turbine sluice at hydropower dams. Eels are prone to entering these channels, where the whirling blades dice them to pieces faster than a sushi chef. Mortality is at or near 100 percent at some turbines. Whereas young salmon, shad, and river herring can be guided toward downstream fishways with barriers, lights, sounds, and attractive currents, none of these strategies have worked on eels, which prefer the stream bottom and use their tight-space maneuvering superpower to slip through barriers intended to herd them to safety.[12]

Motivated by commercial interests, fishery managers are aiming to ensure the population does not dip any further. Scientists are learning more about eel biology and ways to enable safe, timely, and effective eel passage at dams. Similar efforts are underway in Europe and Asia and may also yield innovations in eel fishery management. Certainly the best sign of success thus far is that thanks to eelways, thousands of juvenile eels are migrating up several rivers in the Northeast for the first time in decades.[13] So despite the daunting challenges it still faces, there is hope for the American Eel, fish superhero.

43

Halfway Solutions

WHEN SCIENTISTS AND CONSERVATIONISTS CHAMPION the restoration of fish migrations in the United States, they are thinking about far more than preventing species extinction. They are envisioning the revitalization of ecosystems, economies, and cultures. When schools of spawning fish ascended rivers each spring in the era before dams, they infused rivers with bursts of energy and nutrients that rippled through both freshwater and terrestrial food chains. Imagine, for example, if large shad and river herring schools once again reached the headwater shoals of southeastern rivers. Herons, otters, gars, and other predators would feast on adult fish while the waters downstream bustled with smaller animals snatching eggs and fry, and scavengers stripped carcasses to the bone. Thus, the return of spawning runs would help wildlife populations in the region recover from all-time lows.

The ecological benefits of restoring fish migrations would extend into marine food webs. As river herring, shad, and others rebuilt their populations, so would their marine predators. Populations of red drum, sea trout, mackerel, bluefish, and dolphinfish would rebuild. Dolphin, whale, and seal species would feast on schools of alosines migrating along the Atlantic Coast each summer to the Gulf of Maine and beyond.

The growth of spawning runs would help with the river mussel revival, too. American Shad, Alewife, and Blueback Herring are hosts for the Alewife Floater. Blueback Herring are hosts for the Roanoke Slabshell. Striped Bass are hosts for the Dwarf Wedge Mussel. American Eel are hosts for the Eastern Elliptio, the most common mussel on the East Coast.[1] And as mussel populations rebounded, they would help rivers run clean again by removing nutrients, sediments, and other pollutants.

The restoration of migratory fish populations would help the survival of fishing traditions in the Southeast dating back centuries. Commercial fishers, for example, are a proud, resourceful people who understand our waters in ways that most scientists never will. But it's getting harder to make a living as a fisher because fish populations are declining and regulators are imposing lower catch limits to prevent extinctions.

Subsistence fishing is common in the Southeast. These fishers are usually from impoverished communities, and the fishes they bring home are an affordable and healthy part of their diet. There would be more subsistence fishing opportunities in these communities if migratory fish populations rebounded and were available for harvest. This would also lead to an expansion of recreational fishing. The money sport fishers spend on lodging, guides, gear, and supplies is already vital to the economy of many riverside and coastal communities. The expansion of sport fishing opportunities would support even more prosperity in these communities.

This vision of restoring migratory fishes to their full ecological, economic, and cultural role is alluring, but can fish passage science get us there? Can it develop the technologies and strategies needed to help these fishes travel past dams, reach their spawning grounds, and rebuild their populations?

Many are hoping the answer is yes. Among them are fishery managers, required by law to restore populations of endangered and commercially valuable migratory river fishes. Dam owners also want effective fish passage solutions, since governments require many of them to accommodate fish migrations. And the need for effective fish passage is international. Dams have disrupted fish migrations on every continent except Antarctica, and many countries have watched their fisheries decline due to damming. In the many developing countries whose populations rely on river fisheries for their protein, sustaining or restoring fish populations is critically important.

However, numerous environmentalists and river scientists remain unconvinced that fish passage science can provide effective solutions for sustaining or restoring migratory fish populations. They are skeptical, in large part, due to the widespread failure of fish passage on the East Coast and elsewhere over the past six decades. For example, despite fishways on all dams blocking their migration routes, only 3 percent of American Shad migrating up the Susquehanna, Connecticut, and Merrimack Rivers in the northeastern United States reach spawning habitats. Of the hundreds of endangered Shortnose Sturgeon (a northern anadromous species) crowding the tailwaters of Holyoke Dam on the Connecticut River, an average of four per year ride the fish lift to the other side.[2] And even after hundreds of millions of dollars have been invested to revive Atlantic Salmon populations with fishways and hatcheries, the species is still endangered.

Critics of fish passage also point to ecological problems that fish passage strategies are unlikely to mitigate. Even where fish passage has been more successful (e.g., some of the salmon fishways in the Northwest), fish migrations are a trickle of what they once were and do not fulfill the ecological and economic roles they once did. Furthermore, most successful fishways only assist a targeted few of the many species that once migrated through rivers. Even if scientists find that rock-ramps and bypasses can help more species with migration, it could be prohibitively expensive to build these structures on the thousands of dams in US rivers blocking fish migrations. Finally, critics argue that fishways do little to mitigate other problems that dams create for river ecosystems, such as the transport of sediment to the coast. For these reasons, some have dubbed conventional fish passage strategies as halfway solutions that can only sustain or restore a fraction of a river's full ecological function.[3]

Fish passage scientists acknowledge the field's mediocre record. However, they argue, given the importance of maintaining migratory fish populations on dammed rivers, we must continue seeking effective solutions. These scientists point to progress in their field since the passage of the 1992 National Energy Policy Act, which mandated a more rigorous policy approach to fish passage. They remind critics that the engineers dominating the field for most of the twentieth century used a trial-and-error approach to fishway design. In contrast, today's fish passage scientists are from a range of academic disciplines, and they rigorously apply the scientific method. Modern practitioners have a far better appreciation for the complexities of fish biology and are applying state-of-the-art technologies for studying fish and developing effective fishways. Fish passage scientists also call for more investment in research, especially in rivers of the world where large human populations depend on river fisheries threatened by dams.[4]

The debate over whether fish passage can help migratory fishes rebuild their populations on dammed rivers will not end anytime soon. However, both proponents and critics of fish passage agree on several key points. First, we will not restore migratory fish populations unless we improve river connectivity so fishes can reach their spawning grounds. Second, fish passage is, at best, a partial antidote for the ecological harm that dams impose on free-flowing rivers.[5] Third, and most profoundly, the best way to restore migratory fish populations is to remove dams.[6] Not only would this circumvent the technical challenges of sustaining and rebuilding migratory fish populations, but it would restore many other ecosystem services that free-flowing rivers provide. If dam removal as a solution seems too radical an idea, think again. An age of dam removal is inevitable. In fact, it has already begun.

44

A New Era Begins

THE ELWHA RIVER ORIGINATES HIGH in the glacier-capped mountains of the Olympic Peninsula in Washington State and winds forty-five miles through forest-cloaked hills to the Strait of Juan de Fuca. The river has been the homeland of the Lower Elwha Klallam Tribe for time immemorial. Like with many tribes of the Pacific Northwest, much of its culture centers on the harvest of salmon and trout as these fishes complete their upriver spawning runs. In the nineteenth century, white settlers drawn to the region's wealth of natural resources began to jeopardize Native American lifeways on the Olympic Peninsula.

One of the greatest disruptions for the Lower Elwha Klallam Tribe came in 1912, with the completion of Elwha Dam, a 108-foot-tall hydropower dam built to provide electricity for the nearby town of Port Angeles and its sawmill. The dam restricted fish spawning habitat for ten varieties of salmon and trout to just the final five miles of the river's route to the ocean. The reservoir behind the dam flooded many of the tribe's cultural sites and much of its historical homeland.

After the dam was built the tribe protested because it violated an 1890 Washington State law requiring that dam builders include fish passage technology in any dams on rivers supporting salmon runs. The state ordered the dam owner to build a hatchery to supplement populations of migratory fishes, but the hatchery failed and the owner shut it down in 1922. Five years later the owner built the 210-foot-tall Glines Canyon Dam eight miles upstream from Elwha Dam. The combination of the two dams starved the lower section of the river of sediment, including the gravel over which salmon spawn. This degraded the remaining spawning habitat, and fish populations dipped even further. Over the next forty years, the Lower Elwha Klallam Tribe continued calling for the restoration of salmon runs, but to no avail.[1]

The situation began shifting in favor of the tribe in the early 1970s. The company that owned the dams and the sawmill applied to the Federal Energy Regulatory Commission (FERC) to have its dam operation licenses renewed. The relicensing of Glines Canyon Dam was complicated by its presence inside Olympic National Park, which was established after the dam was built. Congress had prohibited hydropower dams in national parks years earlier. The Lower Elwha Klallam Tribe and park advocates wanted Glines Canyon Dam removed, while the sawmill owner and its powerful allies in the timber and mining industries wanted the dam relicensed. FERC, unsure how to proceed, delayed making a decision. As the commission deliberated, political skirmishes over the future of Glines Canyon Dam dragged on into the 1980s. By then, the Lower Elwha Klallam Tribe had more allies in its battle to free the river. Local and national environmental groups demanded that the Elwha flow freely within the park. Additionally, sport and commercial fishing advocates, concerned about region-wide declines in salmon stocks, wanted both dams removed to restore the river's migratory fish populations. FERC had never faced a relicensing decision as complicated as this, and the commission continued its procrastination.[2]

Congress stepped in to resolve the dispute in 1992 by authorizing the US Department of the Interior to purchase both dams and remove them. By then, the city of Port Angeles was receiving all its electricity from other sources. Furthermore, James River Corporation, the latest owner of the dams, was eager to sell the dams, out of concern that the federal government would eventually force it to remove the dams and pay for the removal itself. The Department of the Interior bought the dams in 2000, Congress approved funding for their removal in 2009, and demolition began in 2011, nearly twenty years after Congress gave the green light for removal, and a full century after the Lower Elwha Klallam Tribe began protesting because its rights had been violated.[3]

The removal of Elwha and Glines Canyon Dams marked the beginning of the largest river restoration project humanity has ever attempted. It may take a century or more for the Elwha River and its watershed to recover, but already there are signs of healing.[4] Within months of removal, Chinook and Coho Salmon and Steelhead Trout

The removal of Elwha and Glines Canyon (*pictured*) Dams in 2012 and 2014, respectively, marked the beginning of the largest river restoration project humanity has attempted. Large-scale restoration like this is needed in the Southeast to restore the full range of ecosystem services rivers can provide. Photo by Jeff Duda, US Geological Survey, Western Fisheries Research Center.

returned to spawning grounds above the former dam sites for the first time in a century. Roots of young trees planted across the now-empty reservoirs are stabilizing sediments and becoming a new forest. Gravel previously trapped behind the dams has washed downstream and refreshed spawning habitats. Finer sediments have reached the coast and rebuilt a quarter-mile-wide delta that had disappeared during the previous century due to sediment starvation.

The Elwha River removals are just two of 1,700 dams that have been removed in the United States as of late 2020. Though nearly all of them were significantly smaller than Elwha and Glines Canyon Dams, these projects have restored connectivity for tens of thousands of miles of streams across the United States. Three-quarters of these projects have occurred this century, and the rate of dam removal is increasing.[5]

Dam removal projects can be motivated by many issues, including ecological, legal, cultural, and financial, but public safety is usually a top concern. Safety played a central role when Congress authorized the removals on the Elwha River. Elwha and Glines Canyon Dams were one hundred years old and eighty-seven years old, respectively, at the time of their removal. Many, from park managers to the dam owners, were concerned about the condition of the dams and what would happen should one of them fail. Though the area downstream from the former dam sites was lightly settled at the time, it was (and still is) popular with tourists.

Glines Canyon and Elwha Dams were not alone in their advanced age—many US dams are quite old. Of the 91,468 large dams in the country, 66 percent were at least fifty years old in 2020, and 19 percent were at least seventy-five years old. The rapid aging of the US dam fleet has many engineers, government officials, and citizens worried about the risk of dam failures, and the removal of unsafe dams is one motivation behind the uptick in dam removal.

Dams do fail, and the consequences are often catastrophic. Since the mid-nineteenth century dam failures have killed over three thousand people in the United States. Federal and state governments didn't systematically address dam safety until a nasty spate of dam failures in the 1970s killed hundreds of people and caused over a billion dollars in damages. Among these tragedies was the 1972 coal slurry dam failure in Buffalo Creek, West Virginia, that we learned about earlier. Another failure, in 1977, killed thirty-nine people, many of whom were college students asleep in their dorms at Toccoa Falls Bible College in Georgia.

In response to these and other failures, the US Army Corps of Engineers tracks the age and condition of all large and potentially hazardous dams in the country. The corps catalogs records from state agencies that inspect dams regularly to find and fix problems before they become disasters. All US states have dam safety inspection programs except one. Alabama's antiregulatory legislature has blocked multiple attempts to establish a dam inspection agency (the corps' database does include other statistics on Alabama's dams).[6]

The corps scores dams in the database based on the degree of risk to people downstream should they fail. In 2020, 60 percent of dams classified as high hazard potential (meaning that failure would probably cause the loss of human life) were at least fifty years old, and 26 percent were at least seventy-five years old. Not only do these percentages rise with every year, but these are underestimates because a fifth of dams in the database are so old that no one knows their age.[7]

Dam failures may increase in the coming years for two reasons related to dam age. The first is that construction materials weaken with time. Concrete deteriorates, metal fatigues and corrodes, and equipment malfunctions. Knowing this, engineers design each dam with a planned operational life span, one typically measured in decades. Troublingly, an overwhelming majority of dams in the United States—about 85 percent—are near the end of their planned operational lives.[8]

A second reason to expect more failures in coming years is that engineers designed most of our dams for a gentler climate that no longer exists. Recall that the warming atmosphere has already caused rainfall events to increase in intensity. This is a climate change trend that will continue well into this century. Heavy rainfall events and flooding cause most of these dam failures. In a process known as overtopping, reservoir levels exceed a dam's capacity to control the release of water through its spillway. The excess water accumulating behind the dam then flows over or around the dam, triggering dam failure.[9]

Extraordinarily heavy rainfall events have caused several recent dam failures that captured national attention. In 2017, heavy rainfall preceded the spectacular failure of Oroville Dam in California, the tallest dam in the United States. Rains waterlogged this earthen dam, causing two spillways to fail and a raging torrent of reservoir water to surge down the dam's foundation. No lives were lost, but repair costs exceeded $1 billion. In 2020, intense rains in Michigan overwhelmed the capacity of Edenville and Sanford Dams. Both dams failed, and reservoir water flooded several towns downstream. Thanks to early evacuations there were no fatalities, but the flood caused over $250 million in property damage. The Southeast has also seen dams fail due to rainstorms intensified by climate change. Perhaps the most dramatic episode was in 2016, when flooding from Hurricane Matthew caused the breaching of forty-two dams in the Carolinas.[10]

Instead of removing aging or unsafe dams, can we repair them as materials degrade and equipment fails? Yes. Can we upgrade dams to handle the high-intensity rainfall events that the new climate is delivering? Absolutely. Dam repairs and upgrades are common, especially at hydroelectric and other large dams providing economically valuable services. However, upgrades and repairs are pricey, and are often prohibitively expensive for owners of smaller dams that generate little or no revenue. These include dams built for mills or factories that no longer exist, or dams that once facilitated commercial transportation that long ago switched to rails and highways. For these reasons, more and more dam owners are opting to remove dams at risk of failure because it is cheaper than repairs, upgrades, or replacement.[11] But as we'll see, even when there are strong incentives like these to remove a dam, the process can be exceptionally complicated.

The dramatic 2017 failure of Oroville Dam in California after exceptionally heavy rains demonstrated how dams built in previous centuries are at risk of failure in this era of rapid climate change. Photo by William Croyle, California Department of Water Resources.

45

The Road to Removal

On April 23, 2006, two teenage boys, Charles and Bryc, walked down to the local swimming hole. The skies were gray, and the area had received two inches of rain the day before, but it was good enough for a Sunday swim on the Maury River at Jordan's Point Park in Lexington, Virginia. Though the water was up about a foot, the river seemed just as enticing as it had always been—flat and smooth. The young men jumped off the floating dock and swam toward the dam, a short distance downstream. They planned to leap from the dam into the plunge hole below, just as they and their friends had done many times in the past.

But when they neared the center of the river, the boys felt the current suddenly accelerate. Because of the previous day's rain, the current was much stronger than the smooth waters implied. Within seconds the river swept Charles and then Bryc over the dam. Bryc found himself tumbling in a hydraulic, a vertical vortex below the dam. After several terrifying seconds Bryc escaped and swam to shore. When he climbed out and looked around, his friend was nowhere in sight. Emergency personnel found Charles's body in the hydraulic nearly twenty-four hours later.[1]

Dams can be unsafe for reasons other than imminent failure. Dams like the one at Jordan's Point Park are known by many as drowning machines. Most are so-called low-head dams, no more than about a dozen feet tall with water spilling over their entire length. Such dams are usually near the center of town at sites of former industry. Though no longer used for their original purpose, these dams are treasured by locals for their recreational, historical, and scenic value. Unfortunately, the water pouring over low-head dams creates a hydraulic, a recirculating wave of tumbling water that can span the length of the dam. If the current is strong enough, the hydraulic can trap even the strongest of swimmers. Low-head dams are also a danger because people walking on them can fall and be injured or killed. In the period from the 1950s to 2015, at least 441 people died at low-head dams across the United States—more than those that died over the same period from dam failures. For example, a small dam in Wilmington, North Carolina, has killed eighteen people since 1984, including several children.[2]

Soon after the tragedy at Jordan's Point Dam, Charles's family and others demanded that the city remove the dam to prevent future tragedies. Their request was honored, but not until thirteen years later. Unfortunately, delays of this length or longer are common in the dam removal process. Why is it so hard to remove a dam, especially one that is a proven danger to people?

It's easy to imagine that removing a dam—especially a small one—could be simple. A few sticks of dynamite or several days of work with heavy equipment could get it done quickly. But dam removal is not so simple. Engineers must determine the safest and most effective way to take the dam down while minimizing collateral risks. And there are many risks. Removal can cause the upstream channel to deepen and threaten infrastructure such as bridges. Sediment from behind the dam can smother habitats downstream when released. Sometimes those sediments hold legacy pollutants that can harm people and wildlife. Some dams are barriers to the migration of non-native invasive species that would disrupt upstream ecosystems. Other dams offer refuge for endangered wildlife needing the swift currents that are found in the tailwaters (eliminated elsewhere in the river by reservoirs and channelization).

With so many concerns, dam removal planners must consult engineers, biologists, and other experts to identify risks and plan accordingly. Surveys are conducted. Samples analyzed. Models developed. Recommendations crafted. Plans drafted, reviewed, and revised. Environmental impact statements approved. Permits obtained. Though the process is complicated, there are numerous guides for dam removal available online for individuals, organizations, or agencies investigating the process.

And then there's the money. The planning process and the physical removal can be expensive, and obtaining the funding is often a challenge. Altogether, it is an exhausting process that can drag on for years, and sometime decades. But as difficult as this is, planning for the physical removal of the dam can be a cakewalk compared to liaising with the public on dam removal.

―⁂―

The process of proposing and planning dam removal usually involves consulting with members of the public to factor their concerns into decision-making. Planners often seek public input because it is required by one or more of the governmental entities involved in the process. This feedback is important because most streams are public property, and dam removal can affect many people. Moreover, the public often brings important issues to the attention of planners.

Conversations about removing dams are often contentious. Advocates for removal are usually motivated by safety or ecological concerns. Opponents often include property owners who live on the reservoir waterfront and people who use the reservoir for recreation. Sometimes a dam owner may be opposed to removal. Because opponents would lose a valued resource, those against dam removal are often the loudest voices in removal conversations. These disputes take time to resolve, and this is another reason why the dam removal process can be so lengthy.

There was substantial opposition to the proposal to remove the dams on the Elwha River. A local coalition, Rescue Elwha Area Lakes (REAL), argued that loss of inexpensive hydropower would threaten the survival of the paper mill in nearby Port Angeles. REAL was also concerned about the small flock of Trumpeter Swans that had begun overwintering on the reservoir above Elwha Dam. Dam removal advocates pointed out that the mill had an arrangement to buy power from Bonneville Dam at a discount, and that the swans had a healthy population in the region and were not endangered. Also fueling the opposition was growing resentment among locals who enjoyed the status quo and were increasingly frustrated with intervention by environmentalists and the federal government in local affairs. Some opponents seemed oblivious to the history of injustices against the Lower Elwha Klallam Tribe or the steep declines in salmon populations.

"They want to take over the whole peninsula and turn it into their version of Jurassic Park," one local retiree complained. "I consider myself an environmentalist. We want to see the Olympic Peninsula stay like it is." He continued, "The dams were a change for the good, despite what they did to the salmon. We've got lots of rivers. We don't have that many lakes."[3]

There was also opposition to removing the dam at Jordan's Point Park in Lexington, though resistance didn't arise until just before the dam was removed. Soon after the 2006 drowning, the city posted signs at the park warning swimmers to stay away from the dam. As the city considered dam removal, it attempted to find the dam owner. To its surprise, the city discovered that it was the owner. Lexington had acquired the dam when it bought the park property in the 1940s. Then, in 2007, the city hired an engineering firm to survey the dam. Inspectors found that the structure was unstable and dangerous. The dam was about a century old and was leaking substantially. One crack spanned the entire upstream face, and the dam was gradually sliding downstream. The engineers warned that dam failure was inevitable unless the city stabilized the structure. However, dam rehabilitation would cost $3 million. This was far more money than the small town of seven thousand residents could easily afford.

For the next ten years nothing happened to the dam. During that interval, the victim's family sued the city for negligence because there had been no warning signs at the park about the dangers of the dam. Virginia's Supreme Court ruled in favor of the family in 2011. After that, talk about the dam and its future subsided. Then, in 2017, a decade after engineers had reported that the dam was unsafe, Virginia's Department of Conservation and Recreation informed the city that it either had to restore the dam or remove it. The cost of removal was over $200,000, less than 7 percent of the cost of dam rehabilitation. As city officials contemplated the choice before them, controversy over the dam's future erupted.

Safety advocates wanted to prevent future accidents and an inevitable dam failure. Environmentalists advocated for reconnecting the downstream section of the Maury River to its network of headwater streams. Though there would be one remaining dam on the Maury, just above its confluence with the James River, removal of the Jordan's Point Dam would connect fifty-six miles of downstream fish habitat to 1,084 miles of upstream headwater habitat. Because of these benefits, the state's Department of Game and Inland Fisheries offered to pay for the removal with funds provided by the US Fish and Wildlife Service and the Virginia Wildlife Grant Program.

Opponents to removal argued that the dam was a valuable link to the city's history and that the dam's rushing waters added beauty to the town's riverfront. Others didn't want to lose the 1.2-mile-long impoundment or the swimming hole at Jordan's Point Park. The Virginia Military Institute in Lexington also opposed removal because it used the pond for cadet training. Altogether, this diversity of viewpoints—both pro and con—illustrates how deeply complex issues of dam removal can be.[4]

After hearing the public's concerns, the Lexington city council voted in 2017 to remove the dam. Money was the deciding factor. Restoration would require an expensive municipal bond that the city would be paying off for decades. Typically, municipal bonds pay for investments that stimulate economic growth and revenue; dam restoration would do none of this. Plus, Lexington would still be legally vulnerable if there was another accident or fatality at the dam. Conversely, removal would cost the city nothing.

Ultimately, fiscal prudence and safety prevailed over convenience, aesthetics, and sentimentality, and the city removed Jordan's Point Dam in late spring 2019. The removal crew spared the millrace and the outer margins of the dam to preserve some of the historical structure.

Social scientists who have studied the dam removal process have found that every project is unique, and, thus far, only a few generalities can be made. One of their findings is that the interaction between removal advocates and opponents during community dialogues often determines whether a bid to remove a dam is successful. Social scientists have also observed that when discussions are contentious, what doesn't work well is a top-down, authoritative approach to dialogue with the public and other stakeholders. Removal advocates, especially those who do not live in the community, can expect significant resistance if they tell dam owners and local stakeholders what is best for them. Such a paternalistic approach can elicit resistance simply as a matter of pride. But social scientists warn that even a more horizontal, inclusive approach to decision-making about the future of a dam is no guarantee for success. Local politics are notoriously volatile, and a few loud and influential oppositional voices can shut down a removal proposal.[5]

When stakeholders have diverse and strongly held opinions about a dam's fate, the success of a dam removal proposal can hinge on the structure of community dialogue. Cooperation and transparency are key to success in such a context. Community leaders, local governments, state and federal agencies, businesses, advocacy groups, and concerned residents—all must be willing to engage in discussion and be forthright about their goals and concerns. Decision makers should integrate diverse viewpoints into their planning. Participants can help by listening more than they speak. Adversaries should interact so they know each other as people, not stereotypes. When leaders promote healthy community dialogue, the process encourages trust, collaboration, and creative problem-solving.[6] Regardless of the outcome, the community comes out stronger.

Despite being the epicenter of aquatic biodiversity in the United States, the Southeast lags behind several other regions in dam removal. From 2000 to 2017, the West Coast and the Northeast removed ten and twenty-nine dams per year, respectively. Over that same period the Southeast averaged only five and a half dams removed per year. At this rate the Southeast will have removed 1 percent of its dams by the year 2062, and just 2.4 percent by the end of this century.[7]

Though southeastern states have removed relatively few dams, we can be hopeful based on how biodiversity has responded where stream connectivity has been restored. For example, in 2004, a partnership of organizations and agencies removed Marvel Slab, one of two small dams on the Cahaba River in central Alabama. In the months that followed, biologists documented five fish species above the dam site for the first time. They even found that snails migrated upstream past the dam site, though, as one might expect, at a much slower pace.[8]

The removal of small dams is vital to the restoration of rivers and aquatic biodiversity in the Southeast. Within months of the 2004 removal of Marvel Slab on the Cahaba River in Alabama, five fish species were documented above the dam site for the first time. Courtesy of Paul L. Freeman

The most noteworthy achievements in the Southeast have been those instances in which all dams on a river have been removed, restoring full connectivity. When Embrey Dam in Virginia was removed in 2004, the Rappahannock became the only river to flow unconstrained between the Appalachian Mountains and Chesapeake Bay. Migrating eel, shad, and river herring numbers in the river have risen ever since.

The 2017 removal of Milburnie Dam in Raleigh fully restored fish passage to North Carolina's Neuse River. It was an accomplishment that was decades in the making and entailed the removal of five other dams on the Neuse and its tributaries. Water quality has improved, paddlers are using the river, and American Shad are rediscovering their spawning grounds. Most importantly, the Neuse River is now safer. Milburnie Dam was a drowning machine that had taken fifteen lives before its removal.[9]

One southeastern community has capitalized on its dam removals. In 2012, demolition teams removed two unsafe dams on the Chattahoochee River, where it passes between Columbus, Georgia, and

366 PART V

Two decrepit milldams on the Chattahoochee River were removed in 2012 and replaced with a whitewater course that attracts thousands of tourists a year to Columbus, Georgia. Courtesy of Explore Georgia.

Phenix City, Alabama. Working with whitewater recreation experts, engineers replaced the dams with a whitewater course that now attracts thousands of tourists a year for tubing, kayaking, and rafting. Business is booming and reinvigorating the local economy.

Dams are not the only stream barriers preventing the movements of aquatic creatures or reducing the ecosystem services that streams can provide. Civil engineers often pipe small streams beneath roadways through concrete or metal pipes known as culverts. Culverts are a much cheaper alternative to bridge building. However, they often discharge stream water many inches above the stream surface. This creates a de facto dam, preventing upstream migration of fishes and other aquatic wildlife. Narrow culverts can create currents too swift and channels too dark for upstream passage. Conversely, oversize culverts can make streams so shallow that fish cannot pass. There are millions of culverts in the Southeast, and they are especially troublesome for aquatic wildlife high in the watershed, where bridges are unnecessary. One study found that culverts blocked access to 33 percent of Brook Trout spawning habitat in one Appalachian Mountain watershed.[10]

Culverts can also cause problems for people due to climate change and urban sprawl. Many culverts installed years ago cannot handle present-day stormwater loads caused by new development higher in the watershed and the increase in heavy rainfall events attributable to climate change. Stormwater can then flood the roadway and, in extreme cases, wash out the road.

In another example of a win-win for people and wildlife, civil engineers can replace dysfunctional culverts with right-sized pipes or broad arches that allow stormwater and fishes to pass with ease. Transportation departments can install these upgrades during routine road improvement projects. Often, state or federal wildlife agencies can share the cost.[11] Culvert reform is a new conservation tool, and most southeastern transportation departments haven't adopted this as a standard practice. But we need efforts like these if we want the benefits a healthy watershed can give.

Because of culvert replacement and removal of small dams, connectivity is gradually improving on streams in the Southeast and elsewhere in the United States. These projects are not just saving lives and protecting infrastructure; they are restoring the ecosystem services that free-flowing rivers can provide, reuniting fragmented populations of aquatic wildlife, improving water quality, and opening spawning habitat for migratory fishes. All are key pieces of the river revival the

Southeast needs. Accelerating the removal of small dams and installing stream-friendly culverts are among the most important conservation actions needed right now in the Southeast.

But what about big dams on big rivers? Restoring connectivity on large southeastern rivers such as the Alabama, Apalachicola, and Savannah is also necessary for reviving the ecological integrity of the region's watersheds. Large dams on the lower sections of these rivers control access to and conditions within the vast majority of aquatic habitats in the region. They block the largest fish migrations. Their reservoirs prevent migration between populations in adjacent headwater streams. Large dams also withhold the bulk of sediments needed for land and marsh building along the coasts, and the fresh water needed to push back the salty tides corroding our river deltas.

Culverts such as this one are as much a barrier to fish migration as dams are. Fortunately, they can be replaced with structures allowing stormwater and fishes to pass with ease. Courtesy of Alan Cressler.

Although we have many ecological reasons to call for the removal of large dams on southeastern rivers, there is one consideration that should give us pause: the large dams of the Southeast produce renewable energy. Given that we have only a decade to significantly reduce our greenhouse gas emissions to avoid a dangerous future climate, is now the time to remove hydropower dams? Restore vital river ecosystems or avert extreme climate change—it is a nasty dilemma, but one for which there may be a solution.

46

The Future of Hydropower

THE SOUTHEASTERN RIVER REVIVAL CANNOT be complete until we remove all dams—especially the largest—from the region's rivers. Only then will the Southeast enjoy all the ecological and economic benefits free-flowing rivers can provide, including the recovery of endangered species, the return of fish migrations, the recharging of our aquifers, and the delivery of fresh water and sediments so dearly needed at the coast.

However, because of climate change we are at a critical moment in the history of our species and life on the planet. Within the next few years, we must substantially reduce greenhouse gas emissions to avoid unleashing the worst that climate change can deliver. Transitioning quickly to a reliance on clean energy will be key to this, and most large dams on southeastern rivers and elsewhere in the United States are hydropower dams producing clean energy. So, what should we do? Save river species and ecosystems by removing large hydropower dams, or keep these dams to help prevent extreme climate change? The hydropower industry believes it has the right answer.

Hydropower is not merely a mode of energy production. It is a full-blown industry consisting of specialized equipment manufacturers, engineering and construction firms, environmental consultants, legal teams supervising everything from land acquisition to licenses, and hundreds of utilities. Like other powerful industries such as coal, petroleum, and agriculture, the hydropower industry has evolved its own culture, complete with a worldview focused on the benefits of hydropower and the expansion of its production. Though hydropower supplies less than 7 percent of the energy consumed in the United States, the industry is eager to expand hydropower production in the country and internationally by building new dams and refurbishing old ones.[1] In the United States, the National Hydropower Association represents those interests. It is a nonprofit group of over 240 member companies. The association aims to create more demand for hydropower in the United States and abroad by influencing legislation and policy at the state and federal level. The association spent over $4.7 million on lobbying for new hydropower projects between 2010 and 2019.

Apparently, this was money well spent. As a result of these and parallel efforts internationally, this century has seen a surge in the planning of new hydropower projects around the world. More new dams are planned for the next few decades than the total count of all the dams built in the twentieth century. Most are in developing countries eager to modernize their economies. For example, 2,200 dams were planned or were under construction in South America in 2006, with most of those in Brazil.[2] Typically, the countries building these hydropower dams lack access to inexpensive fossil fuels for generating power, and they missed out on the previous century's rush of dam building. And, some of these countries or regions within them do not receive enough wind or sunlight to support the development of these alternative renewable energy sources.

Modern dam building in developing countries has a proven record of improving agricultural and economic growth, reducing flooding, and lifting standards of living for millions of people in newly electrified cities. However, these dams have also created big problems, including enormous cost overruns, deforestation, and losses of archaeological sites and biodiversity. Dam construction has accelerated climate change, due to the fossil fuels used in construction and the methane produced by reservoirs.[3] New hydroelectric dams in developing countries have also displaced tens of millions of rural people in recent years. Regrettably, governments have relocated many of the displaced to lands poorly suited for their former lifeways. Others have moved to cities, where they were unwelcome because of their poverty and cultural differences, and the competition for jobs they created. And for each displaced person, reservoirs have disrupted the lives of another eight who lived downstream, often through loss of food and income when river fisheries collapsed. A further injustice is that downstream rural communities often do not receive any of the electricity produced by the dams. Despite international outcry over these impacts, the hydropower industry and national governments are still pushing ahead for new dams. And with 37 percent of the world's largest rivers still undammed, the boom in dam construction could continue for decades.[4]

The growth of international hydropower presents a quandary like the one we confronted at the beginning of this chapter. Developing countries urgently need the economic and technological development that hydropower can facilitate. But, damming the last free-flowing rivers of the world will cause irreparable harm to biodiversity, ecosystems, and rural and Indigenous peoples. Plans for damming in the world's largest watersheds, especially the Amazon, Congo, and Mekong, are particularly worrisome to scientists who study the impact of dam projects. These tropical rivers harbor much of the world's freshwater biodiversity, and the dams planned for these watersheds will drive hundreds of species to extinction. Moreover, damming will displace more people and eliminate fisheries on which millions of people in these regions rely for protein and income.

To address concerns about the expansion of hydropower in developing countries, many river experts are promoting a strategy called sustainable hydropower. The idea is that in watersheds where hydropower is deemed necessary, a limited number of new dams can be placed in well-chosen positions to optimize hydropower production and minimize negative impacts.[5] The approach is possible because dams vary in how much power they can produce and how much ecological damage they cause.

For example, consider a simplistic hypothetical scenario where ten planned dams would capture 100 percent of the hydropower capacity of a watershed. Building all these dams would reduce stream connectivity in the watershed by 100 percent, and thereby cause numerous extinctions and a major loss of ecosystem services. Using sustainable hydropower planning, dams are dropped from the plan that would cause above-average levels of ecological damage and produce below-average levels of hydropower. The resulting sustainable hydropower plan would recommend that only five dams be built. This would limit the loss of stream connectivity to 40 percent, yet 70 percent of the watershed's hydropower capacity would still be captured. The decrease in hydropower capacity could then be replaced by investments in wind or solar energy production and improvements in energy efficiency.

Using the sustainable hydropower approach, river scientists have modeled different numbers and placements of dams to determine the optimal arrangement for new dams in the Amazon and other major watersheds of the world where fleets of new dams are planned. If the countries planning to dam these watersheds adopt the recommendations of these scientists, they would reduce the loss of ecosystem services on which tens of millions of people depend and avoid the extinctions of hundreds, perhaps thousands, of freshwater species. However, it remains to be seen whether the countries planning these hydropower projects will heed these recommendations.[6]

What about the expansion of hydropower here in the United States? Will the National Hydropower Association get its wish to see waves of new dams built on our remaining free-flowing rivers? Hydropower policy experts think not, for several reasons. First, dams already occupy the best hydropower sites in the United States. The remaining sites are remote or lack the capacity to produce much power with existing technology.

Climate change presents another challenge to the feasibility of new hydropower. In California and other mountainous regions of the western United States, rivers fed by annual snowmelt are less dependable for producing hydropower because snowfall is declining. Elsewhere rainfall and surface water availability are becoming less predictable. We've already seen this in the Southeast as droughts rise in frequency and are more intense. Climate change is also causing southeastern summers to become longer and hotter. Demand for electricity peaks in the summer due to air conditioning use, but hydropower production already declines at this time because rainfall is scarce, reservoirs lose water through evaporation, and water use for irrigation increases. This is why several southeastern states import electricity from other regions each summer.[7] Building new hydropower dams in the Southeast would not alleviate these problems.

The most formidable challenge to new hydroelectric dams in the

United States is the boom in wind and solar energy production. Wind and solar farms are much cheaper to build than hydroelectric dams, and utilities can build them faster and closer to where they are needed than they can with dams. Furthermore, the price of energy produced in the United States from new onshore wind farms and new dams is nearly equal, and onshore wind power is expected to undercut the price of energy from new hydropower in the early 2020s.[8] Thus, building new dams is neither a timely nor affordable way to build more clean energy capacity in the United States.

We are now ready to circle back to our original question: should we keep our existing hydropower dams or remove them? The sustainable hydropower approach can help us answer this question. Whereas in developing countries sustainable hydropower is useful for optimizing where to build new hydroelectric dams, here in the United States it can help us identify which hydroelectric dams should be removed. In fact, sustainable hydropower is already saving migratory fish species on the Penobscot River watershed in central Maine.

Earlier, we learned about the Howland Bypass, a fishway built in 2016 to allow Atlantic Salmon and other endangered fishes to circumvent a dam in Howland, Maine. The bypass was one of several major projects completed this century to save migratory fishes in the Penobscot watershed. Fish troubles began a century ago, when nine hydroelectric dams were built in the Penobscot watershed. The dams crippled the spawning migrations of salmon, American Eel, Atlantic and Shortnose Sturgeon, American Shad, and Alewife. Fishways helped a few migrants surmount the lowermost dams, but they were not enough to sustain the fish populations. Consequently, for many decades the dam owners sparred with environmental organizations, the Penobscot Indian Nation, and wildlife agencies about improving fish passage.[9]

All that changed when PPL Corporation bought the Penobscot dams in the early 2000s. Instead of perpetuating adversarial relationships with the other stakeholders, the utility cooperated with them to find a long-term solution. After several years the partners agreed to a plan. To restore fish migrations, they removed Great Works and Veazie

Dams. Fishways on remaining dams, including Howland Dam, were upgraded or installed. To make up for lost power generation capacity, the plan allowed PPL Corporation to discharge more water at the remaining dams in the watershed. Today, fish populations in the Penobscot watershed are on the rise because two thousand miles of stream habitat have been reconnected to the coast. The utility is producing as much electricity as before, and towns along the river are celebrating river restoration with whitewater canoe races and fish festivals.[10]

The Southeast needs similar sustainable hydropower planning for its large river systems. We can begin by identifying dams that are unsafe, inefficient, and too expensive to maintain. Next, we could identify dams whose removal would provide the most ecological benefits. Cross-referencing these lists would reveal which of the region's large hydroelectric dams should come down first. Most of them will be our oldest dams, and there are scores of them to choose from. Of the 380 hydroelectric dams in the Southeast, 30 percent were over a century old as of 2020. By midcentury, this percentage will more than double.[11] Meanwhile, as the region plans for removing these old dams, state governments and utilities should invest rapidly and heavily in solar and wind energy production to replace the lost energy production capacity and, more importantly, help stave off catastrophic levels of climate change.

The closing of a hydropower station in the Florida Panhandle suggests that the renewable energy market is already hastening the transition from existing hydropower to new wind and solar power production. Talquin Dam is a sixty-six-foot-tall structure built in 1928 on the Ochlockonee River. The State of Florida maintains the dam to provide a recreational reservoir. The nearby City of Tallahassee leases the dam and has a multiyear license from the Federal Energy Regulatory Commission to operate the dam's hydropower station. With the 2022 relicensing deadline approaching, Tallahassee opted to surrender its license for purely financial reasons. Producing power from the dam costs the city $85 per megawatt hour, but a new solar power plant the city is building will produce electricity at $50 per megawatt hour. Another consideration in the decision to end hydropower production was that Talquin Dam supplied less than 1 percent of Tallahassee's electricity.[12]

We will see more utilities abandon hydropower production in the coming years as the demand for renewable energy increases, the price of energy produced by other renewable energy technologies declines further, and aging hydropower dams become safety and financial liabilities. Ideally, this transition will be coupled with the removal of these dams for the sake of restoring the region's rivers. No such luck for the Ochlockonee River. For now, the state plans to maintain Talquin Dam to preserve the reservoir. But given that the dam is almost a century old, one wonders how long Florida can afford to do this.

Despite this trend, hydropower will be an economically viable source of power in the United States for years to come, especially at younger dams with low maintenance costs that are still in the prime of their planned operational life spans. Furthermore, engineers are developing and promoting hydropower innovations that do not require new dams. As part of its climate change response, the Obama administration funded investment in increased hydroelectric capacity, but not solely by building dams. Though a handful of dams were built, much more new hydropower capacity came from refurbishing and upgrading existing dams. Another method was to add hydropower turbines to dams built for other purposes or to existing irrigation conduits and canals.[13] Given the need for clean energy like hydropower, it would be good policy to continue similar investments until we can produce enough power from other clean energy sources.

Alternative forms of hydropower are also gaining attention. In pumped-storage hydropower, utilities pump water from a reservoir behind an existing dam to a storage basin at a higher elevation. The pumping is done when demand for electricity is low and dams are producing a surplus of power. When demand is high, utilities release the water from the storage basin to spin hydropower turbines during its return to the reservoir. Thus, a pumped-storage system works like a ginormous rechargeable battery.

Utilities have used pumped-storage hydropower for over a century. The Southeast provides over 50 percent of the nation's pumped-storage capacity, more than any other region. This includes ten pumped-storage facilities: three in South Carolina, four in Georgia, and one each in Tennessee, Virginia, and North Carolina (Virginia has a second facility outside the geographic scope of this book). All are in the uplands of the region, where it is easier to store water nearby at an elevation higher than the turbines.[14]

To meet the rising demand for renewable energy, investors have proposed nearly seventy new pumped-storage facilities across the United States in recent years, including a handful in the Southeast. But new projects are expensive, have large land footprints, and can be incompatible with social and environmental values. Perhaps the worst of the proposed projects is on Navajo Nation lands along the Little Colorado River, upstream from Grand Canyon National Park.[15] The plan calls for damming four side canyons to create storage basins. Not only would this be an injustice to the Navajo, who do not want this facility on their lands, but it would destroy a large expanse of the canyon's ecology and natural beauty.

A completely new form of hydropower production in development is the hydrokinetic turbine. These turbines sit directly in a river and are spun by the river's natural current. In theory, this technology could generate power without interrupting the river's flow. There would be no dam, no reservoir, and fishes could migrate around the turbines. There are technical, economic, and environmental challenges that designers must resolve before these technologies would be ready for deployment. But if successful, hydrokinetic turbines in rivers would augment production of renewable energy.[16]

Given all the trends influencing hydropower, southeastern rivers could look much different by midcentury. We will have removed aging and inefficient hydropower dams on larger rivers. Power production at some remaining dams will be impractical due to the changing climate, and they will be in the queue for removal. Dams built in the late twentieth century that are safer and more efficient will enter their final decades of operation, after having played an important role in the first years of the Southeast's transition to reliance on clean energy. Wind and solar farms in the region will be producing far more energy than hydropower dams ever could have, and they will be doing so with little of the ecological destruction caused by dams.

As a result of these changes—especially the dam removals—there will be much more connectivity within southeastern watersheds. Some of the region's rivers will reach the coast without interruption. Sturgeon, shad, eels, and other migratory fishes will be thriving. Bolstered by the release of more fresh water and sediment from free-flowing rivers, barrier islands, oyster reefs, marshes, and swamps will be expanding at the coast and helping us adjust to sea level rise. And due to widespread culvert replacement and small dam removal, headwater streams in numerous watersheds will be reconnected to one another and the list of endangered aquatic species in the region will shorten. The southeastern river revival will be well underway.

47

Escaping the Flood Trap

THE SOUTHEAST NEEDS A RIVER revival to rebuild wildlife populations, restore fisheries, and help defend our coasts against sea level rise. Given the impact of large hydroelectric dams on watersheds, their removal must be part this effort, preferably through the implementation of a sustainable hydropower strategy. However, not all large dams were built to produce electricity. Flood-control dams account for 9.5 percent (or 1,844) of southeastern dams in the National Inventory of Dams.[1]

Flooding is dangerous and costly. Over the past decade, US fatalities due to flooding averaged ninety-five per year, and the trend is upward. Flood damages in the nation now average $8 billion per year. And due to new rainfall patterns caused by climate change, the number of billion-dollar flooding events in the United States, including the Southeast, is rising.[2]

Based on these statistics one might conclude that we need to invest in more flood-control infrastructure. However, a growing number of flood experts say the opposite is true. They warn us that we have unwittingly built a flood trap for ourselves. By constructing flood dams and other flood-control structures and then moving into flood-prone areas, we have made flood disasters inevitable.

Clearly, we must get out of the flood trap. Flood experts argue that we must do several things that could initially seem radical or counterintuitive: start seeing floods as good, move out of their way or adapt to living with them, and tear down many of our flood-control structures. To understand the flood trap and why these strategies should be part of the solution, it helps to know a little about the history of floods and flood policies in the United States.

European colonists began building the flood trap the moment they stepped off the boat. The yielding, fertile soils next to East Coast rivers

were irresistible. Corn and other crops grew quickly and supplied food security; tobacco and other cash crops provided wealth. Fish and game were abundant, and the level terrain made construction and road building easy. Rivers offered transportation, an endless supply of water, and a means of waste disposal.

But these lands were floodplains, low areas adjacent to the river into which floodwaters spilled when rivers rose and overtopped their banks. In the first years after settlers arrived, floods were rare because adjacent forests absorbed the rain. But as more settlers arrived and deforested more of the landscape, even moderate rains caused rivers to swell. Flooding intensified further as stream valleys filled with sediment eroded from cropland. Floods took lives, inundated or washed away homes, and ruined crops. Despite the threat of floods, those who settled on the floodplains rarely abandoned their property; land ownership was their means to economic security.

The flood trap expanded in the late nineteenth and early twentieth centuries as the US population grew and riverside settlements

Flood events in the Southeast are rising in frequency and intensity due to the new rainfall patterns caused by climate change. The only defensible solution involves changing how we live alongside and manage rivers. Courtesy of Alan Cressler.

increased in number and size. As roofs and streets hardened landscapes, urban stormwater contributed to the frequency and severity of flooding. High ground became valuable real estate where it was available. Economics and discriminatory policies forced the poor, African Americans, and immigrants to settle in the flood-prone neighborhoods of urban areas. When flooding came, hundreds died at a time. Freeport, Texas, 284 fatalities in 1899. Willow Creek, Oregon, 225 fatalities in 1903. Statewide flooding in Ohio, 467 fatalities in 1913.[3]

Communities tried to engineer their way out of the flood trap by building levees, long earthen dams separating rivers from their floodplains. Congress assisted by authorizing the US Army Corps of Engineers to build hundreds of miles of levees along rivers used for interstate commerce. But this only made flood traps deadlier. Emboldened by new levee systems, people moved closer to the river and into places with a history of flooding. However, river currents continuously strip away levee sediments, and river water can force its way through tiny fissures that can expand into full-blown leaks. These erosive processes intensify during floods.

Due to inadequate design and poor maintenance, numerous levees failed when big floods came. One of the nation's worst floods was on the Mississippi River in 1927, after days of intense rainfall across the lower Midwest and Central South. Despite reassurances from the corps that the levees would hold, failures all along the Mississippi and its major tributaries caused flooding that killed 246 people and displaced hundreds of thousands of others. Fatalities were underestimated because the worst flooding occurred in poor African American communities in the Lower Mississippi Valley, which were largely ignored before, during, and after the flood. The flood destroyed vast expanses of agriculture and interrupted the transportation networks distributing food throughout the United States. Several other destructive floods came in the following years, coinciding with the beginning of the Great Depression.[4]

With their disruptions to food supply and commerce, this series of floods convinced Congress that flooding was an economic and security threat to the nation that deserved federal attention. The Flood Control Act of 1936 was one of a series of laws enacted by Congress to help modernize the nation, stimulate the economy, and reduce unemployment during the Depression. The act gave the federal government broad authority to tackle flooding and tasked the corps with getting the job done. In the following decades, the corps constructed newer and more formidable networks of flood protections. The corps raised thousands of miles of levees. It dredged and straightened meandering rivers to move water downstream more quickly. And the corps built hundreds of dams to capture floodwaters and release them gradually.[5]

These projects led to a significant reduction in flooding. To most of the public it seemed the corps had tamed the rivers, and, predictably, more people settled on floodplains, just as had happened decades earlier, when the corps built the first levee systems. The midcentury migration escalated after World War II because cities needed space as the postwar economy boomed and populations surged. Few remembered or knew of the devastating floods from decades earlier. Furthermore, most citizens assumed that local governments would not allow them to live in unsafe areas.[6]

But this assumption was often flawed. Local governments make decisions about where developers can build. Across the United States there is a long history of developers successfully lobbying local politicians to approve their plans for new developments in unsafe areas, especially floodplains. All too often, politicians oblige because building new homes and businesses attracts more customers for local businesses and creates jobs. And city officials have reason to feel secure about their decisions when new levees or flood dams protect these new neighborhoods. As for the developers, after the final sale they move on, with few legal or financial ties to what happens next.[7]

It was inevitable that many of these new neighborhoods would flood. Sometimes levees failed because they were poorly designed or maintained. But even where levees were in good condition, sometimes a flood came that was too powerful for them to restrain. Similarly, flood dams could only safely hold back a limited amount of water before having to release that water downstream.

Eventually it became clear that engineering couldn't provide enough protection from floods. The federal government's response was to pivot from flood prevention to disaster management. Congress created the Federal Emergency Management Agency (FEMA) in 1979 to step in when local and state governments cannot adequately respond to disasters. For FEMA to help, a governor must declare a state of emergency and request federal disaster assistance. If the president approves emergency action, FEMA provides expertise and resources to help with the initial disaster response. FEMA also helps with rebuilding by offering payouts to qualified homeowners and businesses for damages.[8]

FEMA also took over the National Flood Insurance Program (NFIP). Congress created the NFIP in 1968 to pay policyholders for flood damages to buildings and property that private insurance would not cover. The program is only available to participating communities that have agreed to take steps to mitigate flood risk. To administer the program, FEMA maintains maps of flood zone risk for populated areas across the United States. In high-risk zones of participating communities, mortgage lenders require home and business owners to buy federal flood insurance to supplement their private insurance policies. Sometimes lenders also require those living in low- and moderate-risk flood zones to purchase flood insurance. If private insurance is not required or available, they have the option to purchase NFIP coverage.[9]

Today there is broad consensus across the political spectrum that FEMA's flood insurance program and disaster payout system are making the flood trap worse because these policies enable homeowners to reside on high-risk, flood-prone properties. Furthermore, because total federal flood insurance claim awards exceed the collection of insurance premiums, Congress must make up the difference from the federal budget. This means that taxpayers living on high ground are helping homeowners rebuild after floods in areas that are likely to flood again. In many cases, the NFIP has paid for the restoration of the same home after multiple floods, and the sum of these payouts exceeds the value of the home.

Another frequent complaint is that FEMA's flood plain maps are often out of date and do not reflect how flooding risk has changed. Moreover, issues of social equity have also come to light. Researchers have found that FEMA's postdisaster payouts and buyout programs favor wealthy and white communities more than other communities.[10]

Congress should fix FEMA's flood-related programs, but politicians are reluctant to push for reforms that might anger constituents in high-risk areas.[11]

As federal policies prolong occupancy in flood-prone areas, the nation's flood-control infrastructure is deteriorating from insufficient maintenance, and this is making the flood trap more perilous. Though the US Army Corps of Engineers built many modern levees, local and state governments maintain most of them. Unfortunately, local governments have a checkered history of levee inspections and upkeep.[12] The American Society of Civil Engineers periodically evaluates and grades the condition of the nation's infrastructure. In 2021 the society reported on the condition of thirty thousand miles of the nation's levees. While levees protect the lives of seventeen million people and $2.3 trillion in property, the US levee system earned a D on the report card. The society estimated that we need to spend $21 billion to get the levees back into good shape. Similarly, the nation's network of dams tracked in the corps' National Inventory of Dams also received a D. Disturbingly, among the country's high-hazard dams—those whose failure would likely cause fatalities—the society found that over 2,300 were structurally deficient due to lack of maintenance. The number of dangerous dams is rising over time, yet the society reports that we already need to spend $20 billion on high-hazard dams to address existing safety concerns. We need to spend another $73 billion to rehabilitate the nation's remaining dams. The importance of dam and levee maintenance became clear in 2019, when floods broke through dams and levees across the Mississippi River Valley. These floods devastated portions of nineteen states and caused $20.3 billion in damages and a dozen deaths.[13]

Governmental policies promote our vulnerability to flooding, and flood-control infrastructure is deteriorating. This is enough to sound the alarm. But thanks to climate change, the flood trap is getting more dangerous. Our existing flood-control infrastructure is poorly suited for the new weather patterns. Engineers designed our dams and levees for the relatively predictable and gentle climate of the Holocene geological epoch. But the Holocene is over. The warmer atmosphere of the Anthropocene epoch holds more moisture than before, and this is causing storms to drop more rain and deliver it in less time. The result is more flooding now than in the past, and the forecast for this century is that this situation will get worse, not better.

We could attempt to engineer our way out of the trap with more and taller levees and dams. However, history warns us that this will make the flood trap worse. New and refurbished infrastructure would encourage more urbanization in flood zones, putting more people and property at risk. And these projects would cost us hundreds of billions of taxpayer dollars over the next decades, money that could be better spent elsewhere (e.g., by investing in renewable energy production). Furthermore, we'd be building infrastructure for a climate future that is currently unknowable. Though rainfall amounts and storm intensity are increasing, we don't yet know how bad it will get. Thus, we don't know how high new levees and flood dams should be. These are compelling reasons to find an alternative solution to escaping the flood trap. Flood experts say that this begins with us changing how we live in floodplains.

The first solution to the flood trap is so obvious that there shouldn't be a need to say it: we need to move out of floodplains. Floods are only a problem because we put ourselves in their path. Then when floods come—and eventually they do—we describe them as *natural* disasters without any hint of irony.

But because flooding is inevitable and getting worse, more civil engineers, river managers, and planners are advocating that we move out of high-risk, flood-prone areas. Their recommendations have mostly focused on lightly to moderately populated areas. Retreat from densely populated areas with substantial infrastructure is much less practical, affordable, and popular.

Numerous communities across the United States and in the Southeast are heeding this advice and have begun the process of withdrawing from high-risk floodplains. The process is like managed retreat in response to sea level rise along the coast. Willing home and business owners in designated high-risk areas sell their property to the government at fair market values. In situations where flooding has been particularly bad, FEMA has helped fund these buyout programs. However, because local governments must pay 25 percent of the cost, lack of start-up funding has curtailed the participation of small cities and rural counties.[14]

Charlotte, North Carolina, one of the largest cities in the Southeast, has a successful buyout program. After a series of moderate floods in the mid-1990s, the city, adjacent towns, and the county created a joint stormwater utility to reduce flooding and flood damages. In 1999, with funding from FEMA, the city began buying homes, apartments, and businesses in high-risk areas. As of 2019, the city had bought 400 structures and relocated 700 families to higher ground. The city then demolishes structures on the property and revegetates the land to help capture floodwater and allow rainwater infiltration. Charlotte has restored over 185 acres of floodplain, some of which has been converted into greenways and community gardens. Two decades the program began, FEMA supplies only 43 percent of the funding. The remainder comes from local sources, including stormwater service fees assessed on properties based on the amount of impervious surface maintained by a home or business (which contributes to flooding). By 2019 the stormwater utility was spending $4 million a year on buyouts and had saved local governments over $25 million by avoiding emergency rescues, disaster relief, and payments to the NFIP. The city predicts long-term savings to top $300 million for the properties bought thus far.[15] Retreating from the floodplain doesn't just save lives and avert disaster; it is fiscally responsible.

Retreat is not the only option. Some communities and property owners are coping with river flooding by adopting the same strategies that work in coastal communities adjusting to sea level rise. In Charlotte, home and business owners in high-risk areas have elevated their structures, filled basements, and waterproofed below the flood line. The stormwater utility promotes adaptation by monitoring the region's hydrology, offering stormwater education, maintaining and upgrading stormwater infrastructure, funding stream restoration projects, and ensuring that new floodplain developments in high-risk areas are designed with flooding adaptations. One of the towns in the county has become one of the few in the Southeast to adopt low-impact development (LID) into its building ordinances. Recall that LID's goal is to keep rainwater near to where it falls to reduce stormwater flooding.[16]

Moving out of floodplains or adapting to recurrent floods are smart steps, but they are local solutions. Extricating ourselves from the flood trap will also require us to change how we manage water at the scale of the watershed. But to do this, we will need to begin valuing floods for the good things they can do.

Floods are natural events that can benefit us and river ecosystems. While floodwater lingers over the land, much of it seeps belowground to recharge groundwater and aquifers that help forests and farmers during dry periods. Floodwaters bring silt that replenishes soil fertility and offsets subsidence caused by the settling of soft floodplain sediments over time. Flooded wetlands alongside rivers are wildlife habitat and food-rich nurseries for juvenile fishes of commercially valuable species and their prey. Floods refresh river habitats such as sand- and gravel bars, where some fish species lay their eggs. Floods bring fallen trees into the river that become habitat for river wildlife. And importantly, floods transport sediments to the coast to help with building the swamps, marshes, and barrier islands that provide critical protection against sea level rise and storm surges.

If we want to enjoy the ecosystem services that floods can provide, then we must allow rivers more freedom of movement. We can begin by restoring floodplains. Floodplains are like natural dams. They help mitigate floods by slowing down and storing water. Because of these and the other benefits noted above, governments have begun restoring floodplains across the United States. The process involves buying or leasing floodplains from willing owners and breaching or removing levees to reconnect the river to its floodplain. If the floodplains are agriculture, floodplain managers often plant trees because floodplain forests absorb more water and provide more wildlife habitat than do open fields. Another effective approach is to partially restore a floodplain by moving levees back from the river's edge. These levee setbacks reduce maintenance costs by minimizing the levees' exposure to strong river currents. These savings help pay for the restoration. Levee breaching and setbacks create habitat for river wildlife and reduce the risk of

flooding fatalities and property damage because less floodwater flows downstream.[17]

Another way to allow rivers more freedom of movement is to only use dams to withhold floodwater where flooding is severe. Removal may be prudent for some dams because of their age and condition. For example, nearly 40 percent of southeastern flood dams were at least seventy years old in 2020.[18] Because most flood dams can only minimize floods over a relatively small area, the cost of repairing or replacing them as they deteriorate can be high relative to the benefits. It will be much less expensive to fund relocation programs, modify infrastructure to sustain flooding, and restore floodplains. Communities doing this will enjoy cost savings, better safety, and the ecosystem services only a free-flowing river can provide. It's a win-win for people and

The seasonal inundation of floodplains reduces flooding in downstream settlements, recharges groundwater reserves, replenishes soil fertility, and provides wildlife habitat. Floodplain restoration is essential to a secure and sustainable future in the Southeast. Courtesy of Alan Cressler.

biodiversity. These initiatives will not be cheap, but the alternative—enduring floods, upgrading and building new flood-control infrastructure, and paying for rescue and recovery—is even more costly.[19]

River restoration projects at these scales usually involve study, planning, and coordination because any major change to river management affects communities upstream and downstream. Towns along the Mississippi River and its major tributaries have learned this the hard way. In some areas, levees built by upstream communities cause more flooding in downstream communities because levees shunt more water downstream during floods.[20] The smart approach is to address flooding at the watershed level. Flood planning should be part of a comprehensive process involving multiple tiers of government and engaging all stakeholders affected by proposed changes in river management. Such coordination can be challenging, but we need such efforts to restore rivers and help communities escape the flood trap.

Proposals to allow rivers more freedom to flood could seem reckless, given the number of people who die in floods each year or the billions in property damage they cause. But considering the history of flooding, our declining infrastructure, and an era of increasingly destructive storms, this is the smart choice. Sure, there are places where we will need to maintain or build new levees and flood-control dams to protect essential infrastructure and culturally important sites. But moving out of floodplains, adapting to floods, and restoring a river's ability to flood are examples of the proactive strategies we need in this new era of rapid climate change. In combination with flood-reduction approaches such as low-impact development and forest restoration, these strategies will help us escape the flood trap. The longer we delay implementation, the more lives will be lost and the more taxpayer dollars will be spent rebuilding infrastructure that is destined to be flooded again.

48
Lake Life

In July 2020, my wife, Ginger, and I rented a cabin in the woods for a few days near Menlo, Georgia. We had been hunkered down since the COVID-19 pandemic began, and a short vacation with our youngest daughter into the mountains on the border of Alabama and Georgia seemed like a safe way to unwind. The cabin was a short walk from Lake Lahusage, a recreational reservoir at the confluence of the Little River's Middle and East Fork. The Little River is one of a few rivers in the United States perched entirely on a mountain. A few miles downstream from the lake, the river has carved a canyon replete with waterfalls, whitewater, pools for fishing and swimming, massive sandstone cliffs, and lush montane forest.

On one of our morning adventures, Ginger and I slid into a tandem kayak that came with the rental and paddled down the East Fork and onto Lake Lahusage. Though we passed dozens of waterfront houses, it was a Monday morning and the community seemed deserted. Most houses were humble shacks that had been there for decades. Others were newer, larger, and loaded with amenities. A few showed signs of continual occupancy, but most appeared to be for weekend or vacation retreats. Nearly every property had a dock, and many had pontoon boats moored alongside. Yards sported colorful combinations of kayaks, paddleboards, canoes, skiffs, paddleboats, firepits, grills, patio furniture, plastic swans, and pink flamingos. At one cabin a six-foot-tall wooden Sasquatch watched us with suspicion from beneath a shady oak. The only visible resident at another house was a mangy Blue-and-Gold Macaw perched on the deck railing. The only people we saw were two fishers and three teenage swimmers.

Despite the scarcity of people on Lake Lahusage that weekday morning, recreational lakes are very popular in the Southeast. During the warm months, particularly on weekends, these lakes buzz with activity. For many people, lakes provide a more convenient and inexpensive alternative to a coastal vacation. Lakefront homes and lakeside communities are affordable for middle-class retirees, and some can live closer to family and friends than if they were on the coast. In recent years, some coastal retirees have moved inland to lakeside communities, after surviving one too many hurricanes.

The problem with recreational lakes is that nearly every one is a reservoir behind a dam. So, as we prepare for a future with fewer dams, we must also prepare for the day when there are fewer recreational lakes. This makes the discussion of dam removal in the Southeast and elsewhere in the United States more difficult. These conversations would be much simpler if the issues were only about finding alternative means of controlling floods or producing electricity and drinking water. Most people don't know, or even care, how floods are prevented, how their power is generated, or where their water comes from. They simply want safe communities and reliable and affordable utilities.

The issue of recreational dam removal is not trivial. Of the 19,343 southeastern dams in the US Army Corps of Engineers' National Inventory of Dams, 70 percent were built primarily for recreation, and another 10 percent provide recreation as a secondary benefit.[1] The most prominent recreational lakes are large reservoirs bordered by lakeside communities. Nearly all these reservoirs are behind hydropower dams on mainstem rivers or major tributaries. Planning for the communities around these lakes often began when state governments granted or sold land to the utilities and investors who would build the dam. Extra land around the planned reservoir was usually part of the deal. When the reservoir filled, the surrounding land became valuable waterfront property that was sold to the public by the new owners. Lakeshore communities then sprang up in what had been sleepy, rural landscapes. The new concentration of permanent and semipermanent residents and tourists attracted businesses—from bait and tackle shops to boat dealerships to groceries and restaurants. Some lakeside settlements grew into towns, and some towns grew into cities. Given the economic development that now flourishes around recreational lakes, people commonly assume that these lakes are permanent features of the southeastern landscape. However, the future of these lakes is far more tenuous than most realize.

~~~

Our July paddle on Lake Lahusage eventually brought us to the dam that made all the fun possible. The structure was in bad shape. While the dam was originally designed to also serve as a bridge, the roadway across the reservoir had collapsed for a third of the dam's 244-foot length. Concrete slabs littered the tailwaters, where portions of the bridge had fallen and shattered on impact. The remaining dam was showing signs of further deterioration. Chunks of concrete had fallen away to reveal iron reinforcement bars corroding within. Erosion caused by heavy rains was evident around one side of the dam—another sign of instability.

Beneath the remaining bridge was a concrete landing with an opening that faced downstream. We dragged our kayak onto the landing to have a look around and stretch our legs. Hanging from the ceiling were concrete stalactites—signs that water is slowly degrading the concrete. Vandals—probably local children—had painted over the walls with graffiti. A few short steps away from the graffiti was a vertical drop of at least twenty-five feet to rubble below. There was no guardrail. Nor did we see any signs posted on the dam warning boaters or swimmers of danger.

I later did some casual research on the dam. According to the US Army Corps of Engineers, an investor built the dam in 1925 to create a reservoir for recreation. Today, no one owns the structure, but the local homeowner association maintains the dam. But at ninety-five years of age, in an obvious state of deterioration, and with no one legally responsible for its upkeep, how much longer will this dam survive?

The Lahusage Dam's condition illustrates a problem we first confronted a few chapters ago: US dams are aging quickly. Of the southeastern dams in the corps' inventory coded as providing recreation, 54 percent were over fifty years old in 2020, and 10 percent were over seventy-five years old. These tallies do not include several thousand recreation dams for which the construction date is unknown.[2]

The outlook is worse for the large dams built primarily for

hydropower that create the region's largest recreation reservoirs. In 2020, 49 percent of them were at least seventy-five years old, and one in five was at least a century old.[3] Utilities will begin closing many of these dams in the coming years as they require more maintenance. And as we saw earlier with the Talquin Dam near Tallahassee, Florida, the plunging cost of electricity from solar and wind technologies will also force utilities to close hydropower facilities in the coming years.

The awkward issue that no one seems to be discussing is what will happen to lakeside communities when a utility or other dam owner is ready to end its use of the dam. Power utilities will probably not be willing or financially capable of maintaining dams after decommissioning. Could they turn the dams over to local or state governments? Maybe. But should governments accept old dams and assume maintenance costs, liabilities, and the inevitable expense of removal? Plus, considering that relatively few citizens enjoy the benefits of recreational reservoirs and that only a few people use these reservoirs during the winter and on weekdays, this may not be a fair or reasonable way to spend taxpayer dollars.

Could a business, trust, or homeowner association take ownership and pay for upkeep by charging fees to those living near and using the lake? Possibly. But will lake residents and users be willing to pay enough to cover the expenses and liabilities of dam ownership? Who pays the bill if the dam must be removed or replaced for safety reasons?

There is a chance that climate change will decide the future of some lakeside communities. As we learned earlier, dry summers and droughts are a regular feature of the southeastern climate, and climatologists predict both will increase in intensity and frequency in the coming decades. Drone footage from Lake Lahusage just after the 2016–17 drought shows water levels so low that boat ramps and lake house docks were high and dry. Boats rested askew among tall weeds on the bank.[4] Similar scenes played out in reservoirs across the Southeast. Dry reservoirs frustrate lakeside residents and tourists and hurt local businesses. Low water levels also concentrate pollution and trigger harmful algal blooms and fish kills. If these conditions increase in frequency, lakeside residents and visitors may choose to live and play elsewhere, and there may be no point to maintaining a recreation dam.

In the new southeastern climate, low reservoir levels are predicted to increase in frequency. Logan Martin Lake, Alabama (*pictured*), was one of many reservoirs that dried out during the drought of 2016–17. Note the stranded boat at left. Courtesy of Alabama Rivers Alliance.

For all the above reasons, it seems inevitable that we are approaching a time when recreational reservoirs will begin to disappear. When this happens, what do we do then? Most former reservoirs will be inappropriate for building new infrastructure because they might be prone to flooding and because lake bed sediments will be unstable. Instead, we can replace recreational lakes with new, restorative landscapes that revive nature and provide new forms of recreation.

Much of this transformation would happen naturally if we allowed it. Soon after water levels drop, seeds brought by animals and the wind will sprout, and a forest will begin growing. We could speed along the reforestation with tree plantings. Within just a few years a young forest on the former lake bed would be rustling with wildlife. As the forest grows, it will pull carbon dioxide out of the atmosphere and help mitigate climate change. At the same time, a resuscitated river will begin offering the ecosystem services that only a free-flowing river can provide. More fresh water and sediment will flow downstream to restore river and estuarine habitats and help defend coastal communities from sea level rise. Fishes will migrate to new spawning habitats and bring with them young mussels to settle on the river bottom. The recovering landscape could become a public greenway managed by a land trust or a local or state government. Walking and biking trails could wind through the forest and alongside the stream. Canoeists and kayakers could float the river while anglers cast for fishes. On the hills above, former lakefront property will retain value because it borders a nature preserve that attracts new residents to the area.

Two days after Ginger and I explored Lake Lahusage by kayak, we visited Little River Canyon Falls Park, just over the border in Alabama. The park is the gateway to a seventeen-mile-long canyon managed by the US National Park Service. The river flows freely throughout the canyon, beginning with a forty-five-foot waterfall that on this day was roaring due to heavy rains the night before. The rushing water was mesmerizing, so we lingered in the shade of a pine to enjoy the scene. We were not alone. Other visitors watched the river from the boardwalk overlooking the falls. Nearby and at a safe distance upstream from the falls, children played in the shallow rocky flats while adults relaxed in or near the water and chatted.

We then hiked on a cliffside trail to a smaller waterfall less than a mile downstream. There were fewer people here, but there were signs the site was a beloved local hangout. A trio of teenage boys swam in a side channel and scaled large boulders the high waters had transformed into islands. A middle-aged couple arrived for the day with folding chairs and a cooler. A squad of kayakers floated past and, one by one, safely dropped through a small channel around the falls.

Though free-flowing rivers offer many forms of recreation, the swimming, boating, and fishing experiences on a lake are different. And for some people there is a form of serenity that only the still waters of a pond or lake can evoke. But every dam has its own life span, and this includes the Southeast's recreation dams. In a few cases, we will replace these dams so their reservoirs can provide leisure opportunities far into the future. We will remove others to save money, restore rivers, or protect the public. At least a few recreation dams will fail during the strong floods the region is now enduring, while others may become ineffectual if they cannot retain enough water as the climate warms.

Regardless of the reason for its disappearance, the loss of a recreational lake will affect many people. Suffering the most will be those for whom the lake was a regular feature in their lives. While there is no way to fully replace such a loss, perhaps it will help people in affected communities to know that they are not alone. Lifeways and landscapes are changing quickly across much of the planet in this new era of rapid global warming. In times like these, it helps to find satisfaction and joy in new ways as the world around us changes. With time, perhaps those who will mourn the loss of lake life will discover joy in exploring the former reservoir as it fills with forest, birdsong, and flowing streams.

PART VI

49
Where Do We Go from Here?

NEAR THE BEGINNING OF THIS book, we set out to answer several questions. The first was, *how did we get here?* We learned that the environmental problems of our day arose, in part, because until recent decades we lacked an understanding of how we interact with the natural world. We didn't know, for example, that we could harvest enough sturgeon to cause the fishery to collapse, or that if we pour too much waste into a river it will sicken people downstream and destroy vital ecosystems. We didn't understand that marshes and floodplains save lives by absorbing floodwaters. There wasn't even a clear concept of "ecology" or "the environment" until the mid-twentieth century. Since it originated in the 1960s, environmental science has struggled to catch up. Industry heads in a new direction, then science follows to document the consequences.

We now have an empirical and sweeping understanding of how much we have changed the planet and how it is affecting us. Three things are now clear. First, humanity is inescapably dependent on the environment for survival, no matter how far removed from nature our lives may seem to be. We need healthy ecosystems for the food, water, air, and most other resources we use each day. Second, we have been poor stewards of Earth. Through our overconsumption of natural resources and the waste we generate, we are rapidly degrading the very ecosystems that sustain us. Fresh water is overused and polluted. Forests are disappearing, wildlife is vanishing, soils are eroding, ocean chemistry is changing, and the climate is warming. We are committing ecological suicide. Third, the people harmed first and most by the degradation of nature are the ones who benefit the least from the destruction, if they benefit at all. Here in the Southeast and around the world, the poor and people of color are exposed to far more pollution than wealthier and whiter communities.

As a matter of fairness and self-preservation, we must begin practicing new ways of thinking and living. Most importantly, it should become reflex to consider how our choices affect everyone, not just ourselves or those we care about most. We must minimize our role in perpetuating injustices that span social boundaries, geography, and time. And we must extend these concerns to nature, including the other species with whom we share this planet. We rely on them and the ecosystems they create for our day-to-day survival. Thus, caring for other species isn't just the right thing to do; it is the smart thing to do.

If we fail this challenge, our future will be difficult. While this is the same warning issued for decades by environmentalists and scientists, the situation today is different. The science is clear that we are rapidly degrading the planetary systems on which we depend, and we must not procrastinate a response any longer. Most importantly, if we do not cut greenhouse emissions significantly over the next few years, then we'll raise Earth's temperature so much that we will trigger a future of brutal heat, intense droughts and wildfires, violent storms, rapid sea level rise, drowning of coastal cities, declines in agriculture, and waves of species extinction. This will be more than enough to destabilize cultures and governments around the world for centuries to come.

Fortunately, the world is mobilizing a response to the climate change threat. Most countries of the world are signatories to the Paris Agreement, a 2015 accord within the United Nations Framework Convention on Climate Change. Under the agreement, countries are voluntarily reducing greenhouse gas emissions to keep global warming below 3.6°F (2°C), and preferably below 2.7°F (1.5°C), relative to pre-industrial levels. Climate warming beyond these levels will trigger the catastrophic and irreversible changes to the planet noted above. To avoid this outcome, signatory countries aim to cut greenhouse emissions by 45 percent by 2030 (relative to 1990 emission levels) and reach net-zero emissions by 2050.[1]

With our current dependence on fossil fuels, reaching these goals will not be easy. We must reengineer how we live on the planet. This will require us to develop and adopt sustainable practices in all major sectors of the economy, especially energy, transportation, manufacturing, forestry, and food production and distribution. Altogether this will be an unprecedented amount of transformation over a short period of time, but this transition is essential to ensuring that humanity has a hopeful future.

Planetary changes, international agreements, and policy and economic reforms—these topics can seem remote and impersonal. After all, most of us have very little influence on anything at these large scales. But all the changes we need in the coming years must happen locally. It's in our communities where citizens, neighborhoods, businesses, and local governments will take steps to overcome the social and environmental challenges we face. Just as our planetary-scale problems arose from the collective impact of local actions, planetary-scale solutions will arise from the collective influence of our local choices.

Here in the Southeast, much of what we need to do involves the other two questions we set out to answer early in this book: *will the Southeast have enough water in the Anthropocene?* and *can southeastern aquatic biodiversity survive the Anthropocene?* We learned that the answer to each of the above questions is yes. Despite declining water supplies due to climate change this century, the Southeast can have enough water if there is widespread adoption of tried-and-true strategies to conserve fresh water. As for the region's aquatic biodiversity, we learned that much of it can survive if we act quickly. Moreover, it is in our best interest if biodiversity in the region and throughout the world survives because we rely on other species and the ecosystems they create. Fortunately, during our journey we found numerous solutions to the problems of climate change, declining water supply, and biodiversity loss. But given what's at stake and how little time remains to avoid a full ecological meltdown, how should we prioritize these solutions to maximize positive impact?

Without a doubt, our most important priority right now is to slow climate change by ending our reliance on coal, natural gas, and petroleum to generate electricity and fuel transportation. This is the top priority because the world is careening toward a future of runaway global warming with devastating consequences. The solution is to quickly transition to a reliance on solar and wind energy and other sources of clean energy. Every region must do its part, including the Southeast. The United States ranks second among countries for carbon dioxide emissions, and southeastern states produce a fifth of national emissions. Furthermore, the transition to clean energy is in the region's own best interest: climate change is already harming the southeastern economy more than any other US region.[2]

Thus far, investment in wind and solar energy in the Southeast has lagged behind several other US regions because state governments have sheltered the fossil fuel industry and utilities behind protective policies. However, several southeastern states and regional utilities have read the proverbial writing on the wall and begun the transition to renewable energy. Virginia was the first southern state (and the only southern state thus far) to commit to net-zero carbon emissions by 2050. North Carolina and Florida rank second and third among US states for total solar power capacity, respectively, and Georgia ranks thirteenth. Virginia and North Carolina are developing offshore wind farms. While climate change plays a role in these initiatives, thus far the transition to renewable energy in southern states is mainly driven by the low cost of producing renewable energy and the jobs the transition creates.[3]

Ending our use of fossil fuels will also help conserve the region's water supply and protect its aquatic biodiversity. Coal- and gas-fired power plants extract more river water in the Southeast than does any other sector of the economy, including agriculture. Wind and solar (photovoltaic) energy production consumes very little water. So, after we shutter fossil fuel plants, there will be more river water available for using in other ways and for supporting aquatic biodiversity and ecosystems. Abandoning fossil fuels will also end the ecological destruction created by coal, oil, and gas extraction, and the pollution and waste created by transporting and burning these materials. People and ecosystems will be healthier and safer.

The transition to clean energy is an example of multisolving, where one action solves multiple problems. Multisolving is a powerful approach because it boosts the incentive for an action and motivates people with diverse priorities to support that action. Fortunately, as highlighted throughout this book, we have in our tool kit many multisolving strategies for addressing water insecurity, biodiversity decline, and other environmental and social problems.

Stormwater mitigation strategies such as these stepped bioswales should become a commonplace feature of our urban landscape to reduce flooding and pollution, make cities cooler, and improve aesthetics. CC BY NC 2.0 NACTO (National Association of City Transportation Officials).

For example, we need policies to incentivize landowners to preserve existing forests and restore forests where possible. Forested landscapes capture carbon dioxide, reduce flooding, recharge groundwater reserves, and provide habitat for plants and animals. Farmers can multisolve by adopting practices (e.g., the use of conservation tillage) to minimize the need for irrigation, reduce pollution in nearby streams, and return carbon to the soil. We also need multisolving in our urban areas. Stormwater mitigation strategies such as building greenways that double as stormwater basins will reduce flooding and pollution, keep cities cooler, promote healthy lifestyles, improve neighborhood aesthetics, raise property values, and attract new residents. By improving the livability of cities, greenways also reduce urban sprawl, which is a growing threat to watersheds and rivers.

Demographers predict that by 2060, urban sprawl in the Southeast could expand by as much as 200 percent relative to 2010.[4] Sprawl isn't just bad for commuters; it accelerates global warming, pollutes rivers, aggravates flooding, and replaces forests and farms. But cities can curb sprawl by providing the essentials for healthy urban living. This includes affordable housing, high-quality public education, upward mobility for the urban poor, reliable mass transit, bike paths, and walking trails. These services and amenities must be available to all residents, including those in historically underserved urban neighborhoods. To prepare for water shortages and keep the cost of water low for ratepayers, cities should follow Atlanta's example and adopt water efficiency and conservation policies. Initiatives like these will create vibrant cities that attract residents, businesses, and investors. People will be happier, and watersheds will be healthier.

Coastal cities can multisolve to adapt to rising sea level and the rising destructiveness of cyclones. Restoring salt marshes and other coastal ecosystems will help protect communities from flooding, improve water quality, restore wildlife populations, and sustain tourism and fisheries. Where coastal flooding is inevitable, yet armoring and adaptation are impractical, cities can launch managed retreat programs to help residents move out of danger. These initiatives will save lives and money, and abandoned lands can be managed for recreation and wildlife conservation.

Our river management strategies should also adopt the multisolving approach. Rivers and their floodplains can be reconnected to reduce downstream flooding, recharge aquifers, improve water quality and soil fertility, and provide wildlife habitat. We should begin removing small dams, especially those whose removal will help the recovery of endangered species and fish migrations. Where possible, we should remove multiple dams on a single river or within a network of headwater streams to maximize the restoration of ecosystem services provided by free-flowing streams. For example, reconnecting the headwaters will help restore mussel populations so these filter feeders can help clean our streams of pollution. As the benefits of stream connectivity become more evident to the public, the burden of proof will shift from conservationists, who currently must explain why a dam should be removed, to dam owners, who will have to justify why their dams should remain.

It's also time to begin planning the removal of large dams. It will take many years to plan each removal project, given the biological, engineering, social, and financial challenges involved. Thus, we should

begin planning now so that we can remove as many as possible over the next few decades. We can use the sustainable hydropower approach to identify those dams that we should remove first. Near the top of the priority list should be the large dams that pose the greatest safety risk to communities downstream and whose removal will restore endangered fish migrations. These removals will also provide the fresh water and sediment needed by coastal ecosystems to keep up with sea level rise.

Large hydropower dams that are younger, safer, and still useful will likely remain on southeastern rivers well into the second half of this century. These dams will help us reach carbon emission reduction targets by supplying clean energy. Until the day we remove these dams, we should manage them to restore some of the ecosystem services that free-flowing rivers can provide. With environmental flows,

Continued urban sprawl, such as this in metropolitan Atlanta, will hasten global warming, destroy creeks, pollute rivers, aggravate flooding, and replace forests and farms. Courtesy of Alan Cressler.

dam operators can release water to mimic seasonal flow patterns that sustain river ecosystems and wildlife. Additionally, we can install appropriately designed fishways on these dams to rebuild migratory fish populations.

The above solutions and others emphasized in this book should be part of the southeastern river revival. If we commit to the revival, we must abandon the extractive and destructive practices we've used for centuries. In their place we must adopt socially inclusive and ecologically regenerative practices. If we succeed, the Southeast will be safer and more prosperous than it has ever been in the four hundred years since the Euro-American juggernaut began at Jamestown.

For sure, the river revival will require considerable effort and investment from us. But the alternative—attempting to maintain the status quo as the environment becomes more threatening and chaotic—will be far more costly and disruptive. The only reasonable choice is to roll up our sleeves and work together to build a world in which all people and species enjoy a safe and prosperous future.

50
Homecoming

NEARLY EVERY MORNING AND AFTERNOON in Gulf Breeze, my father, Bob Duncan, makes a pilgrimage to The Point, that finger of land jutting deep into Pensacola Bay—the same sacred place of mine where we began our journey to understand the southeastern river crisis. The daily walks keep him fit, but they are mostly about birds. The Point has seen more bird species than most places in North America, and my parents have decades' worth of observations and field notes to prove it.

It is late May, I've submitted final grades, and I'm down for a visit from Birmingham. I'm here for some relaxation, but also to help my parents with repairing some of the damage to the dock caused by the most recent hurricanes. This morning I join Dad for the walk. Spring bird migration is mostly over, but we might find a straggler or, better, a vagrant far from home. But birds are a secondary interest for me. I enjoy the companionship and the visit to a place that's so dear to me.

My mother, Lucy Duncan, tosses aside her morning chore list, grabs her binoculars, and joins us. She walks to The Point with Dad often during peak bird migration, but less often in the warm months. I'm pleased she is with us. It's my final day of the visit and a last chance for the three of us to bird together. Like my father, she's an accomplished birder. With her perseverance and ace observation skills, she would have been a remarkable research scientist. Then again, as a teacher she built a legacy of which she can be proud. Her influence still grows through the workshops and field trips she organizes to teach the community about birds and nature, and through her behind-the-scenes environmental activism.

It's still early morning when we set out for The Point. Though it will be hot within the hour, the air temperature is still pleasant. Our walk is mostly in silence, in part to listen for birds, in part because I've been visiting for several days and we can enjoy one another's company without chitchat. I appreciate the silence because it helps me meditate on thoughts and feelings that are troubling me. Our journey of discovery in this book has been fascinating, but at times it has also been overwhelming. I want resolution, but with so much uncertainty on the horizon, closure is still elusive.

The Point is less than a half mile from the house, but we take a while getting there as we stop for birds and whatever else draws our attention. It's a route we've walked more times than anyone could tally, and through a landscape that has changed dramatically. Most of the oak forest and every one of the old houses that were here when I was a child are gone, replaced by lawns, exotic tropical plants, and immense houses. My childhood paths and secret places in the old landscape are now just memories that fade more each year.

We pass the lot where my grandparents lived when I was a child. It was a drafty wooden structure that was a fragment of the big Duncan family homestead built in the late nineteenth century. My grandmother sold the lot in the early 1990s, and the old house was torn down. I'm grateful I was off at college and not there to see it go. When the new owners built their home, they left most of the moss-draped oaks under which I played as a child. I whisper hello to them each time I pass.

I stoop down to examine soil that erosion has laid bare on the margins of a stormwater catchment basin. I learned as a kid that I could find long-buried treasures in such places, and today I wasn't disappointed. Between a rusted bit of iron and a piece of old glass I find a fragment of Native American pottery. It's small with no markings, but its composition and curvature are diagnostic. I've seen pottery sherds many times elsewhere, but this is the first I've found on The Point. It's a reminder that the full story of human occupancy here, and across the Southeast, is far more enthralling than anyone will ever fathom.

I've fallen behind Mom and Dad. They are up ahead, nearing an empty lot at the street's end that affords us access to the open water. I catch up and show them the sherd. They are as excited about it as I am, but none of us are surprised it is here. We know that Native Americans would have used The Point as a seasonal encampment or as a stopover between the mainland and the barrier island. Mom pockets the sherd so she can give it to a friend who is a local archaeologist.

We scan the open water and shoreline through our binoculars, pointing out birds and other sights to one another. The morning air is still, and the water is glassy. Small waves lap weakly along the narrow beach below the seawall. A pair of Barn Swallows swirl and glide around us, catching insects. A Great Blue Heron stands perfectly still, poised to snatch the next minnow that strays near.

The tide is covering most of the rubble at the tip of The Point, where we began this book, and no patches of sand remain to offer respite for migrating shorebirds. Waters have been continually high during my visit. I reflexively wonder whether this is a sign of sea level rise but remember that a few millimeters per year isn't discernable without precise measurements. More likely, these high tides are from shifting winds and currents pushing Gulf water into the bay.

Unfortunately for us, the lot we are on has recently been sold to a new owner. Surveyors have marked many of its oaks with orange paint for felling. I fear that by my next visit the site will be noisy with construction, and our last direct access to the tip of The Point will be off limits. It's sad, but this was inevitable. Even with sea level rising, this real estate is too valuable to sit idle.

So much has changed here at The Point already. Would the Native Americans who camped here recognize it today? What about my Uncle Dallas, who once strung his sturgeon nets just offshore? I glance over at Dad, who is still scanning the waters. How has all the change he's seen affected him? When he was a kid, The Point extended a half mile into the bay, and most of Gulf Breeze was swamp and scrub thicket. There are children growing up in the new homes here on The Point—what changes will they see in their lifetimes?

Change. It's the only constant. In the next few years everything will change because of global warming. I hope that much of this adjustment will be voluntary as we work together to craft a safe, inclusive, and prosperous future. At least some of this change will be burdensome, especially when it forces us to adjust our lifeways, transforms or destroys the places we love, and harms our friends and loved ones. Many in the Southeast and around the world are already suffering because catastrophes augmented by climate change have struck their communities. But even when we lose the people and places we love, we must somehow find joy in the world—if not, what's the point of sticking around?

Pensacola is certainly changing. The community is politically conservative, and for many here climate change isn't real. Nevertheless, I see a city adapting to sea level rise and stronger hurricanes. Construction crews are replacing the three-mile-long Highway 98 bridge crossing the bay. The new bridge is significantly taller and should be safe from storm surges. The city moved its wastewater treatment plant to higher ground after it was flooded during Hurricane Ivan in 2004 and raw sewage poured into the bay. Investors built a minor league baseball stadium on the waterfront and armored the shoreline with a formidable wall of rock and concrete. Elsewhere, walls of boulders have replaced most of the city's sandy shoreline. The city has protected a mile of waterfront with rock reefs and a restored salt marsh, and the project is a model for using green infrastructure to adapt to sea level rise. Ferries shuttle cyclists and pedestrians from Pensacola to the beach and to historic Fort Pickens, part of Gulf Islands National Seashore. This helps reduce traffic congestion, and the ferries can offer access to the fort when hurricanes wash out the road.

I think about our friend Ann, who lives several miles across the bay in Warrington. The air is clear, and through my binoculars I find her roof nestled among the shoreline pines. I worry about Ann and her neighbors. Like many areas along the southeastern coast, the land is low, riven with bayous, and vulnerable to flooding. Hurricane Ivan's storm surge inflicted terrible damage in Ann's neighborhood. Armoring can help a bit, but floodwater can seep beneath seawalls and rise through the ground and stormwater systems. Although her house is on pilings, just a couple of feet of sea level rise will cut her off from the mainland.

I also worry about my parents and how much more they will see of the changes coming this century. Unlike Ann, they and their neighbors have a lot more freeboard to resist rising bay waters, so daily tides won't threaten the homestead during their lives. But hurricanes are now delivering more rain, bringing stronger winds, and moving slower. It will be a difficult future for my parents and millions of others in the Southeast if the region is regularly pummeled by more Ivans, Florences, Michaels, and Harveys.

My parents, Lucy and Bob Duncan, at The Point, where our river journey began. Photo by R. Scot Duncan.

My brother and I will inherit the homestead someday, and we will eventually face tough decisions about its fate. We may find in our retirement that a future of endless hurricane recovery will be too financially and psychologically exhausting to withstand. But leaving would be wrenching. Much of our identity and history is bound to this land and the bay. We were the sixth generation of Duncans to live here. This is where we learned to walk, swim, and fish; where we've watched thousands of sunsets with family and friends from the dock; and where our parents introduced us to the joys of nature study.

Ultimately, I know that this hand-wringing doesn't deserve a drop of sympathy when the world is brimming with profound suffering, inequity, and injustice. If we abandon the homestead, it will be by choice and we'll still be financially secure. In stark contrast, the World Bank estimates that by 2050 there will be well over a hundred million people displaced from their communities by climate change if carbon emissions go unchecked. Few of them will have a choice, and all will endure years of hardship as they settle elsewhere.[1]

There are many days when I quietly wrestle with my fears about the future of humanity and biodiversity. While some of us understand what needs to happen in these first years of the Anthropocene epoch to ensure a safe and prosperous future, I'm not yet convinced humanity will rise to the challenge in time. Will we stumble ahead clumsily, or will we be agile and strategic? Will ours be a future of social and environmental mayhem, or one of transformation, healing, and renewal? I know I'm not alone with my anxiety. Many are asking these questions right now.

On most days, however, I am optimistic. We humans can be exceptionally compassionate and innovative. We know what needs to be done, and we can muster the resources to do it. And as time passes, more people are engaged on these issues. They are helping their neighbors, working for social and environmental justice, and pressuring their leaders to take action on climate change. On those days I'm convinced that a turning point is just ahead. It seems inescapable. Some social transformations incubate for a long time, then when enough people are on board new perspectives and approaches rapidly become normalized. I hope that day is coming fast. It must.

My parents are ready to go, and they turn toward home. As they walk into the shade of the oaks, I linger for a moment to savor one last look over the bay before turning to join them. It's time to move on. The sun is rising higher, it's getting hotter, and there's much work to do.

Notes

Chapter 1
1. Schwenning, Bruce, and Handley, "Pensacola Bay."
2. US EPA, *Environmental and Recovery Studies of Escambia Bay and the Pensacola Bay System Florida.*
3. Snyder, "Chemical Contaminants in Fish from the Pensacola Bay System and Offshore Northern Gulf of Mexico"; Karouna-Renier et al., "Accumulation of Organic and Inorganic Contaminants in Shellfish Collected in Estuarine Waters near Pensacola, Florida."

Chapter 2
1. Rockström et al., "Planetary Boundaries."
2. Zalasiewicz et al., "The Working Group on the Anthropocene," 56.
3. Díaz et al., *The Global Assessment Report on Biodiversity and Ecosystem Services.*
4. Steffen et al., "The Anthropocene."
5. Zalasiewicz et al., "The Working Group on the Anthropocene."
6. Zalasiewicz et al.
7. Rockström et al., "Planetary Boundaries," 1.
8. Intergovernmental Panel on Climate Change, *Climate Change 2014*, 45.

Chapter 3
1. Jenkins, "Amphibians of the World"; Elkins et al., *The Southeastern Aquatic Biodiversity Conservation Strategy.*
2. Elkins et al., *The Southeastern Aquatic Biodiversity Conservation Strategy.*
3. Díaz et al., *The Global Assessment Report on Biodiversity and Ecosystem Services.*
4. Ruckelshaus was quoted in *Business Week*, June 18, 1990.

Chapter 4
1. Wilson, *The Future of Life.*

Chapter 5
1. Hsiang et al., "Estimating Economic Damage from Climate Change in the United States."

Chapter 6
1. Dunlap and Martin, *Historic Pensacola Photographs.*
2. Oswalt et al., *History and Current Condition of Longleaf Pine in the Southern United States.*
3. NOAA (website), "History of Management of Gulf of Mexico Red Snapper"; Ivanov, "A Look at the History of Red Snapper Fishing in Pensacola."
4. Catarci, *World Markets and Industry of Selected Commercially-Exploited*

Aquatic Species with an International Conservation Profile, section titled "Sturgeons (Acipenseriformes)."

5. Boschung and Mayden, *Fishes of Alabama*.

6. Sulak et al., "Status of Scientific Knowledge, Recovery Progress, and Future Research Directions for the Gulf Sturgeon"; Hilton et al., "Review of the Biology, Fisheries, and Conservation Status of the Atlantic Sturgeon."

7. Cason, Mills, and Tully, *Historical Remembrances of Choctawhatchee Bay*.

8. Sulak et al., "Status of Scientific Knowledge, Recovery Progress, and Future Research Directions for the Gulf Sturgeon."

9. Hilton et al., "Review of the Biology, Fisheries, and Conservation Status of the Atlantic Sturgeon."

10. Saffron, "The Decline of the North American Species."

11. Hilton et al., "Review of the Biology, Fisheries, and Conservation Status of the Atlantic Sturgeon."

12. Hilton et al.

13. Huff, *Life History of Gulf of Mexico Sturgeon*.

14. Sulak et al., "Status of Scientific Knowledge, Recovery Progress, and Future Research Directions for the Gulf Sturgeon."

15. Huff, *Life History of Gulf of Mexico Sturgeon*.

16. Cason, Mills, and Tully, *Historical Remembrances of Choctawhatchee Bay*.

17. Boschung and Mayden, *Fishes of Alabama*.

18. Hilton and Grande, "Review of the Fossil Record of Sturgeons."

Chapter 7

1. Gornitz, "Sea Level Rise, after the Ice Melted and Today."

2. Steponaitis, "Prehistoric Archaeology in the Southeastern United States, 1970–1985."

3. Keel, Cornelison, and Brewer, *Regionwide Archeological Survey Plan*.

4. Keel, Cornelison, and Brewer.

5. Mann, *1491: New Revelations of the Americas Before Columbus*.

6. Steponaitis, "Prehistoric Archaeology in the Southeastern United States, 1970–1985."

7. Holzkamm and Waisberg, "Native American Utilization of Sturgeon."

Chapter 8

1. D. E. Duncan, *Hernando de Soto*.

2. Hudson, *The Southeastern Indians*.

3. D. E. Duncan, *Hernando de Soto*; Rangel, "Account of the Northern Conquest and Discovery of Hernando de Soto"; Bourne and Smith, *True Relation of the Vicissitudes That Attended the Governor Don Hernando de Soto*.

4. Davis, *The Gulf*.

5. Pringle, "How Europeans Brought Sickness to the New World."

6. Mann, *1493: Uncovering the New World Columbus Created*.

7. Saffron, "The Decline of the North American Species"; Garman, "Historical Significance of the Atlantic Sturgeon."

8. Smith, *Capt. John Smith.*
9. Erickson, "Decoding the Mystery of the Jamestown Sturgeon."
10. Mann, *1493: Uncovering the New World Columbus Created.*
11. McCartney, "Starving Time, The."
12. Mann, *1493: Uncovering the New World Columbus Created.*
13. Mann.
14. Mann.
15. M. S. Duncan, "Fall Line"; Manganiello, *Southern Water, Southern Power.*
16. Box and Williams, *Unionid Mussels of the Apalachicola Basin*; Manganiello, *Southern Water, Southern Power.*

Chapter 9

1. Rockström et al., "Planetary Boundaries."
2. Philpott, "A Brief History of Our Deadly Addiction to Nitrogen Fertilizer"; Mann, *1493: Uncovering the New World Columbus Created.*
3. Mann, *1493: Uncovering the New World Columbus Created.*
4. Melillo, "The First Green Revolution."
5. Bastias Saavedra, "Nitrate"; Mann, *1493: Uncovering the New World Columbus Created.*
6. Bastias Saavedra, "Nitrate"; Philpott, "A Brief History of Our Deadly Addiction to Nitrogen Fertilizer."
7. Mann, *1493: Uncovering the New World Columbus Created.*
8. Hergert, Nielsen, and Margheim, "WWII Nitrogen Production Issues in Age of Modern Fertilizers"; Philpott, "A Brief History of Our Deadly Addiction to Nitrogen Fertilizer."
9. Ritter, "The Haber-Bosch Reaction."
10. Cooper, "Historical Aspects of Wastewater Treatment."
11. Cooper.
12. Lemon, "The Great Stink."
13. US EPA (website), "Sanitary Sewer Overflows (SSOs)."
14. Clean Water Atlanta, "Wastewater."
15. USGS (website), "USGS 02337170 Chattahoochee River near Fairburn, GA."
16. Box and Williams, *Unionid Mussels of the Apalachicola Basin.*
17. Wein, "Risk in Red Meat?"
18. US EPA, *Environmental Impacts of Animal Feeding Operations*, 6.
19. S. Rogers and Haines, *Detecting and Mitigating the Environmental Impact of Fecal Pathogens*, 1.
20. Hribar, *Understanding Concentrated Animal Feeding Operations and Their Impact on Communities*; Burkholder et al., "Impacts of Waste from Concentrated Animal Feeding Operations on Water Quality."
21. USDA, *Poultry—Production and Value*; USDA, *2012 Census of Agriculture.*
22. CDC, "Body Measurements."
23. Hribar, *Understanding Concentrated Animal Feeding Operations and Their Impact on Communities.*

24. Harden, *Surface-Water Quality in Agricultural Watersheds of the North Carolina Coastal Plain.*

25. Blake et al., *The Deadliest, Costliest, and Most Intense United States Hurricanes from 1851 to 2004.*

26. Eggert, "Nature Averts a Disaster"; D. Johnson, "Hurricane Matthew Update."

27. Mallin et al., "Hurricane Effects on Water Quality and Benthos in the Cape Fear Watershed."

28. Rundquist, "Exposing Fields of Filth in North Carolina"; National Grain and Feed Association, "North Carolina Estimates Hurricane Florence's Ag Damage."

29. Schramm et al., "Human Health and Climate Change in the Southeast USA"; NOAA (website), "What Is a Harmful Algal Bloom?"

30. Backer, "Cyanobacterial Harmful Algal Blooms (CyanoHABs)."

31. Backer; US EPA (website), "Nutrient Pollution"; Schramm et al., "Human Health and Climate Change in the Southeast USA."

32. Diaz and Rosenberg, "Spreading Dead Zones and Consequences for Marine Ecosystems."

33. Gibbens, "Massive 8,000-Mile 'Dead Zone' Could Be One of the Gulf's Largest."

34. Diaz and Rosenberg, "Spreading Dead Zones and Consequences for Marine Ecosystems."

35. May, "Extensive Oxygen Depletion in Mobile Bay, Alabama"; Loesch, "Sporadic Mass Shoreward Migrations of Demersal Fish and Crustaceans in Mobile Bay, Alabama."

36. Schroeder and Wiseman, "The Mobile Bay Estuary"; May, "Extensive Oxygen Depletion in Mobile Bay, Alabama"; Loesch, "Sporadic Mass Shoreward Migrations of Demersal Fish and Crustaceans in Mobile Bay, Alabama."

37. May, "Extensive Oxygen Depletion in Mobile Bay, Alabama"; Osterman and Smith, "Over 100 Years of Environmental Change Recorded by Foraminifers and Sediments in Mobile Bay, Alabama."

38. NOAA, *NOAA's Estuarine Eutrophication Survey.*

39. Raines, "America's Amazon"; May, "Extensive Oxygen Depletion in Mobile Bay, Alabama"; Osterman and Smith, "Over 100 Years of Environmental Change Recorded by Foraminifers and Sediments in Mobile Bay, Alabama."

Chapter 10

1. US EPA, *Primer for Municipal Wastewater Treatment Systems.*
2. Powers, "Federal Water Pollution Control Act (1948)."
3. Hines, "History of the 1972 Clean Water Act."
4. Powers, "Federal Water Pollution Control Act (1948)."
5. Gallun, "Flushing Out Bubbly Creek."
6. Hines, "History of the 1972 Clean Water Act."
7. Powers, "Federal Water Pollution Control Act (1948)"; McThenia, "An Examination of the Federal Water Pollution Control Act of 1972."

8. Poe, *The Evolution of Federal Water Pollution Control Policies*.

9. Keiser and Shapiro, "Consequences of the Clean Water Act and the Demand for Water Quality"; Inglis et al., *Waterways Restored*; McThenia, "An Examination of the Federal Water Pollution Control Act of 1972."

10. Copeland, *Clean Water Act*; Poe, *The Evolution of Federal Water Pollution Control Policies*.

11. Council on Environmental Quality, *Environmental Quality*.

12. Inglis et al., *Waterways Restored*.

13. US Public Health Service, *Community Water Supply Study*.

14. Oberstar, *The Clean Water Act*, 6.

15. Qtd. in McThenia, "An Examination of the Federal Water Pollution Control Act of 1972," 195n; see also Rinde, "Richard Nixon and the Rise of American Environmentalism."

16. Copeland, *Clean Water Act*.

17. McThenia, "An Examination of the Federal Water Pollution Control Act of 1972."

18. Copeland, *Clean Water Act*; Inglis et al., *Waterways Restored*.

19. US EPA, *National Water Quality Inventory* (1975).

20. Mulligan, *Evolution of the Meaning of "Waters of the United States."*

21. Jenkins et al., "US Protected Lands Mismatch Biodiversity Priorities"; Elkins et al., *The Southeastern Aquatic Biodiversity Conservation Strategy*.

22. Mulligan, *Evolution of the Meaning of "Waters of the United States"*; Stanis, "Drain the Swamps"; US EPA and DOD, "Definition of 'Waters of the United States'"; US EPA, "EPA, Army Announce Intent to Revise Definition of WOTUS."

23. Howe, "Supreme Court Curtails Clean Water Act"; Totenberg, "The Supreme Court Has Narrowed the Scope of the Clean Water Act."

24. US EPA (website), "Wetland Extent, Change, and Sources of Change."

25. Keiser and Shapiro, "Consequences of the Clean Water Act and the Demand for Water Quality."

26. US EPA, *National Water Quality Inventory* (2017), 8.

27. Dodds et al., "Eutrophication of U.S. Freshwaters."

28. US EPA, *National Water Quality Inventory* (2017), 2.

29. Vaughn, Gido, and Spooner, "Ecosystem Processes Performed by Unionid Mussels in Stream Mesocosms"; Jordan, "Stanford Research Shows Value of Clams, Mussels"; Mizoguchi and Valenzuela, "Ecotoxicological Perspectives of Sex Determination."

30. Hu et al., "Detection of Poly- and Perfluoroalkyl Substances (PFASs) in U.S. Drinking Water"; US EPA (website), "Per- and Polyfluoroalkyl Substances (PFAS)"; Andrews and Naidenko, "Population-Wide Exposure to Per- and Polyfluoroalkyl Substances."

31. McThenia, "An Examination of the Federal Water Pollution Control Act of 1972."

32. US EPA, *Report to Congress*; US EPA, "Nonpoint Source Pollution"; US EPA (website), "Basic Information about Nonpoint Source (NPS) Pollution."

33. Copeland, *Clean Water Act*; US Government Accountability Office, *Nonpoint Source Water Pollution*.

34. Dodds et al., "Eutrophication of U.S. Freshwaters."

35. US EPA, "Nonpoint Source Pollution."

36. US EPA, *National Water Quality Inventory* (2017), 8.

37. Keiser and Shapiro, "Consequences of the Clean Water Act and the Demand for Water Quality"; USDA (website), "Agricultural Productivity in the U.S."; Stets, Kelly, and Crawford, "Regional and Temporal Differences in Nitrate Trends."

38. Dodds et al., "Eutrophication of U.S. Freshwaters."

39. Harned et al., *Trends in Water Quality in the Southeastern United States, 1973–2005*.

40. Stets, Kelly, and Crawford, "Regional and Temporal Differences in Nitrate Trends."

41. Harned et al., *Trends in Water Quality in the Southeastern United States, 1973–2005*; Litke, *Review of Phosphorus Control Measures in the United States*; McCoy, "Goodbye, Phosphates."

42. Ramankutty, Heller, and Rhemtulla, "Prevailing Myths about Agricultural Abandonment and Forest Regrowth."

43. Dodds et al., "Eutrophication of U.S. Freshwaters."

44. Sprague and Gronberg, "Relating Management Practices and Nutrient Export in Agricultural Watersheds of the United States"; USDA (website), "Soil Tillage and Crop Rotation."

45. Stavins, "What Can We Learn from the Grand Policy Experiment?"

46. Mehan, "A Symphonic Approach to Water Management."

Chapter 11

1. Vaughn, "Ecosystem Services Provided by Freshwater Mussels."

2. Gatenby, "Mussels Making Moves for Water Quality"; US FWS, "Pearly Mussels of the Susquehanna Basin"; Jordan, "Stanford Research Shows Value of Clams, Mussels"; Stutz, "Natural Filters."

3. Haag, *North American Freshwater Mussels*.

4. Paul Johnson, pers. comm., January 12, 2020.

5. P. D. Johnson and Gagnon, "Freshwater Mollusks"; US FWS (website), Environmental Conservation Online System; Tennessee Wildlife Resources Agency, "Fresh Water Mussels in Tennessee."

6. Haag, *North American Freshwater Mussels*.

7. Haag.

8. Haag.

9. J. D. Williams, Bogan, and Garner, *Freshwater Mussels of Alabama and the Mobile Basin*.

10. Haag, "Past and Future Patterns of Freshwater Mussel Extinctions"; Haag, *North American Freshwater Mussels: Natural History, Ecology, and Conservation*.

11. J. D. Williams, Bogan, and Garner, *Freshwater Mussels of Alabama and the Mobile Basin*.

12. Williams, Bogan, and Garner; Haag, *North American Freshwater Mussels*.
13. Haag, *North American Freshwater Mussels*.
14. Haag.
15. Haag.
16. Haag; Strayer et al., "Distribution, Abundance, and Roles of Freshwater Clams (Bivalvia, Unionidae)."
17. Haag, *North American Freshwater Mussels*.
18. Haag.
19. Haag.
20. J. D. Williams, Bogan, and Garner, *Freshwater Mussels of Alabama and the Mobile Basin*.
21. Haag, *North American Freshwater Mussels*.

Chapter 12
1. Wohl, "Legacy Effects on Sediments in River Corridors"; US EPA, *National Water Quality Inventory* (2017), 8; US EPA, *National Rivers and Streams Assessment 2008–2009*.
2. Mann, *1491: New Revelations of the Americas before Columbus*.
3. Mann, *1493: Uncovering the New World Columbus Created*.
4. Mann.
5. Mann; Manganiello, *Southern Water, Southern Power*.
6. US EPA (website), "What Is Turbidity and Why Is It Important?"
7. Ramankutty, Heller, and Rhemtulla, "Prevailing Myths about Agricultural Abandonment and Forest Regrowth."
8. R. S. Duncan, *Southern Wonder*; Nearing et al., "Natural and Anthropogenic Rates of Soil Erosion."
9. Daniels, "Soil Erosion and Degradation in the Southern Piedmont of the USA"; Nearing et al., "Natural and Anthropogenic Rates of Soil Erosion."
10. Duncan, *Southern Wonder*.
11. Donovan et al., "Sediment Contributions from Floodplains and Legacy Sediments to Piedmont Streams."
12. US EPA, *National Water Quality Inventory* (2017), 8; Natural Resources Conservation Service, *2007 National Resources Inventory*; Neary, Swank, and Riekerk, "An Overview of Nonpoint Source Pollution in the Southern United States."
13. Haag, *North American Freshwater Mussels*.
14. Haag.
15. Nearing et al., "Natural and Anthropogenic Rates of Soil Erosion"; Ramankutty, Heller, and Rhemtulla, "Prevailing Myths about Agricultural Abandonment and Forest Regrowth."

Chapter 13
1. Lacefield, *Lost Worlds in Alabama Rocks*.
2. Waldo, *America's Gibraltar, Muscle Shoals*; Haag, *North American Freshwater Mussels*.

3. City of Muscle Shoals, "The History of the City of Muscle Shoals"; Haag, *North American Freshwater Mussels.*

4. Garner and McGregor, "Current Status of Freshwater Mussels (*Unionidae, Margaritiferidae*) in the Muscle Shoals Area."

5. Billington, Jackson, and Melosi, *The History of Large Federal Dams.*

6. Box and Williams, *Unionid Mussels of the Apalachicola Basin.*

7. International Hydropower Association, "A Brief History of Hydropower"; Billington, Jackson, and Melosi, *The History of Large Federal Dams*; Manganiello, *Southern Water, Southern Power.*

8. Billington, Jackson, and Melosi, *The History of Large Federal Dams*; Manganiello, *Southern Water, Southern Power.*

9. Billington, Jackson, and Melosi, *The History of Large Federal Dams.*

10. Atkins, *Developed for the Service of Alabama.*

11. Billington, Jackson, and Melosi, *The History of Large Federal Dams.*

12. Billington, Jackson, and Melosi.

13. Billington, Jackson, and Melosi; Ezzell, "Wilson Dam and Reservoir."

14. Billington, Jackson, and Melosi, *The History of Large Federal Dams.*

15. Kaetz, "Muscle Shoals."

16. Postel and Richter, *Rivers for Life.*

17. Postel and Richter; Graf, "Dam Nation."

18. Benke, "A Perspective on America's Vanishing Streams."

19. National Inventory of Dams, "Dams of the Nation."

20. Manganiello, *Southern Water, Southern Power.*

21. Southeast Aquatic Resources Partnership, "Southeast Aquatic Barrier Prioritization Tool"; Ignatius and Stallins, "Assessing Spatial Hydrological Data Integration to Characterize Geographic Trends in Small Reservoirs."

22. Haag, *North American Freshwater Mussels.*

23. Billington, Jackson, and Melosi, *The History of Large Federal Dams.*

Chapter 14

1. TVA (website), "Wilson."

2. Haag, *North American Freshwater Mussels.*

3. Garner and McGregor, "Current Status of Freshwater Mussels (*Unionidae, Margaritiferidae*) in the Muscle Shoals Area."

4. Postel and Richter, *Rivers for Life.*

5. Postel and Richter; Haag, *North American Freshwater Mussels.*

6. Postel and Richter, *Rivers for Life.*

7. Poff et al., "Homogenization of Regional River Dynamics by Dams and Global Biodiversity Implications."

Chapter 15

1. TVA (website), "Our History."

2. TVA (website), "TVA at a Glance"; TVA (website), "Hydroelectric"; TVA (website), "Navigation on the Tennessee River."

3. National Archives at Atlanta, *Valley of the Dams.*

4. Gilmer, "In the Shadow of Removal."
5. Rawls, "Forgotten People of the Tellico Dam Fight."
6. Gilmer, "In the Shadow of Removal."
7. Plater, "Tiny Fish/Big Battle."
8. Gilmer, "In the Shadow of Removal."
9. Plater, "Tiny Fish/Big Battle."
10. Billington, Jackson, and Melosi, *The History of Large Federal Dams*.
11. Billington, Jackson, and Melosi.
12. Gilmer, "In the Shadow of Removal."
13. US FWS (website), "History of the Endangered Species Act."
14. Billington, Jackson, and Melosi, *The History of Large Federal Dams*.
15. US Army Corps of Engineers Nashville District, *Duck River Watershed Plan*; TVA, "Use of Lands Acquired for the Columbia Dam Component of the Duck River Project"; Aldrich, "$83 Million Later, Unfinished Dam Being Dismantled."
16. Plater, "Tiny Fish/Big Battle."
17. Gilmer, "In the Shadow of Removal."

Chapter 16

1. Haag, *North American Freshwater Mussels*.
2. Haag.
3. J. D. Williams, Bogan, and Garner, *Freshwater Mussels of Alabama and the Mobile Basin*.
4. Haag, *North American Freshwater Mussels*.
5. Haag.
6. R. S. Duncan, *Southern Wonder*.
7. Haag, *North American Freshwater Mussels*.
8. Box and Williams, *Unionid Mussels of the Apalachicola Basin*.
9. US Army Corps of Engineers Mobile District, "Tennessee-Tombigbee Waterway"; Reeves, "$2B Tenn-Tom Waterway Yet to Yield Promised Boom."
10. Haag, *North American Freshwater Mussels*.
11. Haag.
12. Haag.
13. Haag.

Chapter 17

1. J. D. Williams, Bogan, and Garner, *Freshwater Mussels of Alabama and the Mobile Basin*.
2. Cummings, "*Elliptio crassidens*."
3. McGregor and Garner, *Results of Qualitative Sampling for Protected Mussel Species*.
4. Paul Freeman, pers. comm., March 18, 2020.
5. Randy Haddock, pers. comm., August 16, 2019.
6. J. D. Williams, Bogan, and Garner, *Freshwater Mussels of Alabama and the Mobile Basin*.
7. Williams, Bogan, and Garner.

8. Haag, *North American Freshwater Mussels.*
9. Haag.
10. J. D. Williams, Bogan, and Garner, *Freshwater Mussels of Alabama and the Mobile Basin.*
11. Haag, *North American Freshwater Mussels.*
12. Haag.
13. Haag.
14. J. D. Williams, Bogan, and Garner, *Freshwater Mussels of Alabama and the Mobile Basin*; Hart et al., "Novel Technique to Identify Large River Host Fish."
15. NOAA, "Endangered and Threatened Wildlife and Plants: Notice of 12-Month Finding"; Boschung and Mayden, *Fishes of Alabama.*
16. Freeman et al., "Status and Conservation of the Fish Fauna of the Alabama River System."
17. USDA, *Cahaba River—Alabama*; Onorato, Angus, and Marion, "Historical Changes in the Ichthyofaunal Assemblages of the Upper Cahaba River"; Mettee and O'Neil, "Status of Alabama Shad and Skipjack Herring in Gulf of Mexico Drainages."
18. Boschung and Mayden, *Fishes of Alabama.*
19. NOAA, "Endangered and Threatened Wildlife and Plants: Notice of 12-Month Finding."
20. Haag, *North American Freshwater Mussels.*

Chapter 18
1. Box and Williams, *Unionid Mussels of the Apalachicola Basin.*
2. J. D. Williams, Bogan, and Garner, *Freshwater Mussels of Alabama and the Mobile Basin.*
3. Williams, Bogan, and Garner.
4. Butler et al., *Recovery Plan for Endangered Fat Threeridge.*
5. O'Brien and Williams, "Reproductive Biology of Four Freshwater Mussels."
6. Fritts et al., "Critical Linkage of Imperiled Species"; S. Martin, "UGA Researchers Discover One Endangered Species Depends on Another."
7. Sulak et al., "Status of Scientific Knowledge, Recovery Progress, and Future Research Directions for the Gulf Sturgeon."
8. Fritts et al., "Critical Linkage of Imperiled Species"; S. Martin, "UGA Researchers Discover One Endangered Species Depends on Another."
9. US FWS and Gulf States Marine Fisheries Cornmission, *Gulf Sturgeon Recovery /Management Plan.*
10. Flowers et al., "Spawning Site Selection and Potential Implications of Modified Flow Regimes."
11. Fritts et al., "Critical Linkage of Imperiled Species."

Chapter 19
1. Jenkins, "Amphibians of the World."
2. International Energy Agency, *CO_2</i> Emissions from Fuel Combustion 2017</i>*; US EIA, *Monthly Energy Review: February 2018*; Manganiello, *Southern Water, Southern Power.*

3. J. Fox, "Mountaintop Removal in West Virginia."
4. National Energy Technology Laboratory, "History of U.S. Coal Use."
5. US EPA (website), "Basic Information about Surface Coal Mining in Appalachia."
6. McAuley and Kozar, *Ground-Water Quality in Unmined Areas and Near Reclaimed Surface Coal Mines*.
7. Silverman, "Mechanism of Bacterial Pyrite Oxidation."
8. Ross, McGlynn, and Bernhardt, "Deep Impact."
9. US EPA (website), "What Is Acid Rain?"
10. Bernhardt and Palmer, "The Environmental Costs of Mountaintop Mining Valley Fill Operations."
11. McQuaid, "Mining the Mountains."
12. Bernhardt and Palmer, "The Environmental Costs of Mountaintop Mining Valley Fill Operations"; Ross, McGlynn, and Bernhardt, "Deep Impact."
13. Alliance for Appalachia, "Mountaintop Removal Reclamation Fail."
14. US EPA, *The Effects of Mountaintop Mines and Valley Fills*; A. J. Miller and Zégre, "Mountaintop Removal Mining and Catchment Hydrology"; Ross, McGlynn, and Bernhardt, "Deep Impact."
15. Ross, McGlynn, and Bernhardt, "Deep Impact."
16. US EPA, *The Effects of Mountaintop Mines and Valley Fills*.
17. Bandy, "The Surface Mining Control and Reclamation Act (SMCRA) of 1977."
18. Bernhardt and Palmer, "The Environmental Costs of Mountaintop Mining Valley Fill Operations."
19. Palmer et al., "Mountaintop Mining Consequences."
20. Palmer et al.
21. Bernhardt and Palmer, "The Environmental Costs of Mountaintop Mining Valley Fill Operations."
22. Haag, *North American Freshwater Mussels*.
23. Bernhardt and Palmer, "The Environmental Costs of Mountaintop Mining Valley Fill Operations."
24. Hitt and Chambers, "Temporal Changes in Taxonomic and Functional Diversity of Fish Assemblages."

Chapter 20
1. *Economist*, "Appalachia"; Elam, "Culture, Poverty and Education in Appalachian Kentucky."
2. Partridge, Betz, and Lobao, "Natural Resource Curse and Poverty in Appalachian America"; Lobao et al., "Poverty, Place, and Coal Employment across Appalachia."
3. US MSHA, "Coal Fatalities for 1900 through 2022"; Humphrey, *Historical Summary of Coal-Mine Explosions in the United States*.
4. Boissoneault, "The Coal Mining Massacre America Forgot"; Laurie, "The United States Army and the Return to Normalcy in Labor Dispute Interventions."

5. US Bureau of Labor Statistics, "Injuries, Illnesses, and Fatalities in the Coal Mining Industry."

6. Urbina, "No Survivors Found after West Virginia Mine Disaster."

7. Petsonk, Rose, and Cohen, "Coal Mine Dust Lung Disease."

8. Hamby, "Breathless and Burdened."

9. CDC, "Work-Related Lung Disease Surveillance System (eWoRLD)."

10. Wei-Haas, "Why Black Lung Disease Is Deadlier Than Ever Before."

11. Wei-Haas; Berkes and Lancianese, "Black Lung Study Finds Biggest Cluster Ever of Fatal Coal Miners' Disease."

12. Joy, Colinet, and Landen, "Coal Workers' Pneumoconiosis Prevalence Disparity"; Berkes and Lancianese, "Black Lung Study Finds Biggest Cluster Ever of Fatal Coal Miners' Disease"; US MSHA, "Respirable Dust Rule."

13. Boles, "Federal Watchdog Finds Coal Safety Regulator Not Protecting Miners from Silica Dust."

14. Pickering, *The Buffalo Creek Flood*.

15. Warnick, "I Was the Miracle Baby."

16. National Academies of Science, *Coal Waste Impoundments*.

17. National Academies of Science.

18. Boyles et al., "Systematic Review of Community Health Impacts of Mountaintop Removal Mining."

19. Schiffman, "A Troubling Look at the Human Toll of Mountaintop Removal Mining"; Hendryx and Holland, "Unintended Consequences of the Clean Air Act."

20. Appalachian Voices, "Human Health Impacts."

21. Hendryx and Holland, "Unintended Consequences of the Clean Air Act."

22. Boyles et al., "Systematic Review of Community Health Impacts of Mountaintop Removal Mining."

23. McAuley and Kozar, *Ground-Water Quality in Unmined Areas and Near Reclaimed Surface Coal Mines*.

24. Ray, "Drinking Water Problems Still Plague Eastern Kentucky"; Hendryx, Fulk, and McGinley, "Public Drinking Water Violations in Mountaintop Coal Mining Areas."

25. US EPA (website), "Health and Environmental Effects of Particulate Matter (PM)."

26. Aneja et al., "Particulate Matter Pollution in the Coal-Producing Regions of the Appalachian Mountains"; Aneja, Isherwood, and Morgan, "Characterization of Particulate Matter (PM_{10}) Related to Surface Coal Mining Operations in Appalachia"; Kurth et al., "Atmospheric Particulate Matter Size Distribution and Concentration."

27. Fears, "Trump Administration Halted a Study of Mountaintop Coal Mining's Health Effects"; National Academies of Science, "Statement Regarding National Academies Study on Potential Health Risks of Living in Proximity to Surface Coal Mining Sites."

Chapter 21

1. Satterfield, "TVA Contractor Jacobs Engineering Loses Request to Appeal Lawsuit by Kingston Workers"; J. K. Bourne, "Coal's Other Dark Side."
2. Onyanga-Omara, "Mystery of London Fog That Killed 12,000 Finally Solved"; Murray, "Smog Deaths in 1948 Led to Clean Air Laws."
3. North Carolina Division of Public Health, "Fish Consumption Advisories."
4. Good and VanBriesen, "Current and Potential Future Bromide Loads from Coal-Fired Power Plants."
5. Gutiérrez, "Downstream from Duke Energy Plant"; Fakour and Lo, "Formation of Trihalomethanes as Disinfection Byproducts in Herbal Spa Pools."
6. M. Blake, "Comments on the Draft Title V Permit Modification for Allen Steam Station"; Gutiérrez, "Downstream from Duke Energy Plant."
7. M. Blake, "Comments on the Draft Title V Permit Modification for Allen Steam Station."
8. Blake.
9. US EPA (website), "Coal Ash"; US EPA, "Disposal of Coal Combustion Residuals from Electric Utilities"; Satterfield, "TVA Contractor Jacobs Engineering Loses Request to Appeal Lawsuit by Kingston Workers."
10. American Coal Ash Association, "Production and Use Reports."
11. US EPA (website), "Coal Ash."
12. American Coal Ash Association, "Production and Use Reports."
13. Satterfield, "TVA Contractor Jacobs Engineering Loses Request to Appeal Lawsuit by Kingston Workers."
14. Satterfield; US EPA (website), "EPA's Response to the TVA Kingston Fossil Plant Fly Ash Release."
15. J. K. Bourne, "Coal's Other Dark Side"; Satterfield, "Roane County Lawsuit Asserts TVA and Its Contractor Ran Coal Ash 'Disinformation Campaign'"; Satterfield, "TVA Contractor Jacobs Engineering Loses Request to Appeal Lawsuit by Kingston Workers"; Flessner, "Court Dismisses Lawsuit against TVA, Jacobs Engineering over Coal Ash Cleanup."
16. Washington Post Editorial Board, "The EPA's Move to Regulate 'Coal Ash' Is a Step Forward."
17. Associated Press, "It Will Cost $900 Million to Move Power Plant Ash to Landfill"; US EPA, "EPA Proposes First of Two Rules to Amend Coal Ash Disposal Regulations."
18. Burns and Leslie, "State Orders Duke Energy to Dig Up All Remaining Coal Ash Ponds in NC."
19. Southern Environmental Law Center, "Cleaning Up Coal Ash Pollution."
20. US Department of the Interior, "Dan River Coal Ash Spill"; Dennis, Mufson, and Eilperin, "Dam Breach Sends Toxic Coal Ash Flowing into a Major North Carolina River."
21. Milman, "We're Not a Dump"; Engelman-Lado et al., "Environmental Injustice in Uniontown, Alabama."

22. Black Belt Citizens, "Black Belt Citizens Fighting for Health and Justice."
23. Black Belt Citizens.

Chapter 22

1. Fahrenthold, "Environmental Regulations to Curtail Mountaintop Mining."
2. US MSHA, "Respirable Dust Rule."
3. American Coal Council, "About the American Coal Council."
4. US MSHA, "Coal Fatalities for 1900 through 2022"; Harvey, Larson, and Patel, "History of Power"; Rakes, "Coal Mine Mechanization."
5. Lobao et al., "Poverty, Place, and Coal Employment across Appalachia"; US Census Bureau, "Small Area Income and Poverty Estimates"; CDC, "Statistics."
6. Lobao et al., "Poverty, Place, and Coal Employment across Appalachia"; Partridge, Betz, and Lobao, "Natural Resource Curse and Poverty in Appalachian America."
7. US MSHA, "Coal Fatalities for 1900 through 2022"; US EIA, *Annual Coal Report 2016*; Bernhardt and Palmer, "The Environmental Costs of Mountaintop Mining Valley Fill Operations"; Berkowitz and Meko, "Appalachia Comes Up Small in Era of Giant Coal Mines."
8. Richardson et al., "A Dwindling Role for Coal."
9. Gallegos and Varela, *Trends in Hydraulic Fracturing Distributions and Treatment Fluids, Additives, Proppants, and Water Volumes*.
10. Arena, "Coal Production Using Mountaintop Removal Mining Decreases by 62% since 2008."
11. US EIA, *Monthly Energy Review: February 2018*.
12. Richardson et al., "A Dwindling Role for Coal."
13. Arena, "Coal Production Using Mountaintop Removal Mining Decreases by 62% since 2008"; US EIA, *Annual Coal Report 2016*.
14. US EPA, *Hydraulic Fracturing for Oil and Gas*; Natural Resources Defense Council, "Report: Five Major Health Threats from Fracking-Related Air Pollution"; USGS (website), "Hydraulic Fracturing."

Chapter 23

1. Scripps CO_2 Program, "Keeling Curve Lessons."
2. Global Climate Change, "Carbon Dioxide"; Mooney, "Scientists Say Human Greenhouse Gas Emissions Have Canceled the Next Ice Age."
3. CO_2.Earth, "Daily CO_2."
4. Steffen et al., "Trajectories of the Earth System in the Anthropocene."
5. Steffen et al.
6. K. G. Miller et al., "High Tide of the Warm Pliocene"; Dumitru et al., "Constraints on Global Mean Sea Level during Pliocene Warmth."
7. Hoffman et al., "Regional and Global Sea-Surface Temperatures during the Last Interglaciation."
8. Intergovernmental Panel on Climate Change, *Climate Change 2014*, 16.
9. Intergovernmental Panel on Climate Change, 18–19.

10. Westervelt, *Drilled*; Hall, "Exxon Knew about Climate Change almost 40 Years Ago"; Holden, "How the Oil Industry Has Spent Billions to Control the Climate Change Conversation."

11. Marlon et al., "Yale Climate Opinion Maps 2019."

12. Hsiang et al., "Estimating Economic Damage from Climate Change in the United States."

13. Kunkel et al., *Regional Climate Trends and Scenarios for the U.S. National Climate Assessment*; Konrad et al., "Climate of the Southeast USA."

14. NOAA (website), "National Trends"; Kunkel et al., *Regional Climate Trends and Scenarios for the U.S. National Climate Assessment*; W. Li et al., "Changes to the North Atlantic Subtropical High."

15. W. Li et al., "Changes to the North Atlantic Subtropical High."

16. Konrad et al., "Climate of the Southeast USA"; Kunkel et al., *Regional Climate Trends and Scenarios for the U.S. National Climate Assessment*.

17. L. Li, Li, and Kushnir, "Variation of the North Atlantic Subtropical High Western Ridge."

18. Li, Li, and Kushnir; W. Li et al., "Changes to the North Atlantic Subtropical High."

19. Moss et al., "The Next Generation of Scenarios for Climate Change Research and Assessment"; Wayne, "The Beginner's Guide to Representative Concentration Pathways."

20. Moss et al., "The Next Generation of Scenarios for Climate Change Research and Assessment"; Wayne, "The Beginner's Guide to Representative Concentration Pathways."

21. Moss et al., "The Next Generation of Scenarios for Climate Change Research and Assessment"; Wayne, "The Beginner's Guide to Representative Concentration Pathways."

22. Hayhoe et al., "Our Changing Climate."

23. L. Carter et al., "Southeast"; Konrad et al., "Climate of the Southeast USA."

24. L. Carter et al., "Southeast."

25. Konrad et al., "Climate of the Southeast USA."

26. L. Carter et al., "Southeast."

Chapter 24

1. Hanson, "Evapotranspiration and Droughts."

2. Reidmiller et al., *Impacts, Risks, and Adaptation in the United States*.

3. Vose et al., "Temperature Changes in the United States."

4. Konrad et al., "Climate of the Southeast USA"; Climate Central, "2017 U.S. Temperature Review."

5. Vose et al., "Temperature Changes in the United States."

6. Sun et al., "Impacts of Climate Change and Variability on Water Resources in the Southeast USA"; Marion et al., "Managing Forest Water Quantity and Quality under Climate Change."

7. Vincent and Velkoff, *The Next Four Decades*.

8. L. Carter et al., "Southeast"; L. Li, Li, and Kushnir, "Variation of the North Atlantic Subtropical High Western Ridge."

9. Maupin et al., *Estimated Use of Water in the United States in 2005.*

10. Union of Concerned Scientists (website), "Infographic"; Mitchell et al., "Energy Production, Use, and Vulnerability to Climate Change in the Southeast USA."

11. Wehner et al., "Droughts, Floods, and Wildfires"; Sun et al., "Impacts of Climate Change and Variability on Water Resources in the Southeast USA."

12. Wehner et al., "Droughts, Floods, and Wildfires"; Konrad et al., "Climate of the Southeast USA."

13. Hopkinson et al., "The Effects of Climate Change on Natural Ecosystems of the Southeast USA."

14. Marion et al., "Managing Forest Water Quantity and Quality under Climate Change."

15. Neupane et al., "Hydrologic Responses to Projected Climate Change in Ecologically Diverse Watersheds"; Konrad et al., "Climate of the Southeast USA."

Chapter 25

1. Hopkinson et al., "The Effects of Climate Change on Natural Ecosystems of the Southeast USA."

2. Leuliette and Nerem, "Contributions of Greenland and Antarctica to Global and Regional Sea Level Change."

3. White and Kaplan, "Restore or Retreat?"

4. Brosius et al., "Climate Adaptations in the Southeast USA"; Karegar, Dixon, and Engelhart, "Subsidence along the Atlantic Coast of North America."

5. Leuliette and Nerem, "Contributions of Greenland and Antarctica to Global and Regional Sea Level Change"; Rietbroek et al., "Revisiting the Contemporary Sea-Level Budget on Global and Regional Scales."

6. Frederikse, Riva, and King, "Ocean Bottom Deformation Due to Present-Day Mass Redistribution."

7. Lallensack, "Global Fingerprints of Sea Level Rise Revealed by Satellites."

8. Hayhoe et al., "Our Changing Climate"; Sweet et al., "Sea Level Rise."

9. Karegar, Dixon, and Engelhart, "Subsidence along the Atlantic Coast of North America"; Rietbroek et al., "Revisiting the Contemporary Sea-Level Budget on Global and Regional Scales"; NOAA (website), "Sea Level Trends."

10. Hayhoe et al., "Our Changing Climate."

11. DeConto and Pollard, "Contribution of Antarctica to Past and Future Sea-Level Rise"; Wuebbles et al., *Climate Science Special Report.*

12. DeConto and Pollard, "Contribution of Antarctica to Past and Future Sea-Level Rise"; Sweet et al., "Sea Level Rise."

13. DeConto and Pollard, "Contribution of Antarctica to Past and Future Sea-Level Rise."

14. Sweet et al., "Sea Level Rise."

Chapter 26

1. Sweet and Marra, "Understanding Climate"; Sweet and Marra, *2015 State of U.S. "Nuisance" Tidal Flooding.*

2. Sweet and Marra, "Understanding Climate"; Richardson-Lorente, "Flood City."

3. First Street Foundation, "Sea Level Rise Washes Away $76.4 Million in Texas Home Values."

4. Spanger-Siegfried et al., *When Rising Seas Hit Home*; Sweet et al., *Patterns and Projections of High Tide Flooding along the U.S. Coastline.*

5. Rice, Hong, and Shen, "Assessment of Salinity Intrusion in the James and Chickahominy Rivers."

6. Ferguson and Gleeson, "Vulnerability of Coastal Aquifers to Groundwater Use and Climate Change."

7. Asseng et al., "Agriculture and Climate Change in the Southeast USA"; Sun et al., "Impacts of Climate Change and Variability on Water Resources in the Southeast USA."

8. Roberts et al., "Preventing Local Extinctions of Tidal Marsh Endemic Seaside Sparrows and Saltmarsh Sparrows"; Velasquez-Manoff and Demczuk, "As Sea Levels Rise, So Do Ghost Forests."

9. White and Kaplan, "Restore or Retreat?"

10. Craft et al., "Forecasting the Effects of Accelerated Sea-Level Rise on Tidal Marsh Ecosystem Services."

11. Hopkinson et al., "The Effects of Climate Change on Natural Ecosystems of the Southeast USA."

12. E. S. Blake and Gibney, *The Deadliest, Costliest, and Most Intense United States Tropical Cyclones from 1851 to 2010.*

13. Konrad et al., "Climate of the Southeast USA"; Stewart and Berg, *Tropical Cyclone Report*; Beven, Berg, and Hagen, *Hurricane Michael*; National Hurricane Center, "Costliest U.S. Tropical Cyclones Tables Updated."

14. Kossin et al., "Extreme Storms."

15. Kossin et al.; Walsh et al., "Our Changing Climate."

16. Patricola, "Tropical Cyclones Are Becoming Sluggish."

17. Bender et al., "Modeled Impact of Anthropogenic Warming on the Frequency of Intense Atlantic Hurricanes"; Kossin et al., "Extreme Storms."

Chapter 27

1. United Nations, "World Population Projected to Reach 9.8 Billion in 2050, and 11.2 Billion in 2100."

2. Spratt and Dunlop, *Existential Climate-Related Security Risk.*

3. Wilson, *The Future of Life.*

Chapter 28

1. Stewart, *Tropical Cyclone Report.*

Chapter 29

1. City of Norfolk, *Norfolk Vision 2100*; Green, "Norfolk Forges a Path to a Resilient Future."

2. Union of Concerned Scientists, *Underwater.*

3. Sweet and Marra, *2015 State of U.S. "Nuisance" Tidal Flooding.*

4. City of Norfolk, *Norfolk Vision 2100.*

5. Gittman et al., "Engineering Away Our Natural Defenses."
6. Knabb, Rhome, and Brown, *Tropical Cyclone Report*.
7. Reguero et al., "Comparing the Cost Effectiveness of Nature-Based and Coastal Adaptation"; Narayan et al., "The Effectiveness, Costs and Coastal Protection Benefits of Natural and Nature-Based Defences"; Narayan et al., "Coastal Wetlands and Flood Damage Reduction."
8. Baptiste, "When a Hurricane Takes Your Home"; Honan, "City Had Millions to Buy Out Sandy-Damaged Homes, but Most Didn't Want It."
9. City of Norfolk, *Norfolk Vision 2100*.
10. Parsons, "Climate Gentrification"; Keenan, Hill, and Gumber, "Climate Gentrification."
11. Murphy, "Why Norfolk Is Planning to Tear Down Half of Its Public Housing"; Kusnetz, "Norfolk Wants to Remake Itself as Sea Level Rises."
12. Murphy, "Why Norfolk Is Planning to Tear Down Half of Its Public Housing"; Kusnetz, "Norfolk Wants to Remake Itself as Sea Level Rises."
13. Richardson-Lorente, "Flood City"; Green, "Norfolk Forges a Path to a Resilient Future."
14. Little, "Leaders Sign Norfolk Coastal Storm Risk Management Design Agreement."
15. Richardson-Lorente, "Flood City"; Thornton, "Army Corps of Engineers Turns to Private Sector in Face of Budget Cuts."
16. Davis, *The Gulf*.
17. Evans et al., *Tybee Island Sea-Level Rise Adaptation Plan*.
18. Evans et al.; Sweet et al., "Sea Level Rise."

Chapter 30

1. Gedan et al., "The Present and Future Role of Coastal Wetland Vegetation in Protecting Shorelines"; Costanza et al., "The Value of Coastal Wetlands for Hurricane Protection."
2. Morelle, "Vast Sand Scheme to Protect Norfolk Coast"; Kakissis, "Protecting the Netherlands' Vulnerable Coasts with a 'Sand Motor.'"
3. Osland et al., "Winter Climate Change and Coastal Wetland Foundation Species."
4. Beck et al., "Oyster Reefs at Risk"; Grabowski et al., "Economic Valuation of Ecosystem Services Provided by Oyster Reefs"; Piazza, Banks, and La Peyre, "The Potential for Created Oyster Shell Reefs"; Puckett et al., "Integrating Larval Dispersal, Permitting, and Logistical Factors within a Validated Habitat Suitability Index for Oyster Restoration."
5. Costanza et al., "The Value of Coastal Wetlands for Hurricane Protection"; Gedan et al., "The Present and Future Role of Coastal Wetland Vegetation in Protecting Shorelines."
6. Costanza et al., "The Value of Coastal Wetlands for Hurricane Protection."
7. Reguero et al., "Comparing the Cost Effectiveness of Nature-Based and Coastal Adaptation."

8. Gedan et al., "The Present and Future Role of Coastal Wetland Vegetation in Protecting Shorelines."

9. Kennish, "Coastal Salt Marsh Systems in the U.S."

10. Deegan et al., "Coastal Eutrophication as a Driver of Salt Marsh Loss."

11. He and Silliman, "Biogeographic Consequences of Nutrient Enrichment for Plant-Herbivore Interactions in Coastal Wetlands."

12. Upton, "As Seas Rise, Americans Use Nature to Fight Worsening Erosion."

13. Nature Conservancy, "Public and Private Partnership to Restore and Protect Bayou La Batre Waterfront"; Nature Conservancy (website), "Lightning Point Shoreline Restoration."

14. Beck et al., "Oyster Reefs at Risk"; Eastern Oyster Biological Review Team, *Status Review of the Eastern Oyster*; Woods et al., "Disappearance of the Natural Emergent 3-Dimensional Oyster Reef System of the James River."

15. Eastern Oyster Biological Review Team, *Status Review of the Eastern Oyster*.

16. La Peyre et al., "Oyster Reef Restoration in the Northern Gulf of Mexico"; Puckett et al., "Integrating Larval Dispersal, Permitting, and Logistical Factors within a Validated Habitat Suitability Index for Oyster Restoration."

17. Grabowski et al., "Economic Valuation of Ecosystem Services Provided by Oyster Reefs"; Beck et al., "Oyster Reefs at Risk."

Chapter 31

1. Raines, "Alabama's Rivers Are Crashing Thanks to Water Rules That Favor Industry"; Southeast Regional Climate Center, "The Southeast U.S. Drought of 2016."

2. US Drought Monitor, "Map Archive."

3. US Drought Monitor.

4. Konrad and Knox, *The Southeastern Drought and Wildfires of 2016*.

5. L. Li, Li, and Kushnir, "Variation of the North Atlantic Subtropical High Western Ridge"; A. P. Williams et al., "The 2016 Southeastern U.S. Drought."

6. Baer and Ingle, *Protecting and Restoring Flows in Our Southeastern Rivers*.

7. Baer and Ingle.

8. Georgia Department of Natural Resources, "Water Conservation."

9. Baer and Ingle, *Protecting and Restoring Flows in Our Southeastern Rivers*.

10. Baer and Ingle; North Carolina Division of Water Resources, "Drought Monitor History"; Samuel, "Officials Declare Drought in Atlanta, North Georgia."

11. Raines, "Drought Means a Stunningly Clear Gulf, Happy Fishermen, and Bad News for Oyster Lovers."

12. Douban, "Why Alabama Still Has No Water Management Plan."

13. Konrad and Knox, *The Southeastern Drought and Wildfires of 2016*.

14. Konrad and Knox.

15. Atlanta Regional Commission, "Tri-State Water Wars"; Georgia River Network, "Chattahoochee River."

16. Aaron, "Explore Our Interactive Timeline of the Water Wars."

17. Aaron.

18. Fowler, "New Special Master Finds for Georgia in Most Recent Round of Water Dispute"; Fowler, "No April Fool's Joke for Florida"; Hallerman, "Georgia Is Nowhere Near Last Battle in Tri-State Water Wars."

19. Konrad and Knox, *The Southeastern Drought and Wildfires of 2016*.

20. Konrad and Knox.

21. Konrad and Knox; Carter et al., "Southeast."

22. A. P. Williams et al., "The 2016 Southeastern U.S. Drought"; Konrad and Knox, *The Southeastern Drought and Wildfires of 2016*.

23. A. P. Williams et al., "The 2016 Southeastern U.S. Drought."

Chapter 32

1. National Inventory of Dams, "Dams of the Nation"; Emanuel and Hoffner, *Money Pit*.

2. Emanuel and Hoffner, *Money Pit*; Manganiello, "Excitement Draining for Reservoir Boondoggles"; Billington, Jackson, and Melosi, *The History of Large Federal Dams*.

3. Emanuel and Hoffner, *Money Pit*; Associated Press, "Canton Reservoir Shows Realities of Deal Plan."

4. Emanuel and Hoffner, *Money Pit*.

5. Emanuel and Hoffner; USGS (website), "A Million Gallons of Water—How Much Is It?"; US EPA, "Saving Water in Georgia"; Chapman, "Metro Atlanta Water Use to Increase Much Less Than Earlier Predicted."

6. Tessler et al., "A Model of Water and Sediment Balance as Determinants of Relative Sea Level Rise"; Day et al., "Impacts of Sea-Level Rise on Deltas in the Gulf of Mexico and the Mediterranean."

7. Cornwall, "Hundreds of New Dams Could Mean Trouble for Our Climate."

Chapter 33

1. Herndon, "Energy Makes Up Half of Desalination Plant Costs."

2. Mickley, *Updated and Extended Survey of U.S. Municipal Desalination Plants*; N. T. Carter, *Desalination and Membrane Technologies*.

3. Lone Star Chapter of the Sierra Club, *Desalination*; Bienkowski, "Desalination Is an Expensive Energy Hog"; N. T. Carter, *Desalination and Membrane Technologies*.

4. P. Rogers, "Nation's Largest Ocean Desalination Plant Goes Up near San Diego"; Lone Star Chapter of the Sierra Club, *Desalination*.

5. N. T. Carter, *Desalination and Membrane Technologies*; Lone Star Chapter of the Sierra Club, *Desalination*.

6. USGS, "Groundwater Quality in the Southeastern Coastal Plain Aquifer System"; USGS (website), "Water Use in Georgia, 2010"; Geological Survey of Alabama, "Groundwater Assessment Program"; North Carolina Ground Water Association, home page.

7. J. A. Miller, *Alabama, Florida, Georgia, South Carolina*.

8. Geological Survey of Alabama, "Groundwater Assessment Program"; USGS (website), "Water Use in Georgia, 2010"; US EPA, "North Carolina Water Fact Sheet."

9. Kennedy, "Groundwater Declines across U.S. South over Past Decade."

10. Polycarpou, "US Groundwater Declines More Widespread Than Commonly Thought"; Reilly et al., *Ground-Water Availability in the United States*; Bellino et al., *Hydrogeologic Setting, Conceptual Groundwater Flow System, and Hydrologic Conditions*.

11. Perrone and Jasechko, "Dry Groundwater Wells in the Western United States."

12. Sun et al., "Impacts of Climate Change and Variability on Water Resources in the Southeast USA"; Karegar, Dixon, and Engelhart, "Subsidence along the Atlantic Coast of North America."

13. Bellino et al., *Hydrogeologic Setting, Conceptual Groundwater Flow System, and Hydrologic Conditions*.

14. Chastain, Golladay, and Muenz, "Distribution of Unionid Mussels in Tributaries of the Lower Flint River"; Postel, "How Smarter Irrigation Might Save Rare Mussels and Ease a Water War."

15. Baer and Ingle, *Protecting and Restoring Flows in Our Southeastern Rivers*.

16. US EPA (website), "Aquifer Recharge and Aquifer Storage and Recovery"; Ground Water Protection Council, "Aquifer Storage and Recovery."

Chapter 34

1. Metropolitan North Georgia Water Planning District (website), "About the Metro Water District."

2. Emanuel and Hoffner, *Money Pit*; Hoffner, *Hidden Reservoir*.

3. Hoffner, *Hidden Reservoir*.

4. Hoffner.

5. Metropolitan North Georgia Water Planning District (MNG WPD) (website), "Water Supply and Conservation Plan"; MNG WPD, "Metro Atlanta"; MNG WPD, *2017 Activities & Progress Report*.

6. US EPA, *Potable Reuse Compendium*.

7. National Research Council, *Use of Reclaimed Water and Sludge in Food Crop Production*.

8. Greywater Action, "Greywater Action."

9. US EPA (website), "Water Reuse Resource Hub by End-Use Application"; US EPA, *Potable Reuse Compendium*.

10. US EPA, *Potable Reuse Compendium*.

11. US EPA.

12. US EPA (website), "Water Reuse Resource Hub by End-Use Application."

13. US EPA, *Potable Reuse Compendium*; L. Martin, "Texas Leads the Way with First Direct Potable Reuse Facilities in U.S."

14. US EPA, *Potable Reuse Compendium*.

15. Baer and Ingle, *Protecting and Restoring Flows in Our Southeastern Rivers*.

Chapter 35

1. Haag, *North American Freshwater Mussels*; Haag, "Past and Future Patterns of Freshwater Mussel Extinctions"; Haag and Williams, "Biodiversity on the Brink."

2. Vaughn, "Ecosystem Services Provided by Freshwater Mussels."
3. Haag, *North American Freshwater Mussels*.
4. Egan, "The Cancer of the Great Lakes."
5. Haag, *North American Freshwater Mussels*.
6. Vaughn, Gido, and Spooner, "Ecosystem Processes Performed by Unionid Mussels in Stream Mesocosms."
7. Strayer et al., "Distribution, Abundance, and Roles of Freshwater Clams (Bivalvia, Unionidae)"; Newton et al., "Population Assessment and Potential Functional Roles of Native Mussels in the Upper Mississippi River."

Chapter 36

1. O'Neil et al., *A Survey of Fishes in the Paint Rock River System, Alabama*.
2. Paul Johnson, pers. comm., June 24, 2019.
3. O'Neil et al., *A Survey of Fishes in the Paint Rock River System, Alabama*.
4. Elkins et al., *The Southeastern Aquatic Biodiversity Conservation Strategy*.
5. Paul Johnson, pers. comm., June 24, 2019.
6. Kaetz, "Paint Rock."
7. Nature Conservancy (website), "Who We Are."
8. Steve Northcutt, pers. comm., September 6, 2019.
9. Jason Throneberry, pers. comm., September 5, 2019.
10. Elkins et al., *The Southeastern Aquatic Biodiversity Conservation Strategy*; Jenkins et al., "US Protected Lands Mismatch Biodiversity Priorities."
11. Elkins et al., *The Southeastern Aquatic Biodiversity Conservation Strategy*.
12. Elkins et al.
13. Paul Johnson, pers. comm., June 24, 2019.
14. Johnson.
15. Johnson.

Chapter 37

1. NOAA (website), "Where Is the Largest Estuary in the United States?"; USGS, "The Chesapeake Bay."
2. Chesapeake Bay Foundation (website), "The History of Chesapeake Bay Cleanup Efforts."
3. Chesapeake Bay Program, "Agricultural Runoff."
4. Dance, "Halfway to Cleanup Deadline, Chesapeake Bay"; Chesapeake Bay Foundation, *Agricultural Conservation Practices*; Chesapeake Bay Foundation (website), "What Is Killing the Bay?"
5. Dance, "Halfway to Cleanup Deadline, Chesapeake Bay."
6. Cox, "Bay Scientists Say Stream Restoration Not Delivering as Much as Hoped."
7. Grabar, "How We Built Our Way Into an Urban Flooding Epidemic"; Galloway et al., *The Growing Threat of Urban Flooding*.
8. L. Carter, "Southeast"; Galloway et al., *The Growing Threat of Urban Flooding*.
9. National Research Council, *Urban Stormwater Management in the United States*.

10. Lave, "The Controversy over Natural Channel Design"; Nagle, "Evaluating 'Natural Channel Design' Stream Projects"; Zaffos, "'Restoration Cowboy' Goes against the Flow."

11. Lave, "The Controversy over Natural Channel Design"; Lave, "Bridging Political Ecology and STS."

12. Palmer, Hondula, and Koch, "Ecological Restoration of Streams and Rivers"; Lave, "The Controversy over Natural Channel Design."

13. Cox, "Bay Scientists Say Stream Restoration Not Delivering as Much as Hoped"; Chesapeake Bay Foundation (website), "Chesapeake 2000 EPA Suit."

14. Cox, "Bay Scientists Say Stream Restoration Not Delivering as Much as Hoped"; M. R. Williams et al., "Stream Restoration Performance and Its Contribution to the Chesapeake Bay TMDL."

15. Lave, "The Controversy over Natural Channel Design."

16. Lave.

17. Palmer, Hondula, and Koch, "Ecological Restoration of Streams and Rivers."

18. World Wildlife Fund, *Natural and Nature-Based Flood Management*; US EPA (website), "Urban Runoff."

19. World Wildlife Fund, *Natural and Nature-Based Flood Management*; US EPA (website), "Urban Runoff"; Gibbens, "As the Climate Crisis Worsens, Cities Turn to Parks."

20. US EPA (website), "Climate Change Indicators"; Carter et al., "Southeast."

Chapter 38

1. Tomaszewski, *Tims Ford Dam Release Water Quality Improvements*.

2. Tomaszewski; US Army Corps of Engineers, *Proceedings*.

3. Tomaszewski, *Tims Ford Dam Release Water Quality Improvements*; US Army Corps of Engineers, *Proceedings*.

4. Tomaszewski, *Tims Ford Dam Release Water Quality Improvements*; US Army Corps of Engineers, *Proceedings*.

5. US FWS, *Boulder Darter (Etheostoma wapiti) 5-Year Review*; US FWS, *Boulder Darter Recovery Plan*.

6. US FWS, *Boulder Darter (Etheostoma wapiti) 5-Year Review*.

7. River Network, "Science Module"; Conservation Gateway, "Environmental Flow Components."

8. Postel and Richter, *Rivers for Life*.

9. Federal Energy Regulatory Commission, *Hydropower Primer*; Pittock and Hartmann, "Taking a Second Look."

10. Conservation Gateway, "Environmental Flow Components"; Ward and Meadows, "Adaptive Management of Environmental Flow Restoration in the Savannah River"; Richter et al., "A Collaborative and Adaptive Process for Developing Environmental Flow Recommendations."

11. Poff et al., "The Ecological Limits of Hydrologic Alteration (ELOHA)"; Conservation Gateway, "Ecological Limits of Hydrologic Alteration (ELOHA)."

12. Tomaszewski, *Tims Ford Dam Release Water Quality Improvements*.

13. TVA (website), "Preserving Life on the Elk River"; Pipas and Spaulding, *Management Plan for the Tims Ford Tailwater Trout Fishery*.

14. US FWS, *Boulder Darter (Etheostoma wapiti) 5-Year Review*.

Chapter 39

1. Sulak et al., "Status of Scientific Knowledge, Recovery Progress, and Future Research Directions for the Gulf Sturgeon."

2. Sulak et al.

3. Sulak et al.

4. Sulak et al.

5. Sulak et al.

6. Sulak et al.

7. Sulak et al.

8. Sulak et al.

9. Sulak et al.

10. Hilton et al., "Review of the Biology, Fisheries, and Conservation Status of the Atlantic Sturgeon."

11. Hilton et al.

12. Fears, "A Desperate Try to Restock the Potomac's Sturgeon."

Chapter 40

1. Nickens, "Ode to a Founding Fish"; US FWS (website), "American Shad."

2. Greene et al., *Atlantic Coast Diadromous Fish Habitat*; Manganiello, "Fish Tales and the Conservation State."

3. Watson, "The Common Rights of Mankind."

4. Nickens, "Ode to a Founding Fish."

5. Watson, "The Common Rights of Mankind"; Crable, "Shad Wars"; Manganiello, "Fish Tales and the Conservation State."

6. Watson, "The Common Rights of Mankind."

7. Watson; Crable, "Shad Wars"; Manganiello, "Fish Tales and the Conservation State."

8. Watson, "The Common Rights of Mankind."

9. Watson; Greene et al., *Atlantic Coast Diadromous Fish Habitat*.

10. Watson, "The Common Rights of Mankind."

11. Greene et al., *Atlantic Coast Diadromous Fish Habitat*.

12. Greene et al.

13. Blankenship, "PA Shad Hatchery's 42-Year Run May Be Coming to an End."

14. Atlantic States Marine Fisheries Commission, "Shad & River Herring."

15. NOAA, "Endangered and Threatened Wildlife and Plants; Species Act Listing Determination."

Chapter 41

1. US Army Corps of Engineers, "Columbia River Basin Dams"; American Rivers, "Columbia River"; Billington, Jackson, and Melosi, *The History of Large Federal Dams*.

2. Billington, Jackson, and Melosi, *The History of Large Federal Dams*.

3. Billington, Jackson, and Melosi; Northwest Power and Conservation Council, "Fish Passage at Dams."

4. Billington, Jackson, and Melosi, *The History of Large Federal Dams*.

5. Brownell et al., *Diadromous Fish Passage*.

6. Brownell et al.

7. Brownell et al.

8. US FWS, *Fish Passage Engineering Design Criteria* (2019).

9. Silva et al., "The Future of Fish Passage Science, Engineering, and Practice."

10. Silva et al.

11. Atlantic States Marine Fisheries Commission Fish Passage Working Group, "Upstream Fish Passage Technologies for Managed Species"; Greene et al., *Atlantic Coast Diadromous Fish Habitat*.

12. Atlantic States Marine Fisheries Commission Fish Passage Working Group, "Upstream Fish Passage Technologies for Managed Species"; Greene et al., *Atlantic Coast Diadromous Fish Habitat*.

13. Haro, "Presentation on Fish Passage Concerns for Shad and River Herring, Atlantic (and Shortnose) Sturgeon, and American Eel"; Atlantic States Marine Fisheries Commission Fish Passage Working Group, "Upstream Fish Passage Technologies for Managed Species."

14. Atlantic States Marine Fisheries Commission Fish Passage Working Group, "Upstream Fish Passage Technologies for Managed Species"; Jager et al., "Reconnecting Fragmented Sturgeon Populations in North American Rivers."

15. Mazzocchi, "Cape Fear River."

16. J. A. Smith and Hightower, "Effect of Low-Head Lock-and-Dam Structures on Migration and Spawning of American Shad and Striped Bass"; NOAA, "Endangered and Threatened Wildlife and Plants; Proposed Listings"; Raabe, Ellis, and Hightower, "Evaluation of Fish Passage Following Installation of a Rock Arch Rapids."

17. Aadland, *Reconnecting Rivers*.

18. H. Miller, "River Advocates Work to Add Fish Passages."

19. Raabe et al., "Evaluation of Fish Passage at a Nature-Like Rock Ramp Fishway on a Large Coastal River."

20. US Army Corps of Engineers Savannah District, *Savannah Harbor Expansion Project*; US Army Corps of Engineers Savannah District (website), "SHEP Fish Passage at New Savannah Bluff Lock and Dam."

21. Hodges, "Judge Files Order against Corps of Engineers in Lock and Dam Case"; Associated Press, "Judge Says Dam near Augusta Can't Be Removed if Water Level Will Fall"; Associated Press, "Agency Appeals Order Blocking Savannah River Dam Removal"; Byerly and Allison, "Lock and Dam's Fate in Question after New Ruling from Appeals Court."

22. *Bangor Daily News*, "Three Dams to Get Power Upgrades along Penobscot"; McCarthy, "16-Year Penobscot River Restoration Project Reaches the Finish Line."

23. Silva et al., "The Future of Fish Passage Science, Engineering, and Practice"; US FWS, *Fish Passage Engineering Design Criteria* (2017).

24. Society for Ecological Restoration, "USA: Two Dam Removals"; Natural Resources Council of Maine, "Penobscot River Restoration Trust."

Chapter 42

1. Greene et al., *Atlantic Coast Diadromous Fish Habitat*.
2. Greene et al.
3. Greene et al.
4. Greene et al.
5. Fuller et al., *"Anguilla rostrata."*
6. Fuller et al.
7. US FWS (website), "American Eel (*Anguilla rostrata*)."
8. Sneed, "Glass Eel Gold Rush Casts Maine Fishermen against Scientists."
9. Atlantic States Marine Fisheries Commission, "American Eel"; Trotter, "Price Offered for Maine's Baby Eels Hits Record High"; South Carolina Marine Resources Research Institute, "Diadromous Fish Research"; US Department of Justice, "Maine Men Sentenced for Illegally Trafficking American Eels."; Ebersole, "19 Eel Smugglers Sentenced, but Lucrative Trade Persists."
10. Sneed, "Glass Eel Gold Rush Casts Maine Fishermen against Scientists"; Atlantic States Marine Fisheries Commission Fish Passage Working Group, "Upstream Fish Passage Technologies for Managed Species."
11. Haro, *Proceedings of a Workshop on American Eel Passage Technologies*.
12. Haro; Greene et al., *Atlantic Coast Diadromous Fish Habitat*; Atlantic States Marine Fisheries Commission Fish Passage Working Group, "Upstream Fish Passage Technologies for Managed Species."
13. Haro, *Proceedings of a Workshop on American Eel Passage Technologies*; Greene et al., *Atlantic Coast Diadromous Fish Habitat*; Atlantic States Marine Fisheries Commission Fish Passage Working Group, "Upstream Fish Passage Technologies for Managed Species."

Chapter 43

1. Normandeau, *New Hampshire Wildlife Action Plan*, appendix A (fish); Eads, Price, and Levine, "Fish Hosts of Four Freshwater Mussel Species in the Broad River, South Carolina"; Lellis et al., "Newly Documented Host Fishes for the Eastern Elliptio Mussel *Elliptio complanata*."
2. Silva et al., "The Future of Fish Passage Science, Engineering, and Practice"; Brown et al., "Fish and Hydropower on the U.S. Atlantic Coast"; Holyoke Gas and Electric, "Shortnose Sturgeon."
3. Brown et al., "Fish and Hydropower on the U.S. Atlantic Coast."
4. Silva et al., "The Future of Fish Passage Science, Engineering, and Practice."
5. Silva et al.; Brown et al., "Fish and Hydropower on the U.S. Atlantic Coast."
6. Greene et al., *Atlantic Coast Diadromous Fish Habitat*; Silva et al., "The Future of Fish Passage Science, Engineering, and Practice."

Chapter 44

1. American Rivers, "Elwha River Restoration"; US NPS, "History of the Elwha."

2. American Rivers, "Elwha River Restoration"; US NPS, "History of the Elwha."
3. American Rivers, "Elwha River Restoration"; US NPS, "History of the Elwha."
4. US NPS, "Elwha River Restoration."
5. Bellmore et al., "Status and Trends of Dam Removal Research in the United States"; American Rivers, "American Rivers Dam Removal Database Now Available to Public."
6. Association of State Dam Safety Officials, "Dam Failures and Incidents"; Sheets, "Get This Changed."
7. National Inventory of Dams, "Dams of the Nation."
8. Association of State Dam Safety Officials, "Dam Failures and Incidents"; Doyle et al., "Dam Removal in the United States."
9. Konrad et al., "Climate of the Southeast USA"; Association of State Dam Safety Officials, "Dam Failures and Incidents."
10. Shannon, "Oroville Dam"; Michigan Department of Environment, Great Lakes, and Energy, *Preliminary Report on the Edenville Dam Failure*; FEMA, *Dam Breach Report*.
11. Stanley and Doyle, "Trading Off."

Chapter 45
1. Cox, "Preserve Fish or History?"; Kozlowski, "Hidden Drowning Danger at Riverside Park Dam."
2. US Department of Homeland Security, *Dams Sector*; McFetridge, "Low-Lying Dams Drown Hundreds of Unsuspecting Victims"; Chicago Tribune Editorial Board, "Dismantling Dams Can Protect Rivers, Fish and People."
3. Qtd. in Pryne, "Elwha Dams Won't Be Coming Down Anytime Soon."
4. Cox, "Preserve Fish or History?"
5. C. A. Fox, Magilligan, and Sneddon, "You Kill the Dam, You Are Killing a Part of Me"; Magilligan, Sneddon, and Fox, "The Social, Historical, and Institutional Contingencies of Dam Removal."
6. Gosnell and Kelly, "Peace on the River?"
7. American Rivers, "American Rivers Dam Removal Database Now Available to Public."
8. Bennett et al., "New Upstream Records for Fishes Following Dam Removal in the Cahaba River, Alabama."
9. McCombs, "Neuse River Flows Freely after Milburnie Dam Removed"; Stradling, "We're Going to Have a Different River."
10. US FWS, "Stream Crossings in Georgia"; Neeson et al., "Aging Infrastructure Creates Opportunities for Cost-Efficient Restoration of Aquatic Ecosystem Connectivity"; W. W. Duncan, Bowers, and Frisch, "Missing Compensation"; Poplar-Jeffers et al., "Culvert Replacement and Stream Habitat Restoration."
11. W. W. Duncan, Bowers, and Frisch, "Missing Compensation."

Chapter 46
1. US EIA (website), "What Is U.S. Electricity Generation by Energy Source?"
2. OpenSecrets, Center for Responsive Politics; National Hydropower Association,

"About National Hydropower Association"; Lalasz, "Sustainable Hydropower"; Richter et al., "Lost in Development's Shadow."

3. Richter et al., "Lost in Development's Shadow"; Moran et al., "Sustainable Hydropower in the 21st Century."

4. Grill et al., "Mapping the World's Free-Flowing Rivers."

5. Opperman, "Concentration, Confrontation, Collaboration"; Opperman, Grill, and Hartmann, *The Power of Rivers*.

6. Ziv et al., "Trading-Off Fish Biodiversity, Food Security, and Hydropower in the Mekong River Basin"; Winemiller et al., "Balancing Hydropower and Biodiversity in the Amazon, Congo, and Mekong."

7. US EIA (website), "U.S. States."

8. US EIA, *Levelized Cost and Levelized Avoided Cost of New Generation Resources in the Annual Energy Outlook 2019*.

9. Opperman, "Penobscot River Dam Removal"; Natural Resources Council of Maine, "Penobscot River Restoration Project."

10. Opperman, "Penobscot River Dam Removal"; Natural Resources Council of Maine, "Penobscot River Restoration Project."

11. National Inventory of Dams, "Dams of the Nation."

12. Society for Ecological Restoration, "USA: Two Dam Removals"; Ritchie, "Tallahassee to Shut Down One of Two Hydroelectric Plants in Florida."

13. US Department of Energy, "Obama Administration Announces up to $32 Million Initiative to Expand Hydropower"; Patel, "President Obama Signs Pivotal Hydropower-Boosting Bills into Law"; *Water Power Magazine*, "US President Obama Signs Hydropower Bills into Law"; Uría-Martínez, Johnson, and O'Connor, *2017 Hydropower Market Report*.

14. US EIA (website), "Preliminary Monthly Electric Generator Inventory"; Uría-Martínez, Johnson, and Shan, *U.S. Hydropower Market Report*; DOE Global Energy Storage Database, home page.

15. Uría-Martínez, Johnson, and Shan, *U.S. Hydropower Market Report*; Nelson, "This Land Is All We Have Left."

16. Kirke, "Hydrokinetic and Ultra-Low Head Turbines in Rivers."

Chapter 47

1. National Inventory of Dams, "Dams of the Nation."

2. Nunez, "Floods, Explained"; National Weather Service, "Weather Related Fatality and Injury Statistics"; NOAA (website), "Billion-Dollar Weather and Climate Disasters."

3. Perry, "Significant Floods in the United States during the 20th Century"; Harrington, "In Hurricane Barry's Wake, Here Are the Worst Floods in American History."

4. Harrington, "In Hurricane Barry's Wake, Here Are the Worst Floods in American History"; Ambrose, "Man vs. Nature"; Doyle, *The Source*.

5. Doyle, *The Source*.

6. Doyle.

7. Doyle.
8. FEMA, home page.
9. FEMA.
10. Hersher and Benincasa, "How Federal Disaster Money Favors the Rich."
11. Hersher and Benincasa, "How Federal Disaster Money Favors the Rich"; Doyle, *The Source*.
12. Doyle, *The Source*.
13. NOAA (website), "Billion-Dollar Weather and Climate Disasters."
14. FEMA (website), "Acquisition of Property after a Flood Event."
15. City of Charlotte, "Charlotte-Mecklenburg Storm Water Services."
16. City of Charlotte.
17. World Wildlife Fund, *Natural and Nature-Based Flood Management*.
18. National Inventory of Dams, "Dams of the Nation."
19. Parsons, "Flooding and Natural and Nature-Based Solutions."
20. Parsons; Doyle, *The Source*.

Chapter 48
1. National Inventory of Dams, "Dams of the Nation."
2. National Inventory of Dams.
3. National Inventory of Dams.
4. L. Carter et al., "Southeast"; Toles, "DJI Phantom 3 Standard Lake Lahusage."

Chapter 49
1. Intergovernmental Panel on Climate Change, *IPCC Special Report*.
2. US EIA (website), "Energy-Related CO_2 Emission Data Tables"; Hsiang et al., "Estimating Economic Damage from Climate Change in the United States."
3. Solar Energy Industries Association, "Solar Industry Research Data"; Wagner, "North Carolina Will Work with Maryland, Virginia to Grow Offshore Wind Industry"; Kovaleski, "Virginia Lawmakers Commit to Zero Carbon Emissions by 2050."
4. Terando et al., "The Southern Megalopolis."

Chapter 50
1. Rigaud et al., *Groundswell*.

Bibliography

Aadland, Luther P. *Reconnecting Rivers: Natural Channel Design in Dam Removals and Fish Passage*. Fergus Falls: Minnesota Department of Natural Resources, 2010.

Aaron, Robert. "Explore Our Interactive Timeline of the Water Wars." WABE (website), March 6, 2014.

Aldrich, Marta W. "$83 Million Later, Unfinished Dam Being Dismantled." *Seattle Times*, October 10, 1999.

Alliance for Appalachia. "Mountaintop Removal Reclamation Fail." ILoveMountains (website). Accessed March 23, 2018.

Ambrose, Stephen. "Man vs. Nature: The Great Mississippi Flood of 1927." *National Geographic*, April 30, 2001.

American Coal Ash Association (website). "Production and Use Reports." Accessed August 12, 2020.

American Coal Council (website). "About the American Coal Council." Accessed March 26, 2018.

American Rivers (website). "American Rivers Dam Removal Database Now Available to Public." News release, November 14, 2017.

———. "Columbia River: Washington, Oregon." Accessed August 1, 2020.

———. "Elwha River Restoration." Internet Archive. Accessed August 1, 2020.

Andrews, David Q., and Olga V. Naidenko. "Population-Wide Exposure to Per- and Polyfluoroalkyl Substances from Drinking Water in the United States." *Environmental Science and Technology Letters* 7, no. 12 (2020): 931–36.

Aneja, Viney P., Aaron Isherwood, and Peter Morgan. "Characterization of Particulate Matter (PM_{10}) Related to Surface Coal Mining Operations in Appalachia." *Atmospheric Environment* 54 (2012): 496–501.

Aneja, Viney P., Priya R. Pillai, Aaron Isherwood, Peter Morgan, and Saurabh P. Aneja. "Particulate Matter Pollution in the Coal-Producing Regions of the Appalachian Mountains: Integrated Ground-Based Measurements and Satellite Analysis." *Journal of the Air and Waste Management Association* 67, no. 4 (2017): 421–30.

Appalachian Voices (website). "Human Health Impacts." Accessed March 24, 2018.

Arena, JenAlyse. "Coal Production Using Mountaintop Removal Mining Decreases by 62% since 2008." *Today in Energy* (newsletter), US Energy Information Administration, July 7, 2015.

Asseng, Senthold, Wendy-Lin Bartels, Kenneth J. Boote, Norman E. Breuer, Davide Cammarano, Christine C. Fortuin, and Clyde W. Fraisse, et al. "Agriculture and Climate Change in the Southeast USA." In *Climate of the Southeast United States: Variability, Change, Impacts, and Vulnerability*, edited by Keith T. Ingram, Kirstin Dow, Lynne Carter, and Julie Anderson, 128–64. Washington, DC: Island Press, 2013.

Associated Press. "Agency Appeals Order Blocking Savannah River Dam Removal." AP News (website), January 22, 2021.

———. "Canton Reservoir Shows Realities of Deal Plan." *Chattanooga Times Free Press*, February 14, 2011.

———. "It Will Cost $900 Million to Move Power Plant Ash to Landfill, TVA CEO Says." *Chattanooga Times Free Press*, November 17, 2017.

———. "Judge Says Dam near Augusta Can't Be Removed if Water Level Will Fall." *Atlanta Journal-Constitution*, November 25, 2020.

Association of State Dam Safety Officials (website). "Dam Failures and Incidents." Accessed October 6, 2019.

Atkins, Leah Rawls. *Developed for the Service of Alabama: The Centennial History of the Alabama Power Company, 1906–2006*. Birmingham: Alabama Power Company, 2006.

Atlanta Regional Commission (website). "Tri-State Water Wars—Background and History." Accessed September 29, 2019.

Atlantic States Marine Fisheries Commission (website). "American Eel." Accessed October 6, 2019.

———. "Shad & River Herring." Accessed October 5, 2019.

Atlantic States Marine Fisheries Commission Fish Passage Working Group. "Upstream Fish Passage Technologies for Managed Species." September 2010.

Backer, Lorraine C. "Cyanobacterial Harmful Algal Blooms (CyanoHABs): Developing a Public Health Response." *Lake and Reservoir Management* 18, no. 1 (2002): 20–31.

Baer, Katherine, and April Ingle. *Protecting and Restoring Flows in Our Southeastern Rivers: A Synthesis of State Policies for Water Security and Sustainability*. Boulder, CO: River Network, 2016.

Bandy, Earl. "The Surface Mining Control and Reclamation Act (SMCRA) of 1977: A 30th Anniversary Review." Presentation to the US Department of the Interior Committee on Natural Resources, July 25, 2007.

Bangor Daily News. "Three Dams to Get Power Upgrades along Penobscot." May 31, 2006.

Baptiste, Nathalie. "When a Hurricane Takes Your Home." *Slate*, October 30, 2017.

Bastias Saavedra, Manuel. "Nitrate." In *International Encyclopedia of the First World War*. Last updated October 8, 2014.

Beck, Michael W., Robert D. Brumbaugh, Laura Airoldi, Alvar Carranza, Loren D. Coen, Christine Crawford, and Omar Defeo, et al. "Oyster Reefs at Risk and Recommendations for Conservation, Restoration, and Management." *BioScience* 61, no. 2 (2011): 107–16.

Bellino, Jason C., Eve L. Kuniansky, Andrew M. O'Reilly, and Joann F. Dixon. *Hydrogeologic Setting, Conceptual Groundwater Flow System, and Hydrologic Conditions 1995–2010 in Florida and Parts of Georgia, Alabama, and South Carolina*. Report 2018-5030. Reston, VA: US Geological Survey, 2018.

Bellmore, J. Ryan, Jeffrey J. Duda, Laura S. Craig, Samantha L. Greene, Christian E. Torgersen, Mathias J. Collins, and Katherine Vittum. "Status and Trends of Dam Removal Research in the United States." *WIREs Water* 4, no. 2 (2017): e1164.

Bender, Morris A., Thomas R. Knutson, Robert E. Tuleya, Joseph J. Sirutis, Gabriel A.

Vecchi, Stephen T. Garner, and Isaac M. Held. "Modeled Impact of Anthropogenic Warming on the Frequency of Intense Atlantic Hurricanes." *Science* 327, no. 5964 (January 22, 2010): 454–58.

Benke, Arthur C. "A Perspective on America's Vanishing Streams." *Journal of the North American Benthological Society* 9, no. 1 (1990): 77–88.

Bennett, Micah G., J. Heath Howell, Bernard R. Kuhajda, and Paul L. Freeman. "New Upstream Records for Fishes Following Dam Removal in the Cahaba River, Alabama." *Southeastern Fishes Council Proceedings* 1, no. 55 (2015): 51–61.

Berkes, Howard, and Adelina Lancianese. "Black Lung Study Finds Biggest Cluster Ever of Fatal Coal Miners' Disease." *All Things Considered*. National Public Radio, February 6, 2018.

Berkowitz, Bonnie, and Tim Meko. "Appalachia Comes Up Small in Era of Giant Coal Mines." *Washington Post*, May 5, 2017.

Bernhardt, Emily S., and Margaret A. Palmer. "The Environmental Costs of Mountaintop Mining Valley Fill Operations for Aquatic Ecosystems of the Central Appalachians." *Annals of the New York Academy of Sciences* 1223, no. 1 (2011): 39–57.

Beven, John L., II, Robbie Berg, and Andrew Hagen. *Hurricane Michael (AL142018)*. Miami, FL: National Hurricane Center, 2019.

Bienkowski, Brian. "Desalination Is an Expensive Energy Hog, but Improvements Are on the Way." *World*, PRX, May 15, 2015.

Billington, David P., Donald C. Jackson, and Martin V. Melosi. *The History of Large Federal Dams: Planning, Design, and Construction in the Era of Big Dams*. Denver, CO: US Department of the Interior Bureau of Reclamation, 2005.

Black Belt Citizens (website). "Black Belt Citizens Fighting for Health and Justice." Accessed August 1, 2020.

Blake, Eric S., and Ethan J. Gibney. *The Deadliest, Costliest, and Most Intense United States Tropical Cyclones from 1851 to 2010 (and Other Frequently Requested Hurricane Facts)*. Miami, FL: National Hurricane Center, August 2011.

Blake, Eric S., Jerry D. Jarrell, Edward N. Rappaport, and Christopher W. Landsea. *The Deadliest, Costliest, and Most Intense United States Hurricanes from 1851 to 2004 (and Other Frequently Requested Hurricane Facts)*. Miami, FL: National Hurricane Center, August 2005.

Blake, Myra. "Comments on the Draft Title V Permit Modification for Allen Steam Station." Letter from Southern Environmental Law Center to Edward L. Martin, Division of Air Quality, February 27, 2017.

Blankenship, Karl. "PA Shad Hatchery's 42-Year Run May Be Coming to an End." *Chesapeake Bay Journal*, June 4, 2018.

Boissoneault, Lorraine. "The Coal Mining Massacre America Forgot." *Smithsonian Magazine*, April 25, 2017.

Boles, Sydney. "Federal Watchdog Finds Coal Safety Regulator Not Protecting Miners from Silica Dust." *Ohio Valley ReSource*. Louisville Public Media, November 16, 2020.

Boschung, Herbert T., Jr., and Richard L. Mayden. *Fishes of Alabama*. Washington, DC: Smithsonian Books, 2004.

Bourne, Edward Gaylord, and Buckingham Smith, eds. *True Relation of the Vicissitudes That Attended the Governor Don Hernando de Soto and Some Nobles of Portugal in the Discovery of the Province of Florida Now Just Given by a Fildalgo of Elvas.* New York: A. S. Barnes, 1904.

Bourne, Joel K., Jr. "Coal's Other Dark Side: Toxic Ash That Can Poison Water and People." *National Geographic*, February 19, 2019.

Box, Jayne Brim, and James D. Williams. *Unionid Mussels of the Apalachicola Basin in Alabama, Florida, and Georgia.* Bulletin 21. Tuscaloosa: Alabama Museum of Natural History, April 2000.

Boyles, Abee L., Robyn B. Blain, Johanna R. Rochester, Raghavendhran Avanasi, Susan B. Goldhaber, Sofie McComb, Stephanie D. Holmgren, Scott A. Masten, and Kristina A. Thayer. "Systematic Review of Community Health Impacts of Mountaintop Removal Mining." *Environment International* 107 (2017): 163–72.

Brosius, Ashley, Ernesto Diaz, Rick Durbrow, Rhonda Evans, Stephanie Fauver, Tim Hayden, and Bob Howard, et al. "Climate Adaptations in the Southeast USA." In *Climate of the Southeast United States: Variability, Change, Impacts, and Vulnerability*, edited by Keith T. Ingram, Kirstin Dow, Lynne Carter, and Julie Anderson, 295–320. Washington, DC: Island Press, 2013.

Brown, J. Jed, Karin E. Limburg, John R. Waldman, Kurt Stephenson, Edward P. Glenn, Francis Juanes, and Adrian Jordaan. "Fish and Hydropower on the U.S. Atlantic Coast: Failed Fisheries Policies from Half-Way Technologies." *Conservation Letters* 6, no. 4 (2013): 280–86.

Brownell, Prescott, Alex Haro, Sean McDermott, Al Blott, and Fritz Rohde. *Diadromous Fish Passage: A Primer on Technology, Planning, and Design for the Atlantic and Gulf Coasts.* Washington, DC: National Oceanic and Atmospheric Administration, 2011.

Burkholder, JoAnn, Bob Libra, Peter Weyer, Susan Heathcote, Dana Kolpin, Peter S. Thorne, and Michael Wichman. "Impacts of Waste from Concentrated Animal Feeding Operations on Water Quality." *Environmental Health Perspectives* 115, no. 2 (2007): 308–12.

Burns, Matthew, and Laura Leslie. "State Orders Duke Energy to Dig Up All Remaining Coal Ash Ponds in NC." WRAL News, April 1, 2019.

Butler, Robert S., Jerry Ziewitz, Shawn K. ALam, and Holly N Blalock-Herod. *Recovery Plan for Endangered Fat Threeridge (Amblema neislerii), Shinyrayed Pocketbook (Lampsilis subangulata), Gulf Moccasinshell (Medionidus penicillatus), Ochlockonee Moccasinshell (Medionidus simpsonianus), and Oval Pigtoe (Pleurobema pyriforme), and Threatened Chipola Slabshell (Elliptio chipolaensis), and Purple Bankclimber (Elliptoideus sloatianus).* Atlanta, GA: US Fish and Wildlife Service Southeast Region, 2003.

Byerly, Steve, and Craig Allison. "Lock and Dam's Fate in Question after New Ruling from Appeals Court." WRDW News, April 19, 2023.

Carter, Lynne, Adam Terando, Kirstin Dow, Kevin Hiers, Kenneth E. Kunkel, Aranzazu Lascurain, Doug Marcy, Michael Osland, and Paul Schramm. "Southeast." In *Fourth National Climate Assessment*, vol. 2, *Impacts, Risks, and Adaptation in the*

United States, edited by David R. Reidmiller, Christopher W. Avery, David R. Easterling, Kenneth E. Kunkel, K. L. M. Lewis, Thomas K. Maycock, and Brooke C. Stewart, 743–808. Washington, DC: US Global Change Research Program, 2018.

Carter, Nicole T. *Desalination and Membrane Technologies: Federal Research and Adoption Issues*. R40477. Washington, DC: Congressional Research Service, January 2015.

Cason, James H., Robert A. Mills, and Paula A. Tully. *Historical Remembrances of Choctawhatchee Bay, Northwest Florida*. Havana: Northwest Florida Water Management District, 1985.

Catarci, Camillo. *World Markets and Industry of Selected Commercially-Exploited Aquatic Species with an International Conservation Profile*. Fisheries Circular no. 990. Rome: Food and Agriculture Organization of the United Nations, 2004.

Centers for Disease Control and Prevention (CDC) (website). "Body Measurements." National Center for Health Statistics. Accessed September 22, 2019.

———. "Statistics: Coal Operators." Last updated April 11, 2017.

———. "Work-Related Lung Disease Surveillance System (eWoRLD)." National Institute for Occupational Safety and Health. Accessed March 2, 2023.

Chapman, Dan. "Metro Atlanta Water Use to Increase Much Less Than Earlier Predicted." *Atlanta Journal-Constitution*, August 26, 2015.

Chastain, Charlotte A., Stephen W. Golladay, and Tara K. Muenz. "Distribution of Unionid Mussels in Tributaries of the Lower Flint River, Southwestern Georgia: An Examination of Current and Historical Trends." In *Proceedings of the 2005 Georgia Water Resources Conference*, edited by Kathryn J. Hatcher. Athens: University of Georgia, 2005.

Chesapeake Bay Foundation. *Agricultural Conservation Practices: Clean Water and Climate-Smart Investments*. July 2022.

Chesapeake Bay Foundation (website). "Chesapeake 2000 EPA Suit." Accessed August 1, 2020.

———. "The History of Chesapeake Bay Cleanup Efforts." Accessed August 1, 2020.

———. "What Is Killing the Bay?" Accessed May 15, 2023.

Chesapeake Bay Program (website). "Agricultural Runoff." Accessed May 15, 2023.

Chicago Tribune Editorial Board. "Dismantling Dams Can Protect Rivers, Fish and People." *Chicago Tribune*, August 19, 2016.

City of Charlotte (website). "Charlotte-Mecklenburg Storm Water Services." Accessed August 1, 2020.

City of Muscle Shoals (website). "The History of the City of Muscle Shoals." Accessed March 2, 2023.

City of Norfolk. *Norfolk Vision 2100*. Adopted November 22, 2016.

Clean Water Atlanta (website). "Wastewater." Accessed November 27, 2017.

Climate Central (website). "2017 U.S. Temperature Review." January 9, 2018.

CO_2.Earth (website). "Daily CO_2." Accessed April 23, 2023.

Conservation Gateway (website). "Ecological Limits of Hydrologic Alteration (ELOHA)." Nature Conservancy. Accessed October 4, 2019.

———. "Environmental Flow Components." Nature Conservancy. Accessed October 3, 2019.

Cooper, P. F. "Historical Aspects of Wastewater Treatment." In *Decentralised Sanitation and Reuse: Concepts, Systems and Implementation*, edited by P. Lens, G. Zeeman, and G. Lettinga, 11–38. London: IWA, 2001.

Copeland, Claudia. *Clean Water Act: A Summary of the Law*. RL30030. Washington, DC: Congressional Research Service, October 2016.

Cornwall, Warren. "Hundreds of New Dams Could Mean Trouble for Our Climate." *Science*, September 28, 2016.

Costanza, Robert, Octavio Pérez-Maqueo, M. Luisa Martínez, Paul C. Sutton, Sharolyn Anderson, and Kenneth Mulder. "The Value of Coastal Wetlands for Hurricane Protection." *AMBIO: A Journal of the Human Environment* 37, no. 4 (2008): 241–48.

Council on Environmental Quality. *Environmental Quality: The Second Annual Report of the Council on Environmental Quality*. Washington, DC: Government Printing Office, 1971.

Cox, Jeremy. "Bay Scientists Say Stream Restoration Not Delivering as Much as Hoped." *Maryland Reporter* (blog), November 28, 2018.

———. "Preserve Fish or History? VA Dam Removal Churns Up Debate." *Bay Journal*, January 29, 2019.

Crable, Ad. "Shad Wars: A Long Violent Period in Lancaster County over a Fish in the Susquehanna." *LNP*. Lancaster Online, March 31, 2018.

Craft, Christopher, Jonathan Clough, Jeff Ehman, Samantha Jove, Richard Park, Steve Pennings, Hongyu Guo, and Megan Machmuller. "Forecasting the Effects of Accelerated Sea-Level Rise on Tidal Marsh Ecosystem Services." *Frontiers in Ecology and the Environment* 7, no. 2 (2009): 73–78.

Cummings, K. "*Elliptio crassidens*, Elephantear Mussel." *IUCN Red List of Threatened Species*, 2011, e.T188905A8660568.

Dance, Scott. "Halfway to Cleanup Deadline, Chesapeake Bay Hits Goals for Phosphorus, Sediment, but Misses Nitrogen Target." *Baltimore Sun*, July 17, 2018.

Daniels, R. B. "Soil Erosion and Degradation in the Southern Piedmont of the USA." In *Land Transformation in Agriculture*, edited by M. Gordon Wolman and Frédéric G. A. Fournier, 407–28. Chicester, UK: John Wiley, 1987.

Davis, Jack E. *The Gulf: The Making of an American Sea*. New York: Liveright, 2017.

Day, John W., Didier Pont, Philippe F. Hensel, and Carlès Ibañez. "Impacts of Sea-Level Rise on Deltas in the Gulf of Mexico and the Mediterranean: The Importance of Pulsing Events to Sustainability." *Estuaries* 18, no. 4 (1995): 636–47.

DeConto, Robert M., and David Pollard. "Contribution of Antarctica to Past and Future Sea-Level Rise." *Nature* 531 (2016): 591–97.

Deegan, Linda A., David Samuel Johnson, R. Scott Warren, Bruce J. Peterson, John W. Fleeger, Sergio Fagherazzi, and Wilfred M. Wollheim. "Coastal Eutrophication as a Driver of Salt Marsh Loss." *Nature* 490 (2012): 388–92.

Dennis, Brady, Steven Mufson, and Juliet Eilperin. "Dam Breach Sends Toxic Coal Ash Flowing into a Major North Carolina River." *Washington Post*, September 22, 2018.

Diaz, Robert J., and Rutger Rosenberg. "Spreading Dead Zones and Consequences for Marine Ecosystems." *Science* 321, no. 5891 (2008): 926–29.

Díaz, Sandra, Josef Settele, Eduardo Brondizio, Hien T. Ngo, Maximilien Guèze, John Agard, and Almut Arneth, et al. *The Global Assessment Report on Biodiversity and Ecosystem Services: Summary for Policymakers*. Bonn: Intergovernmental Science-Policy Platform on Biodiversity and Ecosystem Services, 2019.

Dodds, Walter K., Wes W. Bouska, Jeffrey L. Eitzmann, Tyler J. Pilger, Kristen L. Pitts, Alyssa J. Riley, Joshua T. Schloesser, and Darren J. Thornbrugh. "Eutrophication of U.S. Freshwaters: Analysis of Potential Economic Damages." *Environmental Science and Technology* 43, no. 1 (2009): 12–19.

DOE Global Energy Storage Database (website). Home page. US Department of Energy and Sandia National Laboratories. Accessed June 24, 2021.

Donovan, Mitchell, Andrew Miller, Matthew Baker, and Allen Gellis. "Sediment Contributions from Floodplains and Legacy Sediments to Piedmont Streams of Baltimore County, Maryland." *Geomorphology* 235 (2015): 88–105.

Douban, Gigi. "Why Alabama Still Has No Water Management Plan." Public Radio WBHM (website), November 10, 2017.

Doyle, Martin W. *The Source: How Rivers Made America and America Remade Its Rivers*. New York: W. W. Norton, 2018.

Doyle, Martin W., Emily H. Stanley, Jon M. Harbor, and Gordon S. Grant. "Dam Removal in the United States: Emerging Needs for Science and Policy." *Eos* 84, no. 4 (2003): 29–33.

Dumitru, Oana A., Jacqueline Austermann, Victor J. Polyak, Joan J. Fornós, Yemane Asmerom, Joaquín Ginés, Angel Ginés, and Bogdan P. Onac. "Constraints on Global Mean Sea Level during Pliocene Warmth." *Nature* 574 (2019): 233–36.

Duncan, David E. *Hernando de Soto: A Savage Quest in the Americas*. Norman: University of Oklahoma Press, 1996.

Duncan, Mack S. "Fall Line." In *New Georgia Encyclopedia* (online). Last updated July 23, 2018.

Duncan, R. Scot. *Southern Wonder: Alabama's Surprising Biodiversity*. Tuscaloosa: University of Alabama Press, 2013.

Duncan, William W., Kathleen M. Bowers, and John R. Frisch. "Missing Compensation: A Study of Compensatory Mitigation and Fish Passage in Georgia." *Journal of Fish and Wildlife Management* 9, no. 1 (2018): 132–43.

Dunlap, Deborah J., and Tracey L. Martin. *Historic Pensacola Photographs*. Vol. 1. Pensacola, FL: Bayshore, 1999.

Eads, Chris B., Jennifer E. Price, and Jay F. Levine. "Fish Hosts of Four Freshwater Mussel Species in the Broad River, South Carolina." *Southeastern Naturalist* 14, no. 1 (2015): 85–97.

Eastern Oyster Biological Review Team. *Status Review of the Eastern Oyster (Crassostrea virginica)*. Report to the National Marine Fisheries Service, Northeast Regional Office, February 16, 2007.

Ebersole, Rene. "19 Eel Smugglers Sentenced, but Lucrative Trade Persists." *National Geographic*, June 27, 2018.

Economist. "Appalachia: The Fifty Years War." March 19, 2015.

Egan, Dan. "The Cancer of the Great Lakes." *Nautilus*, February 24, 2017.

Eggert, Susan. "Nature Averts a Disaster." *Scientist*, October 2, 2000.

Elam, Constance. "Culture, Poverty and Education in Appalachian Kentucky." *Education and Culture* 18, no. 1 (2002): 10–13.

Elkins, Duncan C., Sarah C. Sweat, Katie S. Hill, Bernard R. Kuhajda, Anna L. George, and Seth J. Wenger. *The Southeastern Aquatic Biodiversity Conservation Strategy*. Athens: University of Georgia River Basin Center, 2016.

Emanuel, Ben, and Jenny Hoffner. *Money Pit: The High Cost and High Risk of Water Supply Reservoirs in the Southeast*. Washington, DC: American Rivers, 2012.

Engelman-Lado, Marianne, Camila Bustos, Haley Leslie-Bole, and Perry Leung. "Environmental Injustice in Uniontown, Alabama, Decades after the Civil Rights Act of 1964: It's Time For Action." *Human Rights Magazine*, May 21, 2021.

Erickson, Mark St. John. "Decoding the Mystery of the Jamestown Sturgeon." *Daily Press* (Newport News, VA), July 19, 2013.

Evans, Jason M., Jill Gambill, Robin J. McDowell, P. Warwick Prichard, and Charles S. Hopkinson. *Tybee Island Sea-Level Rise Adaptation Plan*. Washington, DC: National Oceanic and Atmospheric Administration, 2016.

Ezzell, Patricia Bernard. "Wilson Dam and Reservoir." In *Encyclopedia of Alabama* (online). Last updated June 16, 2020.

Fahrenthold, David A. "Environmental Regulations to Curtail Mountaintop Mining." *Washington Post*, April 2, 2010.

Fakour, Hoda, and Shang-Lien Lo. "Formation of Trihalomethanes as Disinfection Byproducts in Herbal Spa Pools." *Scientific Reports* 8 (2018): article 5709.

Fears, Darryl. "A Desperate Try to Restock the Potomac's Sturgeon." *Washington Post*, January 27, 2013.

———. "Trump Administration Halted a Study of Mountaintop Coal Mining's Health Effects." *Washington Post*, August 21, 2017.

Federal Emergency Management Agency (FEMA). *Dam Breach Report—Hurricane Matthew in North Carolina and South Carolina*. 2018.

Federal Emergency Management Agency (FEMA) (website). "Acquisition of Property after a Flood Event." Fact sheet, November 13, 2018.

———. Home page. Accessed October 9, 2019.

Federal Energy Regulatory Commission. *Hydropower Primer: A Handbook of Hydropower Basics*. February 2017.

Ferguson, Grant, and Tom Gleeson. "Vulnerability of Coastal Aquifers to Groundwater Use and Climate Change." *Nature Climate Change* 2 (2012): 342–45.

First Street Foundation (website). "Sea Level Rise Washes Away $76.4 Million in Texas Home Values." News release, April 23, 2019.

Flessner, Dave. "Court Dismisses Lawsuit against TVA, Jacobs Engineering over Coal Ash Cleanup." *Chattanooga Times Free Press*, October 1, 2020.

Flowers, H. Jared, W. E. Pine III, A. C. Dutterer, K. G. Johnson, J. W. Ziewitz, M. S. Allen, and Frank M. Parauka. "Spawning Site Selection and Potential Implications of Modified Flow Regimes on Viability of Gulf Sturgeon Populations." *Transactions of the American Fisheries Society* 138, no. 6 (2009): 1266–84.

Fowler, Lara. "New Special Master Finds for Georgia in Most Recent Round of Water

Dispute." *SCOTUSblog*, January 9, 2020.

———. "No April Fool's Joke for Florida: Water Rights Case Is Dismissed." *SCOTUSblog*, April 1, 2021.

Fox, Coleen A., Francis J. Magilligan, and Christopher S. Sneddon. "'You Kill the Dam, You Are Killing a Part of Me': Dam Removal and the Environmental Politics of River Restoration." *Geoforum* 70 (2016): 93–104.

Fox, Julia. "Mountaintop Removal in West Virginia: An Environmental Sacrifice Zone." *Organization & Environment* 12, no. 2 (1999): 163–83.

Frederikse, Thomas, Riccardo E. M. Riva, and Matt A. King. "Ocean Bottom Deformation Due to Present-Day Mass Redistribution and Its Impact on Sea Level Observations." *Geophysical Research Letters* 44, no. 24 (2017): 12306–14.

Freeman, Mary C., Elise R. Irwin, Noel M. Burkhead, Byron J. Freeman, and Henry L. Bart Jr. "Status and Conservation of the Fish Fauna of the Alabama River System." *American Fisheries Society Symposium* 45 (2005): 557–85.

Fritts, Andrea K., Mark W. Fritts II, Douglas L. Peterson, Dewayne A. Fox, and Robert B. Bringolf. "Critical Linkage of Imperiled Species: Gulf Sturgeon as Host for Purple Bankclimber Mussels." *Freshwater Science* 31, no. 4 (2012): 1223–32.

Fuller, P., L. Nico, M. Neilson, K. Dettloff, and R. Sturtevant. "*Anguilla rostrata*." US Geological Survey (website), Nonindigenous Aquatic Species Database. Revised September 12, 2019.

Gallegos, Tanya J., and Brian A. Varela. *Trends in Hydraulic Fracturing Distributions and Treatment Fluids, Additives, Proppants, and Water Volumes Applied to Wells Drilled in the United States from 1947 through 2010—Data Analysis and Comparison to the Literature*. Report 2014-5131. Reston, VA: US Geological Survey, 2015.

Galloway, Gerald E, Allison Reilly, Sung Ryoo, Anjanette Riley, Maggie Haslam, Sam Brody, Wesley Highfield, Joshua Gunn, Jayton Rainey, and Sherry Parker. *The Growing Threat of Urban Flooding: A National Challenge*. University of Maryland and Texas A&M University, Galveston. College Park, MD: A. James Clark School of Engineering, 2018.

Gallun, Alby. "Flushing Out Bubbly Creek." *Crain's Chicago Business*, July 25, 2004.

Garman, Greg. "Historical Significance of the Atlantic Sturgeon." VCU News, Virginia Commonwealth University (website), November 2, 2012.

Garner, Jeffrey T., and Stuart W. McGregor. "Current Status of Freshwater Mussels (*Unionidae, Margaritiferidae*) in the Muscle Shoals Area of Tennessee River in Alabama (Muscle Shoals Revisited Again)." *American Malacological Bulletin* 16, no. 1 (2001): 155–70.

Gatenby, Catherine. "Mussels Making Moves for Water Quality." *U.S. Fish and Wildlife Service Northeast* (blog), August 21, 2017.

Gedan, Keryn B., Matthew L. Kirwan, Eric Wolanski, Edward B. Barbier, and Brian R. Silliman. "The Present and Future Role of Coastal Wetland Vegetation in Protecting Shorelines: Answering Recent Challenges to the Paradigm." *Climatic Change* 106, no. 1 (2010): 7–29.

Geological Survey of Alabama (website). "Groundwater Assessment Program." Accessed September 29, 2019.

Georgia Department of Natural Resources: Environmental Protection Division (website). "Water Conservation." Accessed September 29, 2019.

Georgia River Network (website). "Chattahoochee River." Accessed September 29, 2019.

Gibbens, Sarah. "As the Climate Crisis Worsens, Cities Turn to Parks." *National Geographic*, May 21, 2019.

———. "Massive 8,000-Mile 'Dead Zone' Could Be One of the Gulf's Largest." *National Geographic*, June 10, 2019.

Gilmer, Robert A. "In the Shadow of Removal: Historical Memory, Indianness, and the Tellico Dam Project." PhD diss., University of Minnesota, 2011.

Gittman, Rachel K., F. Joel Fodrie, Alyssa M. Popowich, Danielle A. Keller, John F. Bruno, Carolyn A. Currin, Charles H. Peterson, and Michael F Piehler. "Engineering Away Our Natural Defenses: An Analysis of Shoreline Hardening in the US." *Frontiers in Ecology and the Environment* 13, no. 6 (2015): 301–7.

Global Climate Change (website). "Carbon Dioxide: Latest Measurement." National Aeronautics and Space Administration. Accessed November 18, 2021.

Good, Kelly D., and Jeanne M. VanBriesen. "Current and Potential Future Bromide Loads from Coal-Fired Power Plants in the Allegheny River Basin and Their Effects on Downstream Concentrations." *Environmental Science and Technology* 50 (2016): 9078–88.

Gornitz, Vivien. "Sea Level Rise, after the Ice Melted and Today." Science Briefs, NASA Goddard Institute for Space Studies (website), January 2007.

Gosnell, Hannah, and Erin Clover Kelly. "Peace on the River? Social-Ecological Restoration and Large Dam Removal in the Klamath Basin, USA." *Water Alternatives* 3, no. 2 (2010): 361–83.

Grabar, Henry. "How We Built Our Way into an Urban Flooding Epidemic." Slate, November 29, 2018.

Grabowski, Jonathan H., Robert D. Brumbaugh, Robert F. Conrad, Andrew G. Keeler, James J. Opaluch, Charles H. Peterson, Michael F. Piehler, Sean P. Powers, and Ashley R. Smyth. "Economic Valuation of Ecosystem Services Provided by Oyster Reefs." *BioScience* 62, no. 10 (2012): 900–909.

Graf, William L. "Dam Nation: A Geographic Census of American Dams and Their Large-Scale Hydrologic Impacts." *Water Resources Research* 35, no. 4 (1999): 1305–11.

Green, Jared. "Norfolk Forges a Path to a Resilient Future." *The Dirt* (blog), American Society of Landscape Architects, May 14, 2019.

Greene, Karen E., Jennifer L. Zimmerman, R. Wilson Laney, and Jessie C. Thomas-Blate. *Atlantic Coast Diadromous Fish Habitat: A Review of Utilization, Threats, Recommendations for Conservation, and Research Needs*. Habitat Management Series no. 9. Washington, DC: Atlantic States Marine Fisheries Commission, 2009.

Greywater Action (website). "Greywater Action—For a Sustainable Water Culture." Accessed October 1, 2019.

Grill, G., B. Lehner, M. Thieme, B. Geenen, D. Tickner, F. Antonelli, and S. Babu, et al. "Mapping the World's Free-Flowing Rivers." *Nature* 569 (2019): 215–21.

Ground Water Protection Council (website). "Aquifer Storage and Recovery." Accessed September 30, 2019.

Gutiérrez, Bertrand M. "Downstream from Duke Energy Plant, Madison and Eden Continue Costly Search for Cleaner Drinking Water." *Winston-Salem Journal*, March 20, 2017.

Haag, Wendell R. *North American Freshwater Mussels: Natural History, Ecology, and Conservation*. New York: Cambridge University Press, 2012.

———. "Past and Future Patterns of Freshwater Mussel Extinctions in North America during the Holocene." In *Holocene Extinctions*, edited by Samuel T. Turvey, 107–28. Oxford: Oxford University Press, 2009.

Haag, Wendell R., and James D. Williams. "Biodiversity on the Brink: An Assessment of Conservation Strategies for North American Freshwater Mussels." *Hydrobiologia* 735 (2014): 45–60.

Hall, Shannon. "Exxon Knew about Climate Change almost 40 Years Ago." *Scientific American*, October 26, 2015.

Hallerman, Tamar. "Georgia Is Nowhere Near Last Battle in Tri-State Water Wars." *Atlanta Journal-Constitution*, June 28, 2018.

Hamby, Chris. "Breathless and Burdened: A Century of Denial on Black Lung." Center for Public Integrity (website), November 1, 2013.

Hanson, Ronald L. "Evapotranspiration and Droughts." In *National Water Summary 1988–89—Hydrologic Events and Floods and Droughts*, Water-Supply Paper 2375, edited by Richard W. Paulson, Edith B. Chase, Robert S. Roberts, and David W. Moody, 99–104. Denver, CO: US Geological Survey, 1991.

Harden, Stephen L. *Surface-Water Quality in Agricultural Watersheds of the North Carolina Coastal Plain Associated with Concentrated Animal Feeding Operations*. Report 2015-5080. Reston, VA: US Geological Survey, 2015.

Harned, Douglas A., Erik L. Staub, Kelly L. Peak, Kirsten M. Tighe, and Silvia Terziotti. *Trends in Water Quality in the Southeastern United States, 1973–2005*. Report 2009-5268. Reston, VA: US Geological Survey, 2010.

Haro, Alex. "Presentation on Fish Passage Concerns for Shad and River Herring, Atlantic (and Shortnose) Sturgeon, and American Eel." Atlantic States Marine Fisheries Commission, Workshop on Fish Passage Issues Impacting Atlantic Coast States, Jacksonville, FL, April 3 and 4, 2008.

———, ed. *Proceedings of a Workshop on American Eel Passage Technologies*. Gloucester, MA, March 30–31, 2011. Arlington, VA: Atlantic States Marine Fisheries Commission, July 2013.

Harrington, John. "In Hurricane Barry's Wake, Here Are the Worst Floods in American History." *USA Today*, July 17, 2019.

Hart, Michael A., Wendell R. Haag, Robert Bringolf, and James A. Stoeckel. "Novel Technique to Identify Large River Host Fish for Freshwater Mussel Propagation and Conservation." *Aquaculture Reports* 9 (2018): 10–17.

Harvey, Abby, Aaron Larson, and Sonal Patel. "History of Power: The Evolution of the Electric Generation Industry." *Power*, October 2017, last updated October 2022.

Hayhoe, Katharine, Donald J. Wuebbles, David R. Easterling, David W. Fahey, Sarah

Doherty, James P. Kossin, William V. Sweet, Russell S. Vose, and Michael F. Wehner. "Our Changing Climate." In *Fourth National Climate Assessment*, vol. 2, *Impacts, Risks, and Adaptation in the United States*, edited by David R. Reidmiller, Christopher W. Avery, David R. Easterling, Kenneth E. Kunkel, K. L. M. Lewis, Thomas K. Maycock, and Brooke C. Stewart, 72–144. Washington, DC: US Global Change Research Program, 2018.

He, Qiang, and Brian R. Silliman. "Biogeographic Consequences of Nutrient Enrichment for Plant-Herbivore Interactions in Coastal Wetlands." *Ecology Letters* 18, no. 5 (2015): 462–71.

Hendryx, Michael, Florence Fulk, and Andrea McGinley. "Public Drinking Water Violations in Mountaintop Coal Mining Areas of West Virginia, USA." *Water Quality, Exposure and Health* 4 (2012): 169–75.

Hendryx, Michael, and Benjamin Holland. "Unintended Consequences of the Clean Air Act: Mortality Rates in Appalachian Coal Mining Communities." *Environmental Science and Policy* 63 (2016): 1–6.

Hergert, Gary, Rex Nielsen, and Jim Margheim. "WWII Nitrogen Production Issues in Age of Modern Fertilizers." *CropWatch* (newsletter), University of Nebraska-Lincoln, Institute of Agriculture and Natural Resources, April 10, 2015.

Herndon, Andrew. "Energy Makes Up Half of Desalination Plant Costs: Study." Bloomberg, April 30, 2013.

Hersher, Rebecca, and Robert Benincasa. "How Federal Disaster Money Favors the Rich." *All Things Considered*. National Public Radio, March 5, 2019.

Hilton, Eric J., and Lance Grande. "Review of the Fossil Record of Sturgeons, Family Acipenseridae (Actinopterygii: Acipenseriformes), from North America." *Journal of Paleontology* 80, no. 4 (2006): 672–83.

Hilton, Eric J., B. Kynard, M. T. Balazik, A. Z. Horodysky, and C. B. Dillman. "Review of the Biology, Fisheries, and Conservation Status of the Atlantic Sturgeon (*Acipenser oxyrinchus oxyrinchus* Mitchill, 1815)." *Journal of Applied Ichthyology* 32, no. S1 (2016): 30–66.

Hines, N. William. "History of the 1972 Clean Water Act: The Story behind How the 1972 Act Became the Capstone on a Decade of Extraordinary Environmental Reform." *Journal of Energy & Environmental Law* 4, no. 2 (2013): 80–106.

Hitt, Nathaniel P., and Douglas B. Chambers. "Temporal Changes in Taxonomic and Functional Diversity of Fish Assemblages Downstream from Mountaintop Mining." *Freshwater Science* 33, no. 3 (2014): 915–26.

Hodges, Lindsey. "Judge Files Order against Corps of Engineers in Lock and Dam Case." *Post and Courier* (North Augusta, SC), November 24, 2020.

Hoffman, Jeremy S., Peter U. Clark, Andrew C. Parnell, and Feng He. "Regional and Global Sea-Surface Temperatures during the Last Interglaciation." *Science* 355, no. 6322 (2017): 276–79.

Hoffner, Jenny. *Hidden Reservoir: Why Water Efficiency Is the Best Solution for the Southeast*. Washington, DC: American Rivers, October 2008.

Holden, Emily. "How the Oil Industry Has Spent Billions to Control the Climate Change Conversation." *Guardian*, January 8, 2020.

Holyoke Gas and Electric (website). "Shortnose Sturgeon." Accessed October 16, 2020.

Holzkamm, Tim E., and Leo G. Waisberg. "Native American Utilization of Sturgeon." In *Sturgeons and Paddlefish of North America*, edited by Greg T. O. LeBreton, F. William H. Beamish, and R. Scott McKinley, 22–39. Boston: Kluwer Academic, 2004.

Honan, Katie. "City Had Millions to Buy Out Sandy-Damaged Homes, but Most Didn't Want It." *DNA Info*, October 27, 2017.

Hopkinson, Charles S., Alan P. Covich, Christopher B. Craft, Kristine DeLong, Thomas W. Doyle, Neal Flanagan, and Mary C. Freeman, et al. "The Effects of Climate Change on Natural Ecosystems of the Southeast USA." In *Climate of the Southeast United States: Variability, Change, Impacts, and Vulnerability*, edited by Keith T. Ingram, Kirstin Dow, Lynne Carter, and Julie Anderson, 237–70. Washington, DC: Island Press, 2013.

Howe, Amy. "Supreme Court Curtails Clean Water Act." *SCOTUSblog*, May 25, 2023.

Hribar, Carrie. *Understanding Concentrated Animal Feeding Operations and Their Impact on Communities*. Edited by Mark Schultz. Bowling Green, OH: National Association of Local Boards of Health, 2010.

Hsiang, Solomon, Robert Kopp, Amir Jina, James Rising, Michael Delgado, Shashank Mohan, and D. J. Rasmussen, et al. "Estimating Economic Damage from Climate Change in the United States." *Science* 356, no. 6345 (2017): 1362–69.

Hu, Xindi C., David Q. Andrews, Andrew B. Lindstrom, Thomas A. Bruton, Laurel A. Schaider, Philippe Grandjean, and Rainer Lohmann, et al. "Detection of Poly- and Perfluoroalkyl Substances (PFASs) in U.S. Drinking Water Linked to Industrial Sites, Military Fire Training Areas, and Wastewater Treatment Plants." *Environmental Science and Technology Letters* 3, no. 10 (2016): 344–50.

Hudson, Charles. *The Southeastern Indians*. Knoxville: University of Tennessee Press, 1976.

Huff, James A. *Life History of Gulf of Mexico Sturgeon, Acipenser oxyrhynchus desotoi, in Suwannee River, Florida*. St. Petersburg: Florida Department of Natural Resources Marine Research Laboratory, 1975.

Humphrey, Hiram Brown. *Historical Summary of Coal-Mine Explosions in the United States, 1810–1958*. Bulletin no. 586. Washington, DC: US Bureau of Mines, 1960.

Ignatius, Amber, and Jon Anthony Stallins. "Assessing Spatial Hydrological Data Integration to Characterize Geographic Trends in Small Reservoirs in the Apalachicola-Chattahoochee-Flint River Basin." *Southeastern Geographer* 51, no. 3 (2011): 371–93.

Inglis, Jeff, Tom Van Heeke, Gideon Weissman, Lindsey Hallock, and John Rumpler. *Waterways Restored: The Clean Water Act's Impact on 15 American Rivers, Lakes and Bays*. Denver, CO: Environment America Research and Policy Center, October 2014.

Intergovernmental Panel on Climate Change. *Climate Change 2014: Synthesis Report*. Geneva, 2015.

———. *IPCC Special Report: Emissions Scenarios: Summary for Policymakers</i></i>*. Geneva, 2000.

International Energy Agency. *CO₂ Emissions from Fuel Combustion 2017: Overview*. Paris: Organisation for Economic Cooperation and Development, 2018.

International Hydropower Association (website). "A Brief History of Hydropower." Accessed March 2, 2023.

Ivanov, Katya. "A Look at the History of Red Snapper Fishing in Pensacola." WUWF Public Media (website), June 6, 2014.

Jager, Henriette I., Michael J. Parsley, Joseph J. Cech Jr., Robert L. McLaughlin, Patrick S. Forsythe, Robert F. Elliott, and Brenda M. Pracheil. "Reconnecting Fragmented Sturgeon Populations in North American Rivers." *Fisheries* 41, no. 3 (2016): 140–48.

Jenkins, Clinton N. "Amphibians of the World." Biodiversity Mapping (website). Accessed October 11, 2019.

Jenkins, Clinton N., Kyle S. Van Houtan, Stuart L. Pimm, and Joseph O. Sexton. "US Protected Lands Mismatch Biodiversity Priorities." *Proceedings of the National Academy of Sciences of the United States of America* 112, no. 16 (2015): 5081–86.

Johnson, Deborah. "Hurricane Matthew Update: October 14, 2016." North Carolina Pork Council (website), October 14, 2016.

Johnson, Paul D., and Paula Johnson Gagnon. "Freshwater Mollusks." In *New Georgia Encyclopedia* (online). Last updated February 11, 2013.

Jordan, Rob. "Stanford Research Shows Value of Clams, Mussels in Cleaning Dirty Water." Stanford News, Stanford University (website), August 12, 2014.

Joy, G. J., J. F. Colinet, and D. D. Landen. "Coal Workers' Pneumoconiosis Prevalence Disparity between Australia and the United States." *Minerals Engineering* 64, no 7 (2012): 65–71.

Kaetz, James P. "Muscle Shoals." In *Encyclopedia of Alabama* (online). Published March 22, 2011.

———. "Paint Rock." In *Encyclopedia of Alabama* (online). Published July 5, 2017.

Kakissis, Joanna. "Protecting the Netherlands' Vulnerable Coasts with a 'Sand Motor.'" *Weekend Edition Saturday*. National Public Radio, November 25, 2017.

Karegar, Makan A., Timothy H. Dixon, and Simon E. Engelhart. "Subsidence along the Atlantic Coast of North America: Insights from GPS and Late Holocene Relative Sea Level Data." *Geophysical Research Letters* 43, no. 7 (2016): 3126–33.

Karouna-Renier, Natalie K., Richard A. Snyder, Jeffrey G. Allison, Matthew G. Wagner, and K. Ranga Rao. "Accumulation of Organic and Inorganic Contaminants in Shellfish Collected in Estuarine Waters near Pensacola, Florida: Contamination Profiles and Risks to Human Consumers." *Environmental Pollution* 145, no. 2 (2007): 474–88.

Keel, Bennie C., John E. Cornelison Jr., and David M. Brewer. *Regionwide Archeological Survey Plan*. Tallahassee, FL: Southeast Archeological Center, 1996.

Keenan, Jesse M., Thomas Hill, and Anurag Gumber. "Climate Gentrification: From Theory to Empiricism in Miami-Dade County, Florida." *Environmental Research Letters* 13, no. 5 (2018): article 054001.

Keiser, David A., and Joseph S. Shapiro. "Consequences of the Clean Water Act and the Demand for Water Quality." *Quarterly Journal of Economics* 134, no. 1 (2019): 349–96.

Kennedy, Caitlyn. "Groundwater Declines across U.S. South over Past Decade." Climate.gov, October 15, 2014.

Kennish, Michael J. "Coastal Salt Marsh Systems in the U.S.: A Review of Anthropogenic Impacts." *Journal of Coastal Research* 17, no. 3 (2001): 731–48.

Kirke, Brian. "Hydrokinetic and Ultra-Low Head Turbines in Rivers: A Reality Check." *Energy for Sustainable Development* 52 (2019): 1–10.

Knabb, Richard D., Jamie R. Rhome, and Daniel P. Brown. *Tropical Cyclone Report: Hurricane Katrina*. 2006. Miami, FL: National Hurricane Center, 2011.

Konrad, Charles E., II, Christopher M. Fuhrmann, Amanda Billiot, Barry D. Keim, Michael C. Kruk, Kenneth E. Kunkel, Hal Needham, Mark Shafer, and Laura Stevens. "Climate of the Southeast USA: Past, Present, and Future." In *Climate of the Southeast United States: Variability, Change, Impacts, and Vulnerability*, edited by Keith T. Ingram, Kirstin Dow, Lynne Carter, and Julie Anderson, 8–42. Washington, DC: Island Press, 2013.

Konrad, Charles E., II, and Pam Knox. *The Southeastern Drought and Wildfires of 2016*. Chapel Hill, NC: Southeast Regional Climate Center, 2016.

Kossin, James P., Timothy Hall, Thomas Knutson, Kenneth E. Kunkel, Robert J. Trapp, Duane E. Waliser, and Michael F. Wehner. "Extreme Storms." In *Fourth National Climate Assessment*, vol. 1, *Climate Science Special Report*, edited by Donald J. Wuebbles, David W. Fahey, Kathy A. Hibbard, David J. Dokken, Brooke C. Stewart, and Thomas K. Maycock, 257–76. Washington, DC: US Global Change Research Program, 2017.

Kovaleski, Dave. "Virginia Lawmakers Commit to Zero Carbon Emissions by 2050." *Daily Energy Insider* (blog), March 10, 2020.

Kozlowski, James C. "Hidden Drowning Danger at Riverside Park Dam." *Parks & Recreation*, April 2012.

Kunkel, Kenneth E., Laura E. Stevens, Scott E. Stevens, Liqiang Sun, Emily Janssen, Donald Wuebbles, and Charles E. Konrad II, et al. *Regional Climate Trends and Scenarios for the U.S. National Climate Assessment*. Part 2, *Climate of the Southeast U.S.* NOAA Technical Report NESDIS 142-2. Washington, DC: National Oceanic and Atmospheric Administration, 2013.

Kurth, Laura M., Michael McCawley, Michael Hendryx, and Stephanie Lusk. "Atmospheric Particulate Matter Size Distribution and Concentration in West Virginia Coal Mining and Non-Mining Areas." *Modern Pathology* 24 (2014): 405–11.

Kusnetz, Nicholas. "Norfolk Wants to Remake Itself as Sea Level Rises, but Who Will Be Left Behind?," *Inside Climate News*, May 21, 2018.

Lacefield, Jim. *Lost Worlds in Alabama Rocks: A Guide to the State's Ancient Life and Landscapes*. 2nd ed. Tuscaloosa: Alabama Museum of Natural History, 2013.

Lalasz, Bob. "Sustainable Hydropower: Is It Possible?" *Cool Green Science* (blog), Nature Conservancy, August 17, 2015.

Lallensack, Rachael. "Global Fingerprints of Sea Level Rise Revealed by Satellites." *Scientific American*, September 12, 2017.

La Peyre, Megan, Jessica Furlong, Laura A. Brown, Bryan P. Piazza, and Ken Brown. "Oyster Reef Restoration in the Northern Gulf of Mexico: Extent, Methods and Outcomes." *Ocean and Coastal Management* 89 (2014): 20–28.

Laurie, Clayton D. "The United States Army and the Return to Normalcy in Labor Dispute Interventions: The Case of the West Virginia Coal Mine Wars, 1920–1921." *West Virginia History* 50 (1991): 1–24.

Lave, Rebecca. "Bridging Political Ecology and STS: A Field Analysis of the Rosgen Wars." *Annals of the Association of American Geographers* 102, no. 2 (2012): 366–82.

———. "The Controversy over Natural Channel Design: Substantive Explanations and Potential Avenues for Resolution." *Journal of the American Water Resources Association* 45, no. 6 (2009): 1519–32.

Lellis, William A., Barbara St. John White, Jeffrey C. Cole, Connie S. Johnson, Julie L. Devers, Ellen Van Snik Gray, and Heather S. Galbraith. "Newly Documented Host Fishes for the Eastern Elliptio Mussel *Elliptio complanata*." *Journal of Fish and Wildlife Management* 4, no. 1 (2013): 75–85.

Lemon, Johanna. "The Great Stink." Cholera and the Thames (website). Accessed September 22, 2019.

Leuliette, Eric. W, and R. Steven Nerem. "Contributions of Greenland and Antarctica to Global and Regional Sea Level Change." *Oceanography* 29, no. 4 (2016): 154–59.

Li, Laifang, Wenhong Li, and Yochanan Kushnir. "Variation of the North Atlantic Subtropical High Western Ridge and Its Implication to Southeastern US Summer Precipitation." *Climate Dynamics* 39 (2012): 1401–12.

Li, Wenhong, Laifang Li, Rong Fu, Yi Deng, and Hui Wang. "Changes to the North Atlantic Subtropical High and Its Role in the Intensification of Summer Rainfall Variability in the Southeastern United States." *Journal of Climate* 24, no. 5 (2011): 1499–1506.

Litke, David W. *Review of Phosphorus Control Measures in the United States and Their Effects on Water Quality*. Water-Resources Investigations Report no. 99-4007. Denver, CO: US Geological Survey, 1999.

Little, Vince. "Leaders Sign Norfolk Coastal Storm Risk Management Design Agreement." US Army Corps of Engineers: Norfolk District (website), July 1, 2019.

Lobao, Linda, Minyu Zhou, Mark Partridge, and Michael Betz. "Poverty, Place, and Coal Employment across Appalachia and the United States in a New Economic Era." *Rural Sociology* 81, no. 3 (2016): 343–86.

Loesch, Harold. "Sporadic Mass Shoreward Migrations of Demersal Fish and Crustaceans in Mobile Bay, Alabama." *Ecology* 41, no. 2 (1960): 292–98.

Lone Star Chapter of the Sierra Club. *Desalination: Is It Worth Its Salt? A Primer on Brackish and Seawater Desalination*. Austin, TX, 2008, revised 2013.

Magilligan, F. J., C. S. Sneddon, and C. A. Fox. "The Social, Historical, and Institutional Contingencies of Dam Removal." *Environmental Management* 59 (2017): 982–94.

Mallin, Michael A., Martin H. Posey, G. Christopher Shank, Matthew R. McIver, Scott H. Ensign, and Troy D. Alphin. "Hurricane Effects on Water Quality and Benthos in the Cape Fear Watershed: Natural and Anthropogenic Impacts." *Ecological Applications* 9, no. 1 (1999): 350–62.

Manganiello, Christopher J. "Excitement Draining for Reservoir Boondoggles." *Saporta Report* (blog), May 22, 2016.

———. "Fish Tales and the Conservation State." *Southern Cultures* 20, no. 3 (2014): 43–62.

———. *Southern Water, Southern Power: How the Politics of Cheap Energy and Water Scarcity Shaped a Region.* Chapel Hill: University of North Carolina Press, 2015.

Mann, Charles C. *1491: New Revelations of the Americas before Columbus.* New York: Vintage Books, 2006.

———. *1493: Uncovering the New World Columbus Created.* New York: Vintage Books, 2011.

Marion, Daniel A., Ge Sun, Peter V. Caldwell, Chelcy F. Miniat, Ying Ouyang, Devendra M. Amatya, and Barton D. Clinton, et al. "Managing Forest Water Quantity and Quality under Climate Change." In *Climate Change Adaptation and Mitigation Management Options: A Guide for Natural Resource Managers in Southern Forest Ecosystems,* edited by James M. Vose and Kier D. Klepzig, 249–306. Boca Raton, FL: CRC Press, 2014.

Marlon, Jennifer, Peter Howe, Matto Mildenberger, Anthony Leiserowitz, and Xinran Wang. "Yale Climate Opinion Maps 2019." Yale Program on Climate Change Communication (website), September 17, 2019.

Martin, Laura. "Texas Leads the Way with First Direct Potable Reuse Facilities in U.S." *Water Online* (newsletter), September 16, 2014.

Martin, Sandi. "UGA Researchers Discover One Endangered Species Depends on Another." UGA Today, University of Georgia (website), May 7, 2013.

Maupin, Molly A., Joan F. Kenny, Susan S. Hutson, John K. Lovelace, Nancy L. Barber, and Kristin S. Linsey. *Estimated Use of Water in the United States in 2005.* Circular 1344. Reston, VA: US Geological Survey, 2009.

May, Edwin B. "Extensive Oxygen Depletion in Mobile Bay, Alabama." *Limnology and Oceanography* 18, no. 3 (1973): 353–66.

Mazzocchi, Jay. "Cape Fear River." In *Encyclopedia of North Carolina* (online). Published January 1, 2006.

McAuley, Steven D., and Mark D. Kozar. *Ground-Water Quality in Unmined Areas and Near Reclaimed Surface Coal Mines in the Northern and Central Appalachian Coal Regions, Pennsylvania and West Virginia.* Report 2006-5059. Reston, VA: US Geological Survey, 2006.

McCarthy, James. "16-Year Penobscot River Restoration Project Reaches the Finish Line." *Mainebiz,* August 22, 2016.

McCartney, Martha. "Starving Time, The." In *Encyclopedia Virginia* (online). Published September 3, 2021.

McCombs, Erin Singer. "Neuse River Flows Freely after Milburnie Dam Removed." American Rivers (website), November 10, 2017.

McCoy, Michael. "Goodbye, Phosphates." *Chemical and Engineering News* 89, no. 4 (2011): 12–17.

McFetridge, Scott. "Low-Lying Dams Drown Hundreds of Unsuspecting Victims." Associated Press, April 15, 2015.

McGregor, Stuart W., and Jeffrey T. Garner. *Results of Qualitative Sampling for*

Protected Mussel Species at Selected Stations in the Cahaba River System, Alabama, 2005. Tuscaloosa: Geological Survey of Alabama, 2005.

McQuaid, John. "Mining the Mountains." *Smithsonian Magazine*, January 2009.

McThenia, Andrew W., Jr. "An Examination of the Federal Water Pollution Control Act of 1972." *Washington and Lee Law Review* 30, no. 2 (1973): 195–222.

Mehan, G. Tracy, III. "A Symphonic Approach to Water Management: The Quest for New Models of Watershed Governance." *Journal of Land Use and Environmental Law* 26, no. 1 (2010): 1–33.

Melillo, Edward D. "The First Green Revolution: Debt Peonage and the Making of the Nitrogen Fertilizer Trade, 1840–1930." *American Historical Review* 117, no. 4 (2012): 1028–60.

Metropolitan North Georgia Water Planning District. *2017 Activities & Progress Report*. Atlanta, 2017.

———. "Metro Atlanta: Responsible and Efficient Stewards of Our Water Resources." Fact sheet, April 2017.

Metropolitan North Georgia Water Planning District (website). "About the Metro Water District." Accessed September 29, 2019.

———. "Water Supply and Conservation Plan." Accessed October 1, 2019.

Mettee, Maurice F., and Patrick E. O'Neil. "Status of Alabama Shad and Skipjack Herring in Gulf of Mexico Drainages." *American Fisheries Society Symposium* 35 (2003): 157–70.

Michigan Department of Environment, Great Lakes, and Energy. *Preliminary Report on the Edenville Dam Failure, Response Efforts, and Program Reviews*. Lansing, MI, August 2020.

Mickley, Michael. *Updated and Extended Survey of U.S. Municipal Desalination Plants*. Denver, CO: Bureau of Reclamation, December 2018.

Miller, Andrew J., and Nicolas P. Zégre. "Mountaintop Removal Mining and Catchment Hydrology." *Water* 6, no. 3 (2014): 472–99.

Miller, Hannah. "River Advocates Work to Add Fish Passages." *Coastal Review* (online news service), January 9, 2017.

Miller, James A. *Ground Water Atlas of the United States*. Segment 6, *Alabama, Florida, Georgia, South Carolina*. Reston, VA: US Geological Survey, 1990.

Miller, Kenneth G., James D. Wright, James V. Browning, Andrew Kulpecz, Michelle Kominz, Tim R. Naish, Benjamin S. Cramer, Yair Rosenthal, W. Richard Peltier, and Sindia Sosdian. "High Tide of the Warm Pliocene: Implications of Global Sea Level for Antarctic Deglaciation." *Geology* 40, no. 5 (2012): 407–10.

Milman, Oliver. "'We're Not a Dump'—Poor Alabama Towns Struggle under the Stench of Toxic Landfills." *Guardian*, April 15, 2019.

Mitchell, Kenneth L., Marilyn Brown, Ryan Brown, Diana Burk, Dennis Creech, Garry P. Garrett, and Daniel Garver, et al. "Energy Production, Use, and Vulnerability to Climate Change in the Southeast USA." In *Climate of the Southeast United States: Variability, Change, Impacts, and Vulnerability*, edited by Keith T. Ingram, Kirstin Dow, Lynne Carter, and Julie Anderson, 62–85. Washington, DC: Island Press, 2013.

Mizoguchi, Beatriz A., and Nicole Valenzuela. "Ecotoxicological Perspectives of Sex Determination." *Sexual Development* 10, no. 1 (2016): 45–57.

Mooney, Chris. "Scientists Say Human Greenhouse Gas Emissions Have Canceled the Next Ice Age." *Washington Post*, January 13, 2016.

Moran, Emilio F., Maria Claudia Lopez, Nathan Moore, Norbert Müller, and David W. Hyndman. "Sustainable Hydropower in the 21st Century." *Proceedings of the National Academy of Sciences* 115, no. 47 (2018): 11891–98.

Morelle, Rebecca. "Vast Sand Scheme to Protect Norfolk Coast." BBC News, July 18, 2019.

Moss, Richard H., Jae A. Edmonds, Kathy A. Hibbard, Martin R. Manning, Steven K. Rose, Detlef P. Van Vuuren, and Timothy R. Carter, et al. "The Next Generation of Scenarios for Climate Change Research and Assessment." *Nature* 463 (2010): 747–56.

Mulligan, Stephen P. *Evolution of the Meaning of "Waters of the United States" in the Clean Water Act*. R44585. Washington, DC: Congressional Research Service, March 2019.

Murphy, Ryan. "Why Norfolk Is Planning to Tear Down Half of Its Public Housing." *Virginian-Pilot* (Norfolk), February 22, 2018.

Murray, Ann. "Smog Deaths in 1948 Led to Clean Air Laws." *All Things Considered*. National Public Radio, April 22, 2009.

Nagle, Greg. "Evaluating 'Natural Channel Design' Stream Projects." *Hydrological Processes* 21, no. 18 (2007): 2539–45.

Narayan, Siddharth, Michael W. Beck, Borja G. Reguero, Iñigo J. Losada, Bregje Van Wesenbeeck, Nigel Pontee, James N. Sanchirico, Jane Carter Ingram, Glenn Marie Lange, and Kelly A. Burks-Copes. "The Effectiveness, Costs and Coastal Protection Benefits of Natural and Nature-Based Defences." *PLoS ONE* 11, no. 5 (2016): e0154735.

Narayan, Siddharth, Michael W. Beck, Paul Wilson, Christopher Thomas, Alexandra Guerrero, Christine Shepard, Borja G. Reguero, Guillermo Franco, Jane Carter Ingram, and Dania Trespalacios. "Coastal Wetlands and Flood Damage Reduction: Using Risk Industry-Based Models to Assess Natural Defenses in the Northeastern USA." London: Lloyd's Tercentenary Research Foundation, 2016.

National Academies of Science. "Statement Regarding National Academies Study on Potential Health Risks of Living in Proximity to Surface Coal Mining Sites in Central Appalachia." News release, August 21, 2017.

National Archives at Atlanta. *Valley of the Dams: The Impact and Legacy of the Tennessee Valley Authority*. Online exhibit, launched April 2014.

National Energy Technology Laboratory (website). "History of U.S. Coal Use." Internet Archive. Accessed March 2, 2023.

National Grain and Feed Association (website). "North Carolina Estimates Hurricane Florence's Ag Damage." Newsletter, September 21, 2018.

National Hurricane Center. "Costliest U.S. Tropical Cyclones Tables Updated." News release, January 26, 2018.

National Hydropower Association (website). "About National Hydropower Association." Accessed August 1, 2020.

National Inventory of Dams (website). "Dams of the Nation." US Army Corps of Engineers. Accessed March 2, 2023.

National Oceanic and Atmospheric Administration (NOAA). "Endangered and Threatened Wildlife and Plants: Notice of 12-Month Finding on a Petition to List Alabama Shad as Threatened or Endangered under the Endangered Species Act." *Federal Register* 82, no. 8 (2012): 4022–61.

———. "Endangered and Threatened Wildlife and Plants; Endangered Species Act Listing Determination for Alewife and Blueback Herring." *Federal Register* 84, no. 118 (2019): 28630–66.

———. "Endangered and Threatened Wildlife and Plants; Proposed Listings for Two Distinct Population Segments of Atlantic Sturgeon (*Acipenser oxyrinchus oxyrinchus*) in the Southeast." *Federal Register* 75, no. 193 (2010): 61904–29.

———. *NOAA's Estuarine Eutrophication Survey*. Vol. 4, *Gulf of Mexico Region*. Silver Spring, MD: Office of Ocean Resources Conservation and Assessment, 1997.

National Oceanic and Atmospheric Administration (NOAA) (website). "Billion-Dollar Weather and Climate Disasters." National Centers for Environmental Information. Accessed August 1, 2020.

———. "History of Management of Gulf of Mexico Red Snapper." Accessed September 22, 2019.

———. "National Trends." National Centers for Environmental Information. Accessed May 26, 2020.

———. "Sea Level Trends." Tides and Currents. Accessed May 2, 2018.

———. "What Is a Harmful Algal Bloom?" Last updated April 27, 2016.

———. "Where Is the Largest Estuary in the United States?" National Ocean Service. Accessed August 1, 2020.

National Research Council. *Coal Waste Impoundments: Risks, Responses, and Alternatives*. Washington, DC: National Academies Press, 2002.

———. *Urban Stormwater Management in the United States*. Washington, DC: National Academies Press, 2009.

———. *Use of Reclaimed Water and Sludge in Food Crop Production*. Washington, DC: National Academies Press, 1996.

National Weather Service (website). "Weather Related Fatality and Injury Statistics: Weather Fatalities 2018." Accessed October 9, 2019.

Natural Resources Conservation Service. *2007 National Resources Inventory: Soil Erosion on Cropland*. Washington, DC, 2007.

Natural Resources Council of Maine (website). "Penobscot River Restoration Project." Accessed August 1, 2020.

———. "Penobscot River Restoration Trust: Penobscot River Restoration Project Howland Fish Bypass Project Overview." Fact sheet. Accessed August 1, 2020.

Natural Resources Defense Council. "Report: Five Major Health Threats from Fracking-Related Air Pollution." News release, December 16, 2014.

Nature Conservancy. "Public and Private Partnership to Restore and Protect Bayou La Batre Waterfront." News release, April 19, 2019.

Nature Conservancy (website). "Lightning Point Shoreline Restoration." Accessed August 1, 2020.

———. "Who We Are." Accessed October 1, 2019.

Nearing, Mark A., Yun Xie, Baoyuan Liu, and Yu Ye. "Natural and Anthropogenic Rates of Soil Erosion." *International Soil and Water Conservation Research* 5, no. 2 (2017): 77–84.

Neary, D. G., W. T. Swank, and H. Riekerk. "An Overview of Nonpoint Source Pollution in the Southern United States." In *Proceedings of the Symposium: The Forested Wetlands of the Southern United States, July 12–14, 1988, Orlando, FL*, edited by Donal D. Hook and Russ Lea, 1–7. Asheville, NC: US Department of Agriculture, Forest Service, 1989.

Neeson, Thomas M., Allison T. Moody, Jesse R. O'Hanley, Matthew Diebel, Patrick J. Doran, Michael C. Ferris, Timothy Colling, and Peter B. McIntyre. "Aging Infrastructure Creates Opportunities for Cost-Efficient Restoration of Aquatic Ecosystem Connectivity." *Ecological Applications* 28, no. 6 (2018): 1494–1502.

Nelson, Cody. "'This Land Is All We Have Left': Tribes on Edge over Giant Dam Proposal near Grand Canyon." *Guardian*, August 12, 2020.

Neupane, Ram P., Darren L. Ficklin, Jason H. Knouft, Nima Ehsani, and Raj Cibin. "Hydrologic Responses to Projected Climate Change in Ecologically Diverse Watersheds of the Gulf Coast, United States." *International Journal of Climatology* 39, no. 4 (2019): 2227–43.

Newton, Teresa J., Steven J. Zigler, James T. Rogala, Brian R. Gray, and Mike Davis. "Population Assessment and Potential Functional Roles of Native Mussels in the Upper Mississippi River." *Aquatic Conservation: Marine and Freshwater Ecosystems* 21, no. 2 (2011): 122–31.

Nickens, T. Edward. "Ode to a Founding Fish." *National Wildlife*, August 1, 2006.

Normandeau, Glenn. *New Hampshire Wildlife Action Plan*. Concord: New Hampshire Fish and Game Department, 2015.

North Carolina Division of Public Health (website). "Fish Consumption Advisories." Accessed March 2, 2023.

North Carolina Division of Water Resources (website). "Drought Monitor History." Internet Archive. Accessed March 2, 2023.

North Carolina Ground Water Association (website). Home page. Accessed September 29, 2019.

Northwest Power and Conservation Council (website). "Fish Passage at Dams." Accessed October 10, 2020.

Nunez, Christina. "Floods, Explained." *National Geographic*, April 4, 2019.

Oberstar, James L. *The Clean Water Act: 30 Years of Success in Peril*. Washington, DC: Committee on Transportation and Infrastructure, October 2002.

O'Brien, Christine A., and James D. Williams. "Reproductive Biology of Four Freshwater Mussels (Bivalvia: Unionidae) Endemic to Eastern Gulf Coastal Plain Drainages of Alabama, Florida, and Georgia." *American Malacological Bulletin* 17, no. 1–2 (2002): 147–58.

O'Neil, Patrick E., Thomas E. Shepard, Jeffrey R. Powell, Eric W. Spadgenske, Andrew R. Henderson, and Paul L. Freeman. *A Survey of Fishes in the Paint Rock River System, Alabama*. Tuscaloosa: Geological Survey of Alabama, 2013.

Onorato, Dave, Robert A. Angus, and Ken R. Marion. "Historical Changes in the Ichthyofaunal Assemblages of the Upper Cahaba River in Alabama Associated with Extensive Urban Development in the Watershed." *Journal of Freshwater Ecology* 15, no. 1 (2000): 47–63.

Onyanga-Omara, Jane. "Mystery of London Fog That Killed 12,000 Finally Solved." *USA Today*, December 13, 2016.

OpenSecrets (website). Center for Responsive Politics. Accessed August 1, 2020.

Opperman, Jeff. "Concentration, Confrontation, Collaboration: The Future of River Conservation and Sustainable Hydropower." *Cool Green Science* (blog), Nature Conservancy, January 28, 2016.

———. "Penobscot River Dam Removal: Lessons for a World Demanding Energy." *Cool Green Science* (blog), Nature Conservancy, 2013.

Opperman, Jeff, Günther Grill, and Joerg Hartmann. *The Power of Rivers: Finding Balance between Energy and Conservation in Hydropower Development*. Washington, DC: Nature Conservancy, 2015.

Osland, Michael J., Nicholas Enwright, Richard H. Day, and Thomas W. Doyle. "Winter Climate Change and Coastal Wetland Foundation Species: Salt Marshes vs. Mangrove Forests in the Southeastern United States." *Global Change Biology* 19, no. 5 (2013): 1482–94.

Osterman, Lisa E., and Christopher G. Smith. "Over 100 Years of Environmental Change Recorded by Foraminifers and Sediments in Mobile Bay, Alabama, Gulf of Mexico, USA." *Estuarine, Coastal and Shelf Science* 115 (2012): 345–58.

Oswalt, Christopher M., Jason A. Cooper, Dale G. Brockway, Horace W. Brooks, Joan L. Walker, Kristina F. Connor, Sonja N. Oswalt, and Roger C. Conner. *History and Current Condition of Longleaf Pine in the Southern United States*. Asheville, NC: US Department of Agriculture, Forest Service, 2012.

Palmer, Margaret A, Emily S. Bernhardt, William H. Schlesinger, Keith N. Eshleman, Efi Foufoula-Georgiou, Michael S. Hendryx, and A. Dennis Lemly, et al. "Mountaintop Mining Consequences." *Science* 327, no. 5962 (2010): 148–49.

Palmer, Margaret A., Kelly L. Hondula, and Benjamin J. Koch. "Ecological Restoration of Streams and Rivers: Shifting Strategies and Shifting Goals." *Annual Review of Ecology, Evolution, and Systematics* 45, no. 1 (2014): 247–69.

Parsons, Doug. "Climate Gentrification." August 24, 2017. In *America Adapts*, produced by Dan Ackerstein, podcast, 55:08.

———. "Flooding and Natural and Nature-Based Solutions—The Good, the Bad, and the Ugly of Flooding." September 14, 2018. In *America Adapts*, produced by Dan Ackerstein, podcast, 1:10:14.

Partridge, Mark D., Michael R. Betz, and Linda Lobao. "Natural Resource Curse and Poverty in Appalachian America." *American Journal of Agricultural Economics* 95, no. 2 (2013): 449–56.

Patel, Sonal. "President Obama Signs Pivotal Hydropower-Boosting Bills into Law." *Power*, August 15, 2013.

Patricola, Christina M. "Tropical Cyclones Are Becoming Sluggish." *Nature* 558 (2018): 36–37.

Perrone, D., and S. Jasechko. "Dry Groundwater Wells in the Western United States." *Environmental Research Letters* 12 (2017): article 104002.

Perry, Charles A. "Significant Floods in the United States during the 20th Century— USGS Measures a Century of Floods." US Geological Survey. Fact sheet, March 2000.

Petsonk, Edward L., Cecile Rose, and Robert Cohen. "Coal Mine Dust Lung Disease: New Lessons from an Old Exposure." *Concise Clinical Review* 187, no 11 (2013): 1178–85.

Philpott, Tom. "A Brief History of Our Deadly Addiction to Nitrogen Fertilizer." *Mother Jones*, April 19, 2013.

Piazza, Bryan P., Patrick D. Banks, and Megan K. La Peyre. "The Potential for Created Oyster Shell Reefs as a Sustainable Shoreline Protection Strategy in Louisiana." *Restoration Ecology* 13, no. 3 (2005): 499–506.

Pickering, Mimi, dir. *The Buffalo Creek Flood: An Act of Man*. Appalshop, 1975. Film, 39:37.

Pipas, James C., and Justin G. Spaulding. *Management Plan for the Tims Ford Tailwater Trout Fishery (2017–2022)*. Nashville: Tennessee Wildlife Resources Agency, 2017.

Pittock, Jamie, and Joerg Hartmann. "Taking a Second Look: Climate Change, Periodic Relicensing and Improved Management of Dams." *Marine and Freshwater Research* 62, no. 3 (2011): 312–20.

Plater, Zygmunt J. B. "Tiny Fish/Big Battle: 30 Years after TVA and the Snail Darter Clashed, the Case Still Echoes in Caselaw, Politics, and Popular Culture." *Tennessee Bar Journal* 44, no. 4 (2008): 14–20.

Poe, Gregory L. *The Evolution of Federal Water Pollution Control Policies*. Extension Bulletins no. 95-06. Ithaca, NY: Dyson School of Applied Economics and Management, 1995.

Poff, N. LeRoy, Julian D. Olden, David M. Merritt, and David M. Pepin. "Homogenization of Regional River Dynamics by Dams and Global Biodiversity Implications." *Proceedings of the National Academy of Sciences* 104, no. 14 (2007): 5732–37.

Poff, N. LeRoy, Brian D. Richter, Angela H. Arthington, Stuart E. Bunn, Robert J. Naiman, Eloise Kendy, and Mike Acreman, et al. "The Ecological Limits of Hydrologic Alteration (ELOHA): A New Framework for Developing Regional Environmental Flow Standards." *Freshwater Biology* 55, no. 1 (2010): 147–70.

Polycarpou, Lakis. "US Groundwater Declines More Widespread Than Commonly Thought." *State of the Planet* (blog), Columbia Climate School, March 17, 2014.

Poplar-Jeffers, Ira O., J. Todd Petty, James T. Anderson, Steven J. Kite, Michael P. Strager, and Ronald H. Fortney. "Culvert Replacement and Stream Habitat Restoration: Implications from Brook Trout Management in an Appalachian Watershed, U.S.A." *Restoration Ecology* 17, no. 3 (2009): 404–13.

Postel, Sandra. "How Smarter Irrigation Might Save Rare Mussels and Ease a Water War." *Water Currents* (blog), National Geographic, August 19, 2016.

Postel, Sandra, and Brian Richter. *Rivers for Life: Managing Water for People and Nature*. Washington, DC: Island Press, 2003.

Powers, Ann. "Federal Water Pollution Control Act (1948)." Encyclopedia.com. Accessed March 2, 2023.

Pringle, Heather. "How Europeans Brought Sickness to the New World." *Science*, June 4, 2015.

Pryne, Eric. "Elwha Dams Won't Be Coming Down Anytime Soon." *Seattle Times*, May 19, 1994.

Puckett, Brandon J., Seth J. Theuerkauf, David B. Eggleston, Rodney Guajardo, Craig Hardy, Jie Gao, and Richard A. Luettich. "Integrating Larval Dispersal, Permitting, and Logistical Factors within a Validated Habitat Suitability Index for Oyster Restoration." *Frontiers in Marine Science* 5 (2018): article 76.

Raabe, Joshua K., Timothy A. Ellis, and Joseph E. Hightower. "Evaluation of Fish Passage Following Installation of a Rock Arch Rapids at Lock and Dam #1, Cape Fear River, North Carolina." Paper presented at the International Conference on Engineering and Ecohydrology for Fish Passage, University of Wisconsin–Madison, June 9, 2014.

Raabe, Joshua K., Joseph E. Hightower, Timothy A. Ellis, and Joseph J. Facendola. "Evaluation of Fish Passage at a Nature-Like Rock Ramp Fishway on a Large Coastal River." *Transactions of the American Fisheries Society* 148, no. 4 (2019): 798–816.

Raines, Ben. "Alabama's Rivers Are Crashing Thanks to Water Rules That Favor Industry." AL.com, November 15, 2016.

———. "America's Amazon: Old Timers Remember Delta's Pristine Water and Teeming Wildlife." AL.com, December 21, 2014.

———. "Drought Means a Stunningly Clear Gulf, Happy Fishermen, and Bad News for Oyster Lovers." AL.com, November 1, 2016.

Rakes, Paul H. "Coal Mine Mechanization." In *e-WV: The West Virginia Encyclopedia* (online). Last updated June 19, 2012.

Ramankutty, Navin, Elizabeth Heller, and Jeanine Rhemtulla. "Prevailing Myths about Agricultural Abandonment and Forest Regrowth in the United States." *Annals of the Association of American Geographers* 100, no. 3 (2010): 502–12.

Rangel, Rodrigo. "Account of the Northern Conquest and Discovery of Hernando de Soto." In *The De Soto Chronicles: The Expedition of Hernando de Soto to North America in 1539–1543*, edited by Lawrence A. Clayton, Edward C. Moore, and Vernon James Knight, vol. 1, 247–306. Tuscaloosa: University of Alabama Press, 1993.

Rawls, Wendell, Jr. "Forgotten People of the Tellico Dam Fight." *New York Times*, November 11, 1979.

Ray, Tarence. "Drinking Water Problems Still Plague Eastern Kentucky." *Front Porch* (blog), Applachian Voices, May 6, 2016.

Reeves, Jay. "$2B Tenn-Tom Waterway Yet to Yield Promised Boom." Associated Press, September 16, 2019.

Reguero, Borja G., Michael W. Beck, David N. Bresch, Juliano Calil, and Imen Meliane. "Comparing the Cost Effectiveness of Nature-Based and Coastal Adaptation: A Case Study from the Gulf Coast of the United States." *PLoS ONE* 13, no. 4 (2018): e0192132.

Reidmiller, David R., Christopher W. Avery, David R. Easterling, Kenneth E. Kunkel, K. L. M. Lewis, Thomas K. Maycock, and Brooke C. Stewart, eds. *Fourth National Climate Assessment*. Vol. 2, *Impacts, Risks, and Adaptation in the United States*. Washington, DC: US Global Change Research Program, 2018.

Reilly, Thomas E., Kevin F. Dennehy, William M. Alley, and William L. Cunningham. *Ground-Water Availability in the United States*. Circular 1323. Reston, VA: US Geological Survey, 2008.

Rice, Karen C., Bo Hong, and Jian Shen. "Assessment of Salinity Intrusion in the James and Chickahominy Rivers as a Result of Simulated Sea-Level Rise in Chesapeake Bay, East Coast, USA." *Journal of Environmental Management* 111 (2012): 61–69.

Richardson, Jeremy, Sam Gomberg, Julie McNamara, and J. C. Kibbey. "A Dwindling Role for Coal." Union of Concerned Scientists. Fact sheet, September 2017.

Richardson-Lorente, Emily. "Flood City." July 29, 2020. In *Broken Ground*, produced by Southern Environmental Law Center, podcast, 36:33.

Richter, Brian D., Sandra Postel, Carmen Revenga, Thayer Scudder, Bernhard Lehner, Allegra Churchill, and Morgan Chow. "Lost in Development's Shadow: The Downstream Human Consequences of Dams." *Water Alternatives* 3, no. 2 (2010): 14–42.

Richter, Brian D., Andrew T. Warner, Judy L. Meyer, and Kim Lutz. "A Collaborative and Adaptive Process for Developing Environmental Flow Recommendations." *River Research and Applications* 22, no. 3 (2006): 297–318.

Rietbroek, Roelof, Sandra-Esther Brunnabend, Jürgen Kusche, Jens Schröter, and Christoph Dahle. "Revisiting the Contemporary Sea-Level Budget on Global and Regional Scales." *Proceedings of the National Academy of Sciences* 113, no. 6 (2016): 1504–9.

Rigaud, Kanta Kumari, Alex de Sherbinin, Bryan Jones, Jonas Bergmann, Viviane Clement, Kayly Ober, and Jacob Schewe, et al. *Groundswell: Preparing for Internal Climate Migration*. Washington, DC: World Bank, 2018.

Rinde, Meir. "Richard Nixon and the Rise of American Environmentalism." *Distillations* (newsletter), Science History Institute, June 2, 2017.

Ritchie, Bruce. "Tallahassee to Shut Down One of Two Hydroelectric Plants in Florida." *Politico*, July 20, 2017.

Ritter, Steven K. "The Haber-Bosch Reaction: An Early Chemical Impact on Sustainability." *Chemical and Engineering News* 86, no. 33 (2008): n.p.

River Network (website). "Science Module: Environmental Flows and Water Security." Accessed October 3, 2019.

Roberts, Samuel G., Rebecca A. Longenecker, Matthew A. Etterson, Chris S. Elphick, Brian J. Olsen, and W. Gregory Shriver. "Preventing Local Extinctions of Tidal Marsh Endemic Seaside Sparrows and Saltmarsh Sparrows in Eastern North America." *Condor* 121, no. 2 (2019): duy024.

Rockström, Johan, Will Steffen, Kevin Noone, Åsa Persson, F. Stuart Chapin III, Eric

Lambin, and Timothy M. Lenton, et al. "Planetary Boundaries: Exploring the Safe Operating Space for Humanity." *Ecology and Society* 14, no. 2 (2009): article 32.

Rogers, Paul. "Nation's Largest Ocean Desalination Plant Goes Up near San Diego; Future of the California Coast?" *Mercury News*, May 29, 2014.

Rogers, Shane, and John Haines. *Detecting and Mitigating the Environmental Impact of Fecal Pathogens Originating from Confined Animal Feeding Operations: Review*. Cincinnati, OH: Environmental Protection Agency, 2005.

Ross, Matthew R. V., Brian L. McGlynn, and Emily S. Bernhardt. "Deep Impact: Effects of Mountaintop Mining on Surface Topography, Bedrock Structure, and Downstream Waters." *Environmental Science and Technology* 50, no. 4 (2016): 2064–74.

Rundquist, Soren. "Exposing Fields of Filth in North Carolina." Environmental Working Group (website), June 21, 2016.

Saffron, Inga. "The Decline of the North American Species." In *Sturgeons and Paddlefish of North America*, edited by Greg T. O. LeBreton, F. William H. Beamish, and R. Scott McKinley, 1–21. Boston: Kluwer Academic, 2004.

Samuel, Molly. "Officials Declare Drought in Atlanta, North Georgia." WABE, September 9, 2016.

Satterfield, Jamie. "Roane County Lawsuit Asserts TVA and Its Contractor Ran Coal Ash 'Disinformation Campaign.'" *Knox News*, November 12, 2019.

———. "TVA Contractor Jacobs Engineering Loses Request to Appeal Lawsuit by Kingston Workers." *Knox News*, August 16, 2019.

Schiffman, Richard. "A Troubling Look at the Human Toll of Mountaintop Removal Mining." *Yale Environment 360* (online magazine), Yale School of the Environment, November 21, 2017.

Schramm, Paul J., Carina G. M. Blackmore, William Crosson, Christopher T. Emrich, Maury Estes, Erin K. Lipp, and Yang Liu, et al. "Human Health and Climate Change in the Southeast USA." In *Climate of the Southeast United States: Variability, Change, Impacts, and Vulnerability*, edited by Keith T. Ingram, Kristin Dow, Lynne Carter, and Julie Anderson, 43–61. Washington, DC: Island Press, 2013.

Schroeder, William W., and William J. Wiseman Jr. "The Mobile Bay Estuary: Stratification, Oxygen Depletion, and Jubilees." In *Estuarine Case Studies*, edited by Björn Kjerfve, 41–52. Boca Raton, FL: CRC Press, 1988.

Schwenning, Lisa, Traci Bruce, and Lawrence R. Handley. "Pensacola Bay." In *Seagrass Status and Trends in the Northern Gulf of Mexico: 1940–2002*, edited by L. Handley, D. Altsman, and R. DeMay, 128–41. Reston, VA: US Geological Survey, 2007.

Scripps CO_2 Program (website). "Keeling Curve Lessons." Scripps Institute of Oceanography. Accessed April 30, 2018.

Shannon, Chelsea. "Oroville Dam: How the Main Spillway Failed and Was Repaired." ABC10 (website), April 2, 2019.

Sheets, Conner. "'Get This Changed': Alabama Still the Only State That Doesn't Enforce Dam Safety." AL.com, August 3, 2018.

Silva, Ana T., Martyn C. Lucas, Theodore Castro-Santos, Christos Katopodis, Lee J. Baumgartner, Jason D. Thiem, and Kim Aarestrup, et al. "The Future of Fish Passage

Science, Engineering, and Practice." *Fish and Fisheries* 19, no. 2 (2018): 340–62.

Silverman, Melvin P. "Mechanism of Bacterial Pyrite Oxidation." *Journal of Bacteriology* 94, no. 4 (1967): 1046–51.

Smith, John. *Capt. John Smith: Works, 1608–1631*. Part 1. Edited by Edward Arber. Westminster, UK: Archibald Constable, 1895.

Smith, Joseph A., and Joseph E. Hightower. "Effect of Low-Head Lock-and-Dam Structures on Migration and Spawning of American Shad and Striped Bass in the Cape Fear River, North Carolina." *Transactions of the American Fisheries Society* 141, no. 2 (2012): 402–13.

Sneed, Annie. "Glass Eel Gold Rush Casts Maine Fishermen against Scientists." *Scientific American*, August 5, 2014.

Snyder, Richard A. *Chemical Contaminants in Fish from the Pensacola Bay System and Offshore Northern Gulf of Mexico*. Pensacola: University of West Florida, 2009.

Society for Ecological Restoration (website). "USA: Two Dam Removals, One Bypass and One Fish Ladder at the Great Penobscot River Restoration, Penobscot River Watershed, Maine." Accessed May 30, 2023.

Solar Energy Industries Association (website). "Solar Industry Research Data." Accessed December 10, 2020.

South Carolina Marine Resources Research Institute. "Diadromous Fish Research: Eel Sampling." Accessed October 6, 2019.

Southeast Aquatic Resources Partnership (website). "Aquatic Barrier Prioritization Tool." Accessed September 22, 2019.

Southeast Regional Climate Center. "The Southeast U.S. Drought of 2016: Evolution, Climate Perspectives, and Impacts." Fact sheet, October 2016.

Southern Environmental Law Center (website). "Cleaning Up Coal Ash Pollution." Accessed September 22, 2019.

Spanger-Siegfried, Erika, Kristina Dahl, Astrid Caldas, Shana Udvardy, Rachel Cleetus, Pamela Worth, and Nicole Hernandez Hammer. *When Rising Seas Hit Home: Hard Choices Ahead for Hundreds of US Coastal Communities*. Cambridge, MA: Union of Concerned Scientists, July 2017.

Sprague, Lori A., and Jo Ann M. Gronberg. "Relating Management Practices and Nutrient Export in Agricultural Watersheds of the United States." *Journal of Environment Quality* 41, no. 6 (2012): 1939–50.

Spratt, David, and Ian Dunlop. *Existential Climate-Related Security Risk: A Scenario Approach*. Melbourne: National Centre for Climate Restoration, 2019.

Stanis, Tim. "Drain the Swamps: A Brief History of 'Waters of the United States' and the Trump Administration's Attempt at Defining the Term." *Georgetown Environmental Law Review* (blog), March 2, 2017.

Stanley, Emily H., and Martin W. Doyle. "Trading Off: The Ecological Effects of Dam Removal." *Frontiers in Ecology and the Environment* 1, no. 1 (2003): 15–22.

Stavins, Robert N. "What Can We Learn from the Grand Policy Experiment? Lessons from SO_2 Allowance Trading." *Journal of Economic Perspectives* 12, no. 3 (1998): 69–88.

Steffen, Will, Jacques Grinevald, Paul Crutzen, and John McNeill. "The Anthropocene:

Conceptual and Historical Perspectives." *Philosophical Transactions of the Royal Society A: Mathematical, Physical and Engineering Sciences* 369, no. 1938 (2011): 842–67.

Steffen, Will, Johan Rockström, Katherine Richardson, Timothy M. Lenton, Carl Folke, Diana Liverman, and Colin P. Summerhayes, et al. "Trajectories of the Earth System in the Anthropocene." *Proceedings of the National Academy of Sciences of the United States of America* 115, no. 33 (2018): 8252–59.

Steponaitis, Vincas P. "Prehistoric Archaeology in the Southeastern United States, 1970–1985." *Annual Review of Anthropology* 15 (1986): 363–404.

Stets, Edward G., Valerie J. Kelly, and Charles G. Crawford. "Regional and Temporal Differences in Nitrate Trends Discerned from Long-Term Water Quality Monitoring Data." *Journal of the American Water Resources Association* 51, no. 5 (2015): 1394–1407.

Stewart, Stacy R. *Tropical Cyclone Report: Hurricane Ivan.* 2004. Miami, FL: National Hurricane Center, 2011.

Stewart, Stacy R., and Robbie Berg. *Tropical Cyclone Report: Hurricane Florence.* Miami, FL: National Hurricane Center, 2019.

Stradling, Richard. "'We're Going to Have a Different River.' Without Milburnie Dam, the Neuse Comes Alive." *News and Observer* (Raleigh, NC), May 9, 2018.

Strayer, David L., Dean C. Hunter, Lane C. Smith, and Christopher K. Borg. "Distribution, Abundance, and Roles of Freshwater Clams (Bivalvia, Unionidae) in the Freshwater Tidal Hudson River." *Freshwater Biology* 31, no. 2 (1994): 239–48.

Stutz, Bruce. "Natural Filters: Mussels Deployed To Clean Up Polluted Waterways." *Yale Environment 360* (online magazine), Yale School of the Environment, March 31, 2015.

Sulak, Kenneth J., F. Parauka, W. T. Slack, R. T. Ruth, M. T. Randall, K. Luke, M. F. Mettee, and M. E. Price. "Status of Scientific Knowledge, Recovery Progress, and Future Research Directions for the Gulf Sturgeon, *Acipenser oxyrinchus desotoi* Vladykov, 1955." *Journal of Applied Ichthyology* 32, no. S1 (2016): 87–161.

Sun, Ge, Sankar Arumugam, Peter V. Caldwell, Paul A. Conrads, Alan P. Covich, James Cruise, and John Feldt, et al. "Impacts of Climate Change and Variability on Water Resources in the Southeast USA." In *Climate of the Southeast United States: Variability, Change, Impacts, and Vulnerability,* edited by Keith T. Ingram, Kirstin Dow, Lynne Carter, and Julie Anderson, 210–36. Washington, DC: Island Press, 2013.

Sweet, William V., Greg Dusek, Jayantha Obeysekera, and John J. Marra. *Patterns and Projections of High Tide Flooding Along the U.S. Coastline Using a Common Impact Threshold.* Silver Spring, MD: National Oceanic and Atmospheric Administration, 2018.

Sweet, William V., Radley M. Horton, Robert E. Kopp, Allegra Nicole LeGrande, and Anastasia Romanou. "Sea Level Rise." In *Fourth National Climate Assessment,* vol. 1, *Climate Science Special Report,* edited by Donald J. Wuebbles, David W. Fahey, Kathy A. Hibbard, David J. Dokken, Brooke C. Stewart, and Thomas K. Maycock, 333–63. Washington, DC: US Global Change Research Program, 2017.

Sweet, William V., and John J. Marra. *2015 State of U.S. "Nuisance" Tidal Flooding*. Silver Spring, MD: National Oceanic and Atmospheric Administration, June 2016.

———. "Understanding Climate: Billy Sweet and John Marra Explain Nuisance Floods." Climate.gov, September 8, 2015.

Tennessee Valley Authority (TVA). "Use of Lands Acquired for the Columbia Dam Component of the Duck River Project and Future Water Supply Needs in the Upper Duck River Basin." *Federal Register* 66, no. 121 (2001): 33599–601.

Tennessee Valley Authority (TVA) (website). "Hydroelectric." Accessed March 2, 2023.

———. "Navigation on the Tennessee River." Accessed March 2, 2023.

———. "Our History: The 1930s." Accessed March 2, 2023.

———. "Preserving Life on the Elk River." Accessed August 11, 2020.

———. "TVA at a Glance." Accessed March 2, 2023.

———. "Wilson." Accessed March 3, 2018.

Tennessee Wildlife Resources Agency (website). "Fresh Water Mussels in Tennessee." Accessed January 22, 2018.

Terando, Adam J., Jennifer Costanza, Curtis Belyea, Robert R. Dunn, Alexa McKerrow, and Jaime A. Collazo. "The Southern Megalopolis: Using the Past to Predict the Future of Urban Sprawl in the Southeast U.S." *PLoS ONE* 9, no. 7 (2014): e102261.

Tessler, Zachary D., Charles J. Vörösmarty, Irina Overeem, and James P. M. Syvitski. "A Model of Water and Sediment Balance as Determinants of Relative Sea Level Rise in Contemporary and Future Deltas." *Geomorphology* 305 (2017): 209–20.

Thornton, David. "Army Corps of Engineers Turns to Private Sector in Face of Budget Cuts." Federal News Network (website), April 16, 2019.

Toles, Conrad. "DJI Phantom 3 Standard Lake Lahusage." YouTube video, posted September 17, 2016. 5:46.

Tomaszewski, Tina M. *Tims Ford Dam Release Water Quality Improvements*. Chattanooga: Tennessee Valley Authority, April 2008.

Totenberg, Nina. "The Supreme Court Has Narrowed the Scope of the Clean Water Act." *All Things Considered*. National Public Radio, May 25, 2023.

Trotter, Bill. "Price Offered for Maine's Baby Eels Hits Record High." *Bangor Daily News*, March 27, 2018.

Union of Concerned Scientists. *Underwater: Rising Seas, Chronic Floods, and the Implications for US Coastal Real Estate*. Cambridge, MA, 2018.

Union of Concerned Scientists (website). "Infographic: The Energy Water Collision." July 27, 2012.

United Nations (website). "World Population Projected to Reach 9.8 Billion in 2050, and 11.2 Billion in 2100." June 21, 2017.

Upton, John. "As Seas Rise, Americans Use Nature to Fight Worsening Erosion." Climate Central (website), July 14, 2018.

Urbina, Ian. "No Survivors Found after West Virginia Mine Disaster." *New York Times*, April 9, 2010.

Uría-Martínez, Rocío, Megan M. Johnson, and Patrick W. O'Connor. *2017 Hydropower Market Report*. Oak Ridge, TN: Oak Ridge National Laboratory, 2018.

Uría-Martínez, Rocío, Megan M. Johnson, and Rui Shan. *U.S. Hydropower Market Report: January 2021*. Oak Ridge, TN: US Department of Energy, 2021.

US Army Corps of Engineers. "Columbia River Basin Dams." Fact sheet, May 4, 2012.

———. *Proceedings: CE Workshop on Reservoir Releases*. Washington, DC, July 1987.

US Army Corps of Engineers Mobile District (website). "Tennessee-Tombigbee Waterway." Accessed July 31, 2020.

US Army Corps of Engineers Nashville District. *Duck River Watershed Plan: Final Watershed Assessment*. Nashville, TN, May 2018.

US Army Corps of Engineers Savannah District. *Savannah Harbor Expansion Project, Georgia and South Carolina: Fish Passage at New Savannah Bluff Lock and Dam: Integrated Post-Authorization Analysis Report and Supplemental Environmental Assessment*. Savannah, GA, August 2019.

US Army Corps of Engineers Savannah District (website). "SHEP Fish Passage at New Savannah Bluff Lock and Dam: Overview." Accessed October 7, 2019.

US Bureau of Labor Statistics. "Injuries, Illnesses, and Fatalities in the Coal Mining Industry." Fact sheet, April 2010.

US Census Bureau (website). "Small Area Income and Poverty Estimates (SAIPE) Program." Accessed March 26, 2018.

US Department of Agriculture (USDA). *2012 Census of Agriculture: United States Summary and State Data*. Vol. 1, *Geographic Area Series*, Part 51. AC-12-A-51. Washington, DC, May 2014.

———. *Cahaba River—Alabama: Wild and Scenic River Study*. 1979.

———. *Poultry—Production and Value: 2016 Summary*. April 2017.

US Department of Agriculture (USDA) (website). "Agricultural Productivity in the U.S." Economic Research Service. Last updated January 6, 2022.

———. "Soil Tillage and Crop Rotation." Economic Research Service. Last updated April 28, 2020.

US Department of Energy. "Obama Administration Announces up to $32 Million Initiative to Expand Hydropower." News release, July 1, 2009.

US Department of Homeland Security. *Dams Sector: Estimating Loss of Life for Dam Failure Scenarios*. Washington, DC, September 2011.

US Department of Justice. "Maine Men Sentenced for Illegally Trafficking American Eels." News release, May 3, 2018.

US Department of the Interior (website). "Dan River Coal Ash Spill." Natural Resource Damage Assessment and Restoration Program. Accessed March 25, 2018.

US Drought Monitor (website). "Map Archive." National Drought Mitigation Center. Data for March 29, 2016.

US Energy Information Administration (US EIA). *Annual Coal Report 2016*. Washington, DC, 2017.

———. *Levelized Cost and Levelized Avoided Cost of New Generation Resources in the Annual Energy Outlook 2019*. February 2019.

———. *Monthly Energy Review: February 2018*. Washington, DC: Office of Energy Statistics, 2018.

US Energy Information Administration (US EIA) (website). "Energy-Related CO_2 Emission Data Tables." Accessed December 27, 2020.

———. "Preliminary Monthly Electric Generator Inventory." Accessed March 7, 2021.

———. "U.S. States: State Profiles and Energy Estimates." Accessed October 8, 2019.

———. "What Is U.S. Electricity Generation by Energy Source?" Accessed August 1, 2020.

US Environmental Protection Agency (US EPA). "Disposal of Coal Combustion Residuals from Electric Utilities; Final Rule." *Federal Register* 80, no. 74 (2015): 21302–501.

———. *The Effects of Mountaintop Mines and Valley Fills on Aquatic Ecosystems of the Central Appalachian Coalfields*. Washington, DC: National Center for Environmental Assessment, March 2011.

———. *Environmental and Recovery Studies of Escambia Bay and the Pensacola Bay System Florida*. Atlanta, GA, July 1975.

———. *Environmental Impacts of Animal Feeding Operations*. Office of Water, 1998.

———. "EPA, Army Announce Intent to Revise Definition of WOTUS." News release, June 9, 2021.

———. "EPA Proposes First of Two Rules to Amend Coal Ash Disposal Regulations, Saving Up To $100M Per Year in Compliance Costs." News release, March 1, 2018.

———. *Hydraulic Fracturing for Oil and Gas: Impacts from the Hydraulic Fracturing Water Cycle on Drinking Water Resources in the United States*. Washington, DC: Office of Research and Development, December 2016.

———. *National Rivers and Streams Assessment 2008–2009: A Collaborative Survey*. Office of Water, Washington, DC, 2016.

———. *National Water Quality Inventory: 1975 Report to Congress*. Washington, DC: Office of Water Planning and Standards, 1975.

———. *National Water Quality Inventory: Report to Congress*. Washington, DC, 2017.

———. "Nonpoint Source Pollution: The Nation's Largest Water Quality Problem." Fact sheet, March 1996.

———. "North Carolina Water Fact Sheet." July 2010.

———. *Potable Reuse Compendium*. Washington, DC: Office of Ground Water and Drinking Water, 2017.

———. *Primer for Municipal Wastewater Treatment Systems*. Washington, DC: Office of Water, September 2004.

———. *Report to Congress: Nonpoint Source Pollution in the U.S.* Washington, DC: Office of Water Program Operations, January 1984.

———. "Saving Water in Georgia." Fact sheet, June 2013.

US Environmental Protection Agency (US EPA) (website). "Aquifer Recharge and Aquifer Storage and Recovery." Accessed September 30, 2019.

———. "Basic Information about Nonpoint Source (NPS) Pollution." Last updated December 22, 2022.

———. "Basic Information about Surface Coal Mining in Appalachia." Accessed March 23, 2018.

———. "Climate Change Indicators: Heat-Related Deaths." Accessed October 3, 2019.
———. "Coal Ash (Coal Combustion Residuals, or CCR)." Accessed March 25, 2018.
———. "EPA's Response to the TVA Kingston Fossil Plant Fly Ash Release." Last updated January 9, 2009.
———. "Health and Environmental Effects of Particulate Matter (PM)." Accessed August 1, 2020.
———. "Nutrient Pollution: The Effects." Accessed November 29, 2017.
———. "Per- and Polyfluoroalkyl Substances (PFAS)." Accessed June 16, 2021.
———. "Sanitary Sewer Overflows (SSOs)." Accessed August 1, 2020.
———. "Urban Runoff: Low Impact Development." Accessed October 3, 2019.
———. "Water Reuse Resource Hub by End-Use Application." Accessed March 2, 2023.
———. "Wetland Extent, Change, and Sources of Change." EPA's Report on the Environment (ROE). Accessed March 2, 2022.
———. "What Is Acid Rain?" Accessed March 23, 2018.
———. "What Is Turbidity and Why Is It Important?" Accessed January 4, 2020.
US Environmental Protection Agency and Department of Defense (US EPA and DOD). "Definition of 'Waters of the United States'—Recodification of Pre-Existing Rules." *Federal Register* 84, no. 204 (2019): 56626–71.
US Fish and Wildlife Service (US FWS). *Boulder Darter (Etheostoma wapiti) 5-Year Review: Summary and Evaluation.* Cookeville, TN: Cookeville Ecological Services Field Office, 2017.
———. *Boulder Darter Recovery Plan.* Atlanta, 1989.
———. *Fish Passage Engineering Design Criteria.* Hadley, MA: Northeast Region R5, February 2017.
———. *Fish Passage Engineering Design Criteria.* Hadley, MA: Northeast Region R5, 2019.
———. "Pearly Mussels of the Susquehanna Basin." Fact sheet, updated March 15, 2020.
US Fish and Wildlife Service (US FWS) (website). "American Eel (*Anguilla rostrata*)." Accessed March 2, 2023.
———. "American Shad (*Alosa sapidissima*)." Accessed October 4, 2019.
———. Environmental Conservation Online System. Accessed May 15, 2023.
———. "History of the Endangered Species Act: Principal Amendments." Accessed May 15, 2023.
———. "Stream Crossings in Georgia." Accessed May 15, 2023.
US Fish and Wildlife Service (US FWS) and Gulf States Marine Fisheries Cornmission. *Gulf Sturgeon Recovery/Management Plan.* Atlanta, GA, 1995.
US Geological Survey (USGS). "The Chesapeake Bay: Geologic Product of Rising Sea Level." Fact sheet 102-98, October 1998.
———. "Groundwater Quality in the Southeastern Coastal Plain Aquifer System, Southeastern United States." Fact sheet 2016-3076, September 2016.
US Geological Survey (USGS) (website). "Hydraulic Fracturing." Water Resources Mission Area. March 2, 2019.

———. "A Million Gallons of Water—How Much Is It?" Water Science School. June 8, 2018.

———. "USGS 02337170 Chattahoochee River near Fairburn, GA." National Water Information System: Web Interface. Accessed November 27, 2017.

———. "Water Use in Georgia, 2010." South Atlantic Water Science Center. January 17, 2017.

US Government Accountability Office. *Nonpoint Source Water Pollution: Greater Oversight and Additional Data Needed for Key EPA Water Program*. Washington, DC, 2012.

US Mine Safety and Health Administration (US MSHA) (website). "Coal Fatalities for 1900 through 2022." Accessed March 2, 2023.

———. "Respirable Dust Rule: A Historic Step Forward in the Effort to End Black Lung Disease." Accessed March 25, 2018.

US National Park Service (US NPS) (website). "Elwha River Restoration." Accessed October 7, 2019.

———. "History of the Elwha." Accessed August 1, 2020.

US Public Health Service. *Community Water Supply Study*. Washington, DC, 1970.

Vaughn, Caryn C. "Ecosystem Services Provided by Freshwater Mussels." *Hydrobiologia* 810 (2017): 15–27.

Vaughn, Caryn C., Keith B. Gido, and Daniel E. Spooner. "Ecosystem Processes Performed by Unionid Mussels in Stream Mesocosms: Species Roles and Effects of Abundance." *Hydrobiologia* 527 (2004): 35–47.

Velasquez-Manoff, Moises, and Gabriella Demczuk. "As Sea Levels Rise, So Do Ghost Forests." *New York Times*, October 8, 2019.

Vincent, Grayson K., and Victoria A. Velkoff. *The Next Four Decades—The Older Population in the United States: 2010 to 2050*. Washington, DC: US Census Bureau, May 2010.

Vose, Russell S., David R. Easterling, Kenneth E. Kunkel, Allegra N. LeGrande, and Michael F. Wehner. "Temperature Changes in the United States." In *Fourth National Climate Assessment*, vol. 1, *Climate Science Special Report*, edited by Donald J. Wuebbles, David W. Fahey, Kathy A. Hibbard, David J. Dokken, Brooke C. Stewart, and Thomas K. Maycock, 185–206. Washington, DC: US Global Change Research Program, 2017.

Wagner, Adam. "North Carolina Will Work with Maryland, Virginia to Grow Offshore Wind Industry." *News and Observer* (Raleigh, NC), November 1, 2020.

Waldo, W. G. *America's Gibraltar, Muscle Shoals: A Brief for the Establishment of Our National Nitrate Plant at Muscle Shoals on the Tennessee River*. Nashville, TN: Muscle Shoals Association, 1916.

Walsh, John, Donald Wuebbles, Katharine Hayhoe, James Kossin, Kenneth Kunkel, Graeme Stephens, and Peter Thorne, et al. "Our Changing Climate." In *Climate Change Impacts in the United States: The Third National Climate Assessment*, edited by J. M. Melillo, Terese Richmond, and G. W. Yohe, 19–67. Washington, DC: US Global Change Research Program, 2014.

Ward, J. M., and A. W. Meadows. "Adaptive Management of Environmental Flow

Restoration in the Savannah River." Paper presented at the International Conference on Implementing Environmental Water Allocations, Port Elizabeth, South Africa, February 23–26, 2009.

Warnick, Melody. "'I Was the Miracle Baby': How Kerry Albright Survived a Deadly Flood at 9 Months Old." *Reader's Digest*, January 5, 2016.

Washington Post Editorial Board. "The EPA's Move to Regulate 'Coal Ash' Is a Step Forward." *Washington Post*, January 1, 2015.

Water Power Magazine. "US President Obama Signs Hydropower Bills into Law." August 12, 2013.

Watson, Harry L. "'The Common Rights of Mankind': Subsistence, Shad, and Commerce in the Early Republican South." *Journal of American History* 83, no. 1 (1996): 13–43.

Wayne, Graham P. "The Beginner's Guide to Representative Concentration Pathways." Skeptical Science (website), August 2013.

Wehner, Michael F., Jeff R. Arnold, Thomas Knutson, Kenneth E. Kunkel, and Allegra N. LeGrande. "Droughts, Floods, and Wildfires." In *Fourth National Climate Assessment*, vol. 1, *Climate Science Special Report*, edited by Donald J. Wuebbles, David W. Fahey, Kathy A. Hibbard, David J. Dokken, Brooke C. Stewart, and Thomas K. Maycock, 231–56. Washington, DC: US Global Change Research Program, 2017.

Wei-Haas, Maya. "Why Black Lung Disease Is Deadlier Than Ever Before." *Smithsonian Magazine*, May 15, 2017.

Wein, Harrison. "Risk in Red Meat?" NIH Research Matters, National Institutes of Health (website), March 26, 2012.

Westervelt, Amy. *Drilled*. Critical Frequency Podcast, 2018.

White, Elliott, Jr., and David Kaplan. "Restore or Retreat? Saltwater Intrusion and Water Management in Coastal Wetlands." *Ecosystem Health and Sustainability* 3, no. 1 (2017): article 11879065.

Williams, A. Park, Benjamin I. Cook, Jason E. Smerdon, Daniel A. Bishop, Richard Seager, and Justin S. Mankin. "The 2016 Southeastern U.S. Drought: An Extreme Departure from Centennial Wetting and Cooling." *Journal of Geophysical Research: Atmospheres* 122, no. 20 (2017): 10888–905.

Williams, James D., Arthur E. Bogan, and Jeffrey T. Garner. *Freshwater Mussels of Alabama and the Mobile Basin in Georgia, Mississippi, and Tennessee*. Tuscaloosa: University of Alabama Press, 2008.

Williams, Michael R., Gopal Bhatt, Solange Filoso, and Guido Yactayo. "Stream Restoration Performance and Its Contribution to the Chesapeake Bay TMDL: Challenges Posed by Climate Change in Urban Areas." *Estuaries and Coasts* 40, no. 5 (2017): 1227–46.

Wilson, Edward O. *The Future of Life*. New York: Vintage Books, 2002.

Winemiller, K. O., P. B. McIntyre, L. Castello, E. Fluet-Chouinard, T. Giarrizzo, S. Nam, and I. G. Baird, et al. "Balancing Hydropower and Biodiversity in the Amazon, Congo, and Mekong." *Science* 351, no. 6269 (2016): 128–29.

Wohl, Ellen. "Legacy Effects on Sediments in River Corridors." *Earth-Science Reviews* 147 (2015): 30–53.

Woods, Helen, William J. Hargis, Carl H. Hershner, and Pam Mason. "Disappearance of the Natural Emergent 3-Dimensional Oyster Reef System of the James River, Virginia, 1871–1948." *Journal of Shellfish Research* 24, no. 1 (2005): 139–42.

World Wildlife Fund. *Natural and Nature-Based Flood Management: A Green Guide*. Washington, DC, May 2017.

Wuebbles, Donald J., David W. Fahey, Kathy A. Hibbard, David J. Dokken, Brooke C. Stewart, and Thomas K. Maycock, eds. *Fourth National Climate Assessment*. Vol. 1, *Climate Science Special Report*. Washington, DC: US Global Change Research Program, 2017.

Zaffos, Joshua. "'Restoration Cowboy' Goes against the Flow." *High Country News*, November 10, 2003.

Zalasiewicz, Jan, Colin N. Waters, Colin P. Summerhayes, Alexander P. Wolfe, Anthony D. Barnosky, Alejandro Cearreta, and Paul Crutzen, et al. "The Working Group on the Anthropocene: Summary of Evidence and Interim Recommendations." *Anthropocene* 19 (2017): 55–60.

Ziv, Guy, Eric Baran, So Nam, Ignacio Rodríguez-Iturbe, and Simon A. Levin. "Trading-Off Fish Biodiversity, Food Security, and Hydropower in the Mekong River Basin." *Proceedings of the National Academy of Sciences* 109, no. 15 (2012): 5609–14.

Further Readings

Billington, David P., Donald C. Jackson, and Martin V. Melosi. *The History of Large Federal Dams: Planning, Design, and Construction in the Era of Big Dams*. Denver, CO: US Department of the Interior Bureau of Reclamation, 2005.

Cushing, Colbert E., and J. David Allen. *Streams: Their Ecology and Life*. San Diego, CA: Academic Press, 2001.

Doyle, Martin W. *The Source: How Rivers Made America and America Remade Its Rivers*. New York: W. W. Norton, 2018.

Haag, Wendell R. *North American Freshwater Mussels: Natural History, Ecology, and Conservation*. New York: Cambridge University Press, 2012.

Manganiello, Christopher J. *Southern Water, Southern Power: How the Politics of Cheap Energy and Water Scarcity Shaped a Region*. Chapel Hill: University of North Carolina Press, 2015.

Postel, Sandra, and Brian Richter. *Rivers for Life: Managing Water for People and Nature*. Washington, DC: Island Press, 2003.

Reidmiller, David R., Christopher W. Avery, David R. Easterling, Kenneth E. Kunkel, K. L. M. Lewis, Thomas K. Maycock, and Brooke C. Stewart, eds. *Fourth National Climate Assessment*. Vol. 2, *Impacts, Risks, and Adaptation in the United States*. Washington, DC: US Global Change Research Program, 2018.

Richardson-Lorente, Emily. "Flood City." July 29, 2020. In *Broken Ground*, produced by Southern Environmental Law Center, podcast, 36:33.

World Wildlife Fund. *Natural and Nature-Based Flood Management: A Green Guide*. Washington, DC: World Wildlife Fund, 2017.

Index

Page numbers in italics refer to figures.

acid rain, 157, 176, 187
Acornshell, 130
adaptation: climate, 230, 235; proactive, 229, 230–34; reactive, 229
African Americans, 180, 237, 240, 382
agriculture, 6, 7, 9, 20, 23, 31, 33, 42, 46–52, 55–58, 62–64, 66, 68–69, 72, 74, 80–81, 83–86, 96–98, 100–105, *101*, 109, 111–12, 120–22, 125–26, 145, 179, 188–89, 200–201, 213, 219–21, 228, 249, 251–56, 258, 260, 262, 267, 269–71, 285, 291–92, 294, 303–5, 310, 324–26, 371–72, 382, 388, 402, 404–5, 407
Alabama Aquatic Biodiversity Center, 17, 87, 284–85, 289–90, *289*, 314
Alabama Lampmussel, 33, 284–85, 289–90, *289*, 314
Alabama Power Company, 110–11, 131
Alabama River, 137, 142
Alabama Shad, 142–44, 147
Albright, Kerry, 168
Albright, Sylvia, 168
Alewife, 323–24, 328, 350, 375. *See also* river herring
algae, 64–65, *65*, 70, 74, 86, 99, 131, 201, 220, 251, 256, 282. *See also* harmful algal bloom
Alliance for Appalachia, 158
Altamaha River, 112, 324
Amazon River, 373–74
American Eel, 60, 342–50, *348*, 366, 375, 379
American Shad, 323–28, *324*, *327*, 332, *335*, 336–37, 350, 352, 366, 375
amphibians, 13, 17, 139, 154

Antarctica, 26, 205, 207–8, 351
Anthropocene, 9, 11–12, 20–21, 32–34, 150, 190, 219–20, 223, 226–27, 229, 279, 385, 403, 414
Apalachicola Bay, 6, 71, 258
Apalachicola-Chattahoochee-Flint River Basin (ACF), 145–47, 149, 257
Apalachicola River, *18*, 101, 145–46, 148–49, 258, 319, 369
Apalachicola River Basin, 202
Appalachian Mountains, 7, 13, 15–17, *19*, 100, 153–58, *159*, 163, 172, 176, 192, 197, 203, 366
Appleton Edison Electric Light Company, 109
aquifers, 6, 199, 201, 205, 213, 216, 220, 266–72, 276, 371, 388, 406
Atlanta, Georgia, 33, 68, 108, 145, 199, 257–58, 263, 267, 271–73, 406
Atlantic Coast, 38, 43, 53, 205–6, 210, 212, 243, 245, 250, 259, 268, 270, 315, 324, 326–28, 337, 344, 350
Atlantic Ocean, 14, 192–93, 215, 230, 254
Atlantic Salmon, 332, 340–41, *341*, 352, 375
Atlantic Shad, 323–24, 326–28, 336–37, 341, 350, 366
Atlantic Sturgeon, 38–45, *39*, 53, 58, 89, 148, 150, 316, 318, 321–22, 336, 343
Augusta, Georgia, 339

barrier islands, 20, 222, 225–26, 242, 244–45, 248–49, 252, 379, 388, 410
Battle of Blair Mountain, 165
beach nourishment, 245

483

Belews Creek Steam Station, 174
Bermuda High, 192–93, 200, 254, 259
biodiversity, 4, 6, 8, 12–13, 15, 17, 20–21, 27–28, 30–31, 33, 58, 71–72, 80–81, 84–85, 99, 102, 118, 126, 191, 201–2, 214, 220–23, 230, 244, 246, 261, 278–79, 283, 285, 287, 290, 293, 298, 308–10, 314, 365–66, 372–73, 390, 403–4, 414
birding, 24–25, *25*, 27, 114
Birmingham, Alabama, 13, 18, 27, 29, 30, 33, 137, 257, 274, 296, 302, 409
Birmingham-Southern College, 13, 60, *296*
Black Warrior River, 57, 295
Blackwater River, 320
Blohm, Bendix, 42, 89
Blueback Herring, 323–24, 328, 350. *See also* river herring
Blue Crab, 5, 23, *24*, 60, 251
Boepple, Johann, 89–90, *90*
Bonneville Dam, 330–32, *330*, 336, 362
Bosch, Carl, 63, 110
Boulder Darter, 306–9, 313–14
bromide, 174–76
Buford Dam, 257
bypass, fish, 328, 340–41, *341*, 352, 375

Cahaba River, 29, 136–38, *137*, *138*, 142–44, 146, 285, 365–66, *366*
Cahaba River Society, 30, 31
canals, 25, 132
cancer, 5–6, 68, 103, 163, 169, 172, 174, 182
Cape Fear River, 179, 337–38, *338*
carbon dioxide, 11, 12, 31, 155, 186, 188–90, 196, 209, 214, 246, 264, 279, 395, 403, 405
Caribbean Sea, 14, 52, 54, 193, 254
Carson, Rachel, 77
Carter, Jimmy, 160
caviar, 41–42, 54, 89, 145, 148, 150
Centers for Disease Control and Prevention, 69

channelization, 132, 137, 150, 284, 287, 361
Chapman, Colin, 27
Charleston, South Carolina, 6, 71, 210
Charlotte, North Carolina, 33, 175, 199, 267, 272, 387
Chattahoochee River, 68, 131, 146, 148, *149*, 257–58, 367, *367*
Cherokee, 52, 55, 105, 122, 125–26
Chesapeake Bay, 6, 71, 212, 233, 291–94, 300, 303–4, 313, 366
Chesapeake Watershed, 292, 294, 300
Chickasaw, 52, 55
Choctaw, 52, 55
Choctawhatchee Bay, 43
Choctawhatchee River, 320
Claiborne Lock and Dam, 142
clay, 95, 98–99, 102–3, 143, 153–54, 293–94
Clean Air Act, 78, 157, 176
Clean Water Act (CWA), 78–86, 160, 274, 280, 282, 291, 297
climate change, 10–12, 31–33, 85, 155, 188–97, 201–2, 204–8, 214–16, 219–24, 226–30, 233–35, 239, 241, 245–46, 252, 258, 261, 264, 266, 269, 270–71, 278, 297, 302, 311, 321, 358–59, 368, 370, 371–72, 374, 376–78, 380, *381*, 385, 390, 394–95, 397, 402–5, 407, *407*, 411–14
climate gentrification, 239, 240
Clinch River, 130, 134, 177, 285
coal, 7, 11–12, 37, 155–89, 200, 303, 357, 371, 403–4
coal ash, 11, 172, 175–82, 185, 187
Coal Mine Health and Safety Act, 166
coal smog, 173, 187
Coal Wars, 165
coal worker's pneumoconiosis (CWP), 166–67
coastal armoring, 5, 236–39, 244, 248, 406, 412
Coastal Plain, 13, 16–18, *18*, 57,

100–101, 153–54, 259, 267, 270–71, 324, 326
coastal retreat, 238, 386, 406
colonization, European, 49, 52, 249, 323
Colorado River, 378
Columbia Dam, 125
Columbia River, 329–32, *330*
community planning, 234
Conasauga River, 285
concentrated animal feeding operations (CAFOs), 69
Congaree River, 57
Cooper River, 54, 346
crayfish, 14, 16–17, *16*, *17*, 99, 137, 141, 153, 288
Creek, 52, 55
culverts, 368–69, *369*, 379
Cumberland River, 46, 130, 155, 178
Cuyahoga River, 25, 77

dam failure, 357–60, 363, 385
dam relicensing, 311–12, 355, 376
dam removal, 337, 353, 355–57, *356*, 361–70, *366*, *367*, 376–79, 392, 394, 406–7
dams, 7, 57–58, 107–34, 137, 142–43, 145, 148–50, 166–69, 176–77, 184, 208, 212–13, 249, 257, 261–64, 271–72, 275, 277–78, 283–85, 305–14, 316, 318–26, 328–41, 343, 347–80, 382, 383, 385–86, 388–90, 392–94, 396, 406–8; flood control, 110, 241, 329, 380, 383–84, 386–87, 389; hydroelectric, 109–10, 116–17, 120–21, 184, 329, 340, 359, 372, 374–77, 380; irrigation, 109, 111; milldams, 325–26, 367; navigation, 58, 111, 118; recreational, 7, 307, 339–40, 360, 362, 376, 391–96
Dan River, 174, 179
Dan River Steam Station, 179
dead zones, 5, 6, 71–73, 81, 291
deforestation, 10, 27, 46–47, 55, 97–100, *101*, 150, 158, 189, 194, 202, 219, 284, 326, 372, 381
delta, 7, 57, 59, 118, 206, 213, 220, 256, 264, 312, 356, 369
density-dependent foraging, 93
desalination, 265–66, 270–72, 275
de Soto, Hernando, 50, 88
disease, 20, 281; human, 20, 51–52, 54–56, 62, 66–68, 89, 134, 163, 166, 169, 170–73, 221, 274, 279, 302; mussel, 93, 133, *134*
dissolved oxygen, 61, 65, 81, 99, 103, 117, 201, 251, 256, 282, 319
Dodds, Walter, 83
dredge, 6, 7, 57, 73, 78, 91, 93, 109, 131–33, 145, 248–50, 383
drinking water, 30, 58, 174, 265
drought, 19, 31, 52–53, 68, 104, 109, 118, 155, 191, 200–202, 219–20, 226, 229–30, 235, 253–61, *260*, 263, 265–66, 268–69, 272, 275–77, 283, 309, 374, 394–95, *395*, 402
drought plans, 254–55, 258
Duck River, 125–26, 263
Duck River Dam, 263
Duke Energy, 174–75, 178–79
Duncan, Bob, 24, 26, 203, 409, 413, *413*
Duncan, Dallas, 37–40, 43, 411
Duncan, Ginger, 13, 26–27, 391, 396
Duncan, Lucy, 22–26, 59–60, 74, 203, 409, 413, *413*
Duncan, Mary Ann McElheran, 23
Duncan, Will, 22, 24, 29–30, 225, 312
Duncan family history, 413
Dust Bowl, 104, 198

ecological limits of hydrologic alteration (ELOHA), 312–13
ecological succession, 227
ecosystem services, 20–22, 28, 71, 100, 131, 214, 223, 228, 237, 244, 249, 251, 262, 279, 287, 301, 310, 312, 353, 356, 368, 373, 388, 395, 406
Edenville Dam, 358
Edisto River, 324

INDEX 485

Elephantear, 136–39, *138*, 141–47, 283
Elk River, 290, 306–9, 313–14
El Niño/Southern Oscillation, 192, 259
Elwha Dam, 354, *356*, 357, 362
Elwha River, 354–57, *356*, 362
Embrey Dam, 366
Emory River, 177
endangered species, *16*, *17*, 18, 30–32, 40, 44, 48, 58, 83, 87, 92, 124–25, 127, 130, 136, 139, 143, 145–46, 148, 150, 219, 227, 269, 278, 280, 283, 288–90, 305–6, 308, 313, 316–17, 321, 338, 346, 351–52, 361–62, 371, 375, 379, 406–7
Endangered Species Act, 78, 88, 123–25, 127–28, 288, 319, *319*, 328
endemic species, 17, 33, 132, 146, 268, 284, 288
enslavement, 23, 41, 47, 50, 52, 55–56, 95, 179, 326
environmental flows, 309–14, 321, 407
environmentalists, 22, 26, 113, 123, 126, 161, 177, 333, 351, 363, 402
environmental justice, 8, 414
Environmental Protection Agency (EPA), 68, 78–84, 158–60, 170–71, 174–75, 178, 182, 213, 291, 297, 300, 303–4
erosion, 6, 20, 31, 85, 96, 98, 100–101, *101*, 104, 160, 203, 237, 244–45, 247, 252, 279, 287, 292–93, 340, 410
Escambia River, 26, 204, 319
Etnier, David, 123
Etowah River, 47, 285
eutrophication, 67–68, 70–71, 74, 80, 83, 86, 95, 104, 251, 293
evaporation, 198–201, 220, 260, 263, 268, 303, 307, 321, 374
evolution, biological, 17, 27
evolution, stream, 293–94
explosive production, 63–64, 110
extinction, 12, 15, *15*, *17*, 18–21, 27–28, 31, 33, 40, 43, 45, 55–56, 58, 87–88, 93–94, 123, 128–33, *133*, *134*, 145–46, 219, 220, 223, 228, 262, 278–80, 283–84, 289–90, 305, 314, 328, 350–51, 373, 402
extinction crisis, 4, 28, 150, 228
extinction debt, 130, 145

Fair Point, *2*, 3, 22–24, 37, 40, 43
Fall Line, 57
Federal Emergency Management Agency (FEMA), 238, 384–85, 387
Federal Energy Regulatory Commission (FERC), 311–12, 355, 376
fertilizers, 62–64, 69, 71–73, 84, 95, 251, 292
fish kills, 70, 74, 394
fish passage, 329, *330*, 332–39, 342–43, 351–54, 366, 375
fishways, 325–42, *338*, 347–48, 352–53, 375–76, 408
fishways, nature-like, 337, 340–41. *See also* bypass, fish; rock-ramps
Flint River, 146, 148–49, *149*, 258, 268
floating infrastructure, 237
flood control, 122, 238
Flood Control Act, 382
flooding adaptation, 235, 237–38, 244–45, 247, 387
floodplains, 7, 20, 46–47, 57, 69–70, 95, 114, 204, 287, 297, 302, 310, 381, 383, 386–90, *389*, 401, 406
floods, 219, 220, 229, 236, 240; coastal, 31, 206, 210–16, *211*, 222, 224, 233–50, 252, 264, 406, 412; river, 7, 20, 31, 47, 70, 80, 95, 98–99, 102–3, 108–11, 114–15, 118, 120, 122, 126, 130, 156, 167–69, 179, 191, 196–97, 205, 226, 233, 235, 253, 259, 264, 283, 285, 293–97, 301, 303, 307, 309–11, 313, 317, 319, 339, 358, 368, 372, 380–90, *381*, *389*, 392, 395, 397, 405–7, *405*, 412
forestry, 20, 80, 124, 402
Forster, Ann, 3, 222, 225, 269, 412
fossil fuels, 12, 20, 189, 191, 200, 209,

264–66, 372, 402, 404. *See also* coal; methane
Fox River, 109
fracking, 185–86
fugitive dust, 171, 177, 180–81

generalists, 33, 100, 117, 147–48, 223
ghost forests, 213–14, *214*
glaciers, 16, 188, 194, 204–6, 208, 345, 354
Glines Canyon Dam, 354–57, *356*
global circulation models (GCMs), 193–96, 199
global warming. *See* climate change
glochidia, 139–41, 144, 146–48
golden age of dam building, 112–13, 128–30, 142–43, 314, 328, 345
gray infrastructure, 244, 246, 248
graywater, 274–75
Great Depression, 92, 111, 329, 382
Great Lakes, 74, 92, 281
Great Works Dam, 375
greenhouse gasses, 12, 31, 186, 188–90, 194–96, 199, 206–8, *207*, 216, 221–22, 226, 229, 264, 370–71, 402
green infrastructure, 244–48, 282, 302, 412
Green Revolution, 64
Grisham, Charles, 25, 114
groundwater, 69, 79, 97, 157, 161, 177, 179, 186, 200–201, 256, 263, 266–70, 276–77, 279, 293, 302, 388–89, *389*, 405
guano, 63
Gulf, The: The Making of An American Sea, 51
Gulf Breeze, Florida, 2, 3, 22–24, *24*, 203, 224, 226, 409, 411
Gulf Coast, 13, 71, 132–33, 145, 202, 247–48, 250, 259, 316, 335–36
Gulf Islands National Seashore, 4, 226, 412
Gulf of Mexico, 14, 16, 24, 38, 68, 71, 106, 142, 149, 192–93, 206, 246, 250
Gulf Sturgeon, 38–45, *39*, 51, 58, 147–50, 316–22, *319*, *320*, 336, 343

Haag, Wendell, 102–3, 129–32, 134–35, 142–43
Haber, Fritz, 63, 110
Haddock, Randy, 30
Hard Labor Creek Dam, 263
harmful algal bloom, 6, 70–72, 246, 303, 394
hatcheries, 251, 308, 314, 328, 331, 352, 354
herbicides, 95
herring, 142
Hickory Log Creek Dam, 263
Hill, Hiram, 123–25
hogs, 42, 50–51, 56, 69, 70
Holliman, Dan, 29
Holocene, 9, 150, 188–90, 204, 219, 227, 385
Holyoke Dam, 352
Hothouse Earth, 188–89
Howland, Maine, 340–41, *341*, 375–76
Hurricane Dorian, 215
Hurricane Elena, 203
Hurricane Florence, 70, 179, 215
Hurricane Hermine, 259
Hurricane Ivan, 224, 412
Hurricane Katrina, 237
Hurricane Matthew, 70, 259, 358
Hurricane Michael, 215
Hurricane Sandy, 234, 238, 241
hurricanes, 20, 23, 31, 69–70, 192, 214–16, 222, 225–26, 229, 242, 247–48, 319, 392, 409, 412–13
hydrokinetic turbines, 378
hydropower, 20, 111, 155, 257, 306, 311, 333, 348, 362, 370–79, 394, 407
hydropower, sustainable, 373–76, 380, 407
hypoxia, 65, 70–73

INDEX 487

ice ages. *See* Pleistocene
ice sheets, 188–89, 205–6, 208
industrialization, 25; landscape, 54–58, 219, 220, 227, 325; planetary, 9–10, 219, 222, 228; river, 7, 9, 32, 58, 73, 87, 150, 220, 285–86, *286*, 318, 323–24
Industrial Revolution, 66, 108, 163, 189
invasive species, 6, 281, 361
irrigation, 20, 32, 58, 67, 74, 111, 200–201, 220, 254–56, 258, 266–69, 273–75, 278, 310, 318, 374, 377, 405

Jacobs Engineering, 172, 177–78
James River, 52–53, 97–99, 155, 178, 212–13, 233, 250, 322, 324, 355, 364
Jamestown, Virginia, 52–55, 58, 94–97, 100, 201, 268, 322, 408
Jim Woodruff Lock and Dam, 148–49, *149*
Johnson, Paul, 284
jubilee, 59–61, 65, 72–74
Jungle, The, 75

Keiser, David, 80
Kingston Steam Plant, 172, 175, 177–80, 182

Lahusage Dam, 393
Lake Lahusage, 391, 393–94, 396
Lake Lanier, 68, 257–58, 272
landfills, 178–82
Largemouth Bass, 33, 173
Lave, Rebecca, 300–301
lead, 80–81, 156, 175
levees, 7, 205, 236–37, 241, 244, 247–48, 382–83, 385–86, 388, 390
Lexington, Virginia, 363–64
Li, Wenhong, 193
Lightning Point, 250
Little Ice Age, 46
Little River, 391, 396

Little Tennessee River, 120, 123, 125–27
Lower Elwha Klallam Tribe, 354–55, 362
low impact development (LID), 301–4, 388

maize, 46–48, 51
Malthus, Thomas, 62
mangroves, 27, 245–46, 249, 252, 279
manure, 34, 62, 64, 66–67, 69–70, 85, 292
Marvel Slab, 365–66, *366*
Maury River, 360, 363
McPhee, John, 324
megafauna, 45–46
Mekong River, 373
mercury, 59, 81, 156, 173–76, 187
Merrimack River, 352
methane, 31, 75, 164, 170, 185–86, 188–89, 200, 214, 264, 372, 403
Meyer, Judy, 30
Milburnie Dam, 366
Millers Ferry Lock and Dam, 142
Mine Safety and Health Administration, 167, 183
mining: mountaintop, 158–62, *159*, 169–70, 176, 182, 186
Mississippi River, 51, 71, 90, 120, 132, 136, 142, 205, 214, 282, 382, 385, 390
Mobile Bay, 58, 61, 71–74, 256
Mobile River, *16*, 59
Mobile River Basin, 17, 101, 155
Mobile River Delta, 13, 253, 256
mud, 95, 99, 100, 131, 168, 185, 256–57, 262, 264, 308, 315, 339. *See also* polllution
mullet, 3, 5, 6, 22, 37, 252
Muscle Shoals, 105–7, *106*, 110–11, 113, 116, 118–19, 130, 262, 306
mussel die-offs, 132
mussels, 14, 16–17, *16*, 33, 46, 48, 86–94, *87*, 98–99, 102–7, *106*, 109,

113, 116–18, 125–26, 128–38, *134*, 145–48, 150, 161, 219–20, 256, 262, 268, 280–85, 288–90, 293, 306, 309, 313–14, 346, 350, 396, 406; reproduction, 139–44

National Energy Policy Act, 333, 352
National Environmental Policy Act, 78, 122–23, 125
National Flood Insurance Program (NFIP), 384, 387
National Hydropower Association, 371, 374
National Inventory of Dams, 112, 380, 385
National Labor Relations Act, 165
National Marine Fisheries Service, 143, 328, 346
Native Americans, 41, 46–57, *47*, 88–89, 92, 94–95, 105, 122, 129, 219, 250, 262, 277, 323, 329–30, 334, 354, 410–11; Archaic period, 46, 154; Mississippian cultural period, 47–48, 51–52, 88; Woodland period, 46, 51, 154. *See also* Cherokee; Chickasaw; Choctaw; Creek; Lower Elwha Klallam Tribe; Navajo Nation; Penobscot Indian Nation; Powhatan; Seminole
natural channel design, 297–301
natural gas. *See* methane
Nature Conservancy, *18*, *247*, 250, 286–87, 289, 311
Navajo Nation, 378
navigation, river, 32, 57, 98, 110, 118, 120–21, 131, 133, 216, 249, 325, 337, 339
Neuse River, 324, 366
New Orleans, Louisiana, 205, 236–37, 264
New Savannah Bluff Lock and Dam, 338
nitrogen, 61–64, 66–67, 81, 83–84, 292–93, 304

Nixon, Richard, 77–78
non-governmental organizations, 286–88
Norfolk, Virginia, 233–35, 238, 239–43, 264, 291
nuclear energy, 7, 20, 200, 265

Obama, Barak H., 167, 178, 182, 185, 377
Ochlockonee River, 146, 319, 376–77
Ocmulgee River, 57
Office of Surface Mining, 171
Oroville Dam, 358–59, *359*
oyster farming, 93, 251–52
oyster reef, 244–46, 248–52, 291, 293, 379
oysters, 20, 23, *24*, 58, 70, 92, 246, 248–52

Paint Rock River, 284–90, *286*
Paleo-Indians, 45–46
Pamlico Sound, 6, 70–71
Paris Agreement, 402
Pascagoula River, 319
pathogens, 7, 66–67, 81, 86, 133
Patuxent River, 291
PCBs. *See* polychlorinated biphenyls
Pearl River, 43
pearls, 48, 88–89, 92–94, 129, 150
Pee Dee River, 112, 324
Penobscot Indian Nation, 340
Penobscot River, 340, 375
Pensacola, Florida, 3–6, 22–24, 26, 37, 39–40, 43, 60, 71, 203–4, 209, 222, 224–25, 409, 412
Pensacola Bay, 3, 5, 6, 22–23, *23*, *24*, 26, 37, 39, 40, 43, 60, 204, 209, 222, 225, 409
per- and polyfluoroalkyl substances (PFAS), 82
permafrost, 189, 191
pesticides, 82–83, 95, 103
PFAS. *See* per- and polyfluoroalkyl substances

INDEX 489

pharmaceuticals, 68, 82, 86
phosphate, 62, 84
phosphorus, 61–63, 66–67, 81, 83–84, 292–93, 304
Pickwick Dam, 107
Piedmont, 100–101, 112, 199, 267, 271, 277, 324
Piscataquis River, 340–41, *341*
planetary boundaries, 12
planetary systems, 10–12, 20, 61, 72, 190, 219, 279, 402
plankton, 5, 65, 71–72, 252, 256, 281–82
Plater, Zygmunt, 123, 126
Pleistocene, 4, 9, 16, 45, 154, 188, 204–5, 227, 291
Point, The, *2*, 3–5, 8, 37, 222, 225, 409–11, 413, *413*
pollution, 5–6, 9–10, 25–26, 32–33, 64, 73–87, 93, 95–96, 100, 102–3, 109, 128, 133, 137, 139, 143, 150, 157, 159, *159*, 173–76, 180, 186–87, 202, 211, 219–20, 228, 251, 277, 280, 282, 291–92, 300–304, 311, 318, 320, 322, 350, 361, 394, 401, 404–6, *405*; mud, 34, 95–104, *96*, 117, 128–29, 133, 143, 156, 168, 251, 256, 282, 287, 293–94, 300, 308, 315, 326, 339; nonpoint sources, 82–84, 86, 96, 292, 304; nutrient, 7, 12, *65*, 70–71, 84, 291–92; point sources, 82, 280, 292, 304
polychlorinated biphenyls (PCBs), 5–6, 81
potable reuse, 275–77
Potomac River, 291, 313
poultry, 56, 68–69
power: solar, 266, 375–76, 394, 403–4; wind, 183, 372, 374–76, 378, 394, 403–4
Powhatan, 53–54, 97
Purple Bankclimber, 145–50, 289

Quackenbush, Jacob, 89
Quagga Mussel, 281

rainfall patterns, 12, 118, 192–93, 196–202, 264, 297, 358, 368, 374, 380–81, 386
Rappahannock River, 291, 366
Reagan, Ronald, 26
refugium river, 285–86, *286*, 288
Refuse Act, 75
representative concentration pathways (RCPs), 195–96
reservoirs, 70, 82, 112, 114–18, 121–22, 126–28, 130, 173, 200, 220, 253, 256–58, 261–64, 270–72, 276, 278, 280, 303, 306–9, 311, 314, 321–22, 331–32, 337–41, 354, 356, 358, 361–62, 372, 374, 377–78, 392–97
resource curse, 185
restoration, marsh, *247*, 249–50
restoration, river. *See* stream restoration
Rio Grande River, 318
river herring, 323, 326–28, 334–36, *335*, *338*, 350
rock-ramps, 337–40, *338*
Rolfe, John, 53
Roosevelt, Franklin, 111, 165
Roosevelt Dam, 111
Rosgen, David, 297–301
Ruckelshaus, William, 21
Rutland, Judy, 30, 59, 60, 74
Rutland, Lucy Wilkinson, 59
Rutland, Richard, 59

sacrifice zones, 155, 162, 171
Safe Drinking Water Act, 78, 274, 276
salamanders, 154, 159, 161
salmon, 324, 328–32, 341, 345, 348, 352, 354–55, 362–63, 375
salt marshes, 204, 213–14, 244–52, 279, 406, 412
saltpeter, 63
saltwater intrusion, 212–13, 270, 276
sand engines, 245
Sanford Dam, 358
Santa Barbara oil spill, 25, 77

Santee River, 324
Savannah Process, 312–13
Savannah River, 46, 311–12, 324, 338
seagrass, 5, 291–93, 304
sea level, 5, 12, 16, 31, 45, 156, 188–90, 203–16, *207*, 222, 227, 229–30, 233–36, 238, 241–49, 252, 256, 261, 264, 268, 294, 321, 379–80, 386–88, 396, 402, 406, 407, 411–12
sediment pollution. *See* pollution: mud
seed pearl, 92–93
selenium, 156, 161, 176
Seminole, 55
Sequatchie River, 290
sewage, 7, 58, 66–67, 73, 75, 77, 82, 95, 109, 180, 225, 234, 251, 274, 412
sewer overflows, 68, 72
sewer systems, 66–67, 211
shad, *47*, 330, 335, *338*, 343, 348, 350, 366
Shapiro, Joseph, 80
shell buttons, 90–93, *90*, 109, 136
shellfish, 5, 20, 70, 80, 252
shelling, 90–94, *91*, 136
Shoal Creek, 309, 314
Shortnose Sturgeon, 338, 352
shrimp, 38, 60, 214, 252, 256
Silent Spring, 77
silt, 7, 20, 86, 95, 98–99, 117, 143, 256, 264, 286, 293, 310, 388
sinkholes, 268–69, *269*
Skipjack Herring, 142–44
Smith, Captain John, 53
Snail Darter, 121, *121*, 123–27, *124*
snails, 14, 17, *17*, 99, 107, 126, 131, 137, 139, 220, 249, 262, 282, 289–90, 365
Snake River, 329
snapper, 37, 214, 251–52
social inequity, 235, 239, 241–42, 297, 413
Southern Environmental Law Center, 178, 287
specialists, 33, 140, 147

Stewart, Beth, 30
stormwater, 7, 30, 98, 100, 112, 211, 237, 239, 242–43, 270, 285, 292–97, *294, 295, 296,* 300–305, 368–69, *369,* 382, 387–88, 405, *405,* 410, 412
stream restoration, 287–88, 292–94, 297–301, *299, 356,* 387
streams: ephemeral, 153–54, 160–61; intermittant, 154, 160–61
Striped Bass, 334–38, *338,* 350
sturgeon, *39,* 45, *47,* 48–49, 87, 99, 219, 252, 262, 314–15, 334–35, *338,* 401, 411. *See also* Atlantic Sturgeon; Gulf Sturgeon; Shortnose Sturgeon
subsidence, 205, 264, 268, 270, 276, 294, 388
Surface Mining Control and Reclamation Act (SMCRA), 160
surge barriers, 236, 241, 244
Susquehanna River, 291, 352
Suwannee River, 43, 112, 202, 318, 320–21, *320,* 335
systematic habitat destruction, era of, 132

Talquin Dam, 376–77, 394
Tellico Dam, 120–27, *121*
Tennessee River, *16,* 46, 88, 105–10, *106, 108,* 115–17, 119–20, 130, 132, 262, 284–85, 306, 308
Tennessee River Basin, 17, 127
Tennessee Valley Authority (TVA), 120–27, 133, 172, 175, 177–78, 180, 257, 284, 306–9, 313–14
Tenn-Tom Waterway, 132, *133*
Thames River, 67
tidal floods, 210–13, 216, 234, 238, 242
Tims Ford Dam, 306–11, 313–14
tipping points, 189
tobacco, 53–54, 56, 58, 95, 97–98, 100, 381
Tombigbee River, 132–33, 319
tourism, 70, 80, 113, 126, 229, 244, 252, 329, 406

INDEX 491

toxins, 5–6, 8, 69–70, 77, 84, 175–77, 179, 181, 185, 251, 280, 292, 297
trihalomethanes (THMs), 174, 187
Tri-State Water Wars, 257–58, 271
Tropical Storm Julia, 259
trout, 5, 37, 308–9, 311, 313–14, 354–55, 368
Trump, Donald J., 171, 178, 183
turtles, 13, 17, 27, 146
Tybee Island, Georgia, 242–43

Uniontown, Alabama, 180–81
United Nations, 20, 190, 402
Upton Sinclair, 75
urban heat island, 302
urban sprawl, 7, 33, 137, 223, 252, 257, 293, 368, 386, 405, 407, *407*
US Army Corps of Engineers, 78, 111, 134, 149, 169, 241, 258, 264, 311, 330–31, 339, 357, 382, 385, 392–93
US Fish and Wildlife Service, 30, 39, 134, 288, 308, 313, 316–17, 321, 335, 341, 346, 364
US Geological Survey, 146, 170, 348, 356
US Supreme Court, 124, 160, 258

Vaughn, Caryn, 282
Veazie Dam, 375
Vision 2100, 238–39, 241

Wando River, 54
wastewater, 67–68, 137, 168, 174, 256, 274–76, 304–5
wastewater treatment plants, 67–68, 76, 79–82, 84, 225, 274–75, 292, 300, 304, 412

water budgets. *See* water plans
water conservation, 200, 255, 271–73, 277–78
water efficiency, 271–78
water plans, 213, 258, 260, 277–78
Water Pollution Control Act, 75–76, 78–79
water quality, 5, 31, 73–76, 78, 80, 82–85, 87, 94, 96, 100, 124, 201, 251–52, 256, 258, 280–83, 287, 290, 298, 306–8, 314, 366, 368, 406
Water Quality Act, 76, 83
water recycling, 274–77
water shortages, 31, 197, 199–202, 220, 229, 254–55, 262, 267, 270, 274–77, 406
waters of the United States (WOTUS), 78–80, 82, 84, 160
water yield, 199
wetlands, 7, 10, 55–56, 69, 79–81, 98, 157, 201, 214, 247–48, 252, 266, 268–69, 303, 388
Wheeler Dam, 308
wildfire, 31, 56, 190–91, 220, 226, 229, 259, 402
Wilson Dam, 107–8, *108*, 110–20, *115*, 130, 285, 314
Wilson, Edward O., 27, 223
Wilson, Woodrow, 110
World War I, 63–64, 71, 75, 92, 110–11, 186, 383
World War II, 64, 71, 75, 92, 186, 383

Yadkin River, 324
Yellow River, 320

Zebra Mussel, 92–93, 132, 281